T0214538

Communications in Computer and Information Science 1018

Commenced Publication in 2007
Founding and Former Series Editors:
Phoebe Chen, Alfredo Cuzzocrea, Xiaoyong Du, Orhun Kara, Ting Liu,
Krishna M. Sivalingam, Dominik Ślęzak, Takashi Washio, and Xiaokang Yang

Editorial Board Members

More information about this series at http://www.springer.com/series/7899

Stanisław Kozielski · Dariusz Mrozek ·
Paweł Kasprowski · Bożena Małysiak-Mrozek ·
Daniel Kostrzewa (Eds.)

Beyond Databases, Architectures and Structures

Paving the Road to Smart Data Processing and Analysis

15th International Conference, BDAS 2019
Ustroń, Poland, May 28–31, 2019
Proceedings

 Springer

Editors
Stanisław Kozielski
Institute of Informatics
Silesian University of Technology
Gliwice, Poland

Dariusz Mrozek (ID)
Institute of Informatics
Silesian University of Technology
Gliwice, Poland

Paweł Kasprowski (ID)
Institute of Informatics
Silesian University of Technology
Gliwice, Poland

Bożena Małysiak-Mrozek (ID)
Institute of Informatics
Silesian University of Technology
Gliwice, Poland

Daniel Kostrzewa (ID)
Institute of Informatics
Silesian University of Technology
Gliwice, Poland

ISSN 1865-0929   ISSN 1865-0937   (electronic)
Communications in Computer and Information Science
ISBN 978-3-030-19092-7   ISBN 978-3-030-19093-4   (eBook)
https://doi.org/10.1007/978-3-030-19093-4

This Springer imprint is published by the registered company Springer Nature Switzerland AG
The registered company address is: Gewerbestrasse 11, 6330 Cham, Switzerland

Preface

Collecting, processing, and analyzing data have become important branches of computer science. Many areas of our existence generate a wealth of information that must be stored in a structured manner and processed appropriately in order to gain knowledge from the inside. Databases have become a ubiquitous way of collecting and storing data. They are used to hold data describing many areas of human life and activity, and as a consequence, they are also present in almost every IT system. Today's databases and data stores have to face the problem of data proliferation and growing variety. More efficient methods for data processing are needed more than ever. New areas of interests that deliver data require innovative algorithms for data analysis.

Beyond Databases, Architectures and Structures (BDAS) is a series of conferences located in Central Europe and are significant for this geographic region. The conference intends to give the state of the research that satisfies the needs of modern, widely understood database systems, architectures, models, structures, and algorithms focused on processing various types of data. The conference aims to reflect the most recent developments of databases and allied techniques used for solving problems in a variety of areas related to database systems or even go one step forward – beyond the horizon of existing databases, architectures, and data structures.

The 15th International BDAS Scientific Conference (BDAS 2019), held in Ustroń Poland during May 28–31, 2019, was a continuation of the highly successful BDAS conference series started in 2005. For many years BDAS was organized under the technical co-sponsorship of the Institute of Electrical and Electronics Engineers (IEEE). The 14th edition of the BDAS conference (BDAS 2018) was organized under the technical co-sponsorship of the International Federation for Information Processing (IFIP) within the IFIP World Computer Congress (IFIP WCC 2018), which attracted hundreds of participants. For many years, BDAS has attracted thousands of researchers and professionals working in the field of databases and data analysis. Among the attendees at our conference were scientists and representatives of IT companies. Several editions of BDAS were supported by our commercial, world-renowned partners, developing solutions for the database domain, such as IBM, Microsoft, Sybase, Oracle, and others. BDAS annual meetings have become an arena for exchanging information on the widely understood database systems and data processing algorithms.

BDAS 2019 was the 15th edition of the conference, organized under the technical co-sponsorship of the IEEE Poland Section. We also continued our successful cooperation with Springer, which resulted in the publication of this book. The conference attracted participants from 15 countries, who made this conference a successful and memorable event. There were several keynotes and invited talks given during BDAS by leading scientists. The keynote speeches, tutorials, and plenary sessions allowed participants to gain insight into new areas of data analysis and data processing.

BDAS is intended to have a broad scope related to collecting, storing, processing, and analyzing data. Proceedings of the conference were published in this book and the *Studia Informatica* journal. Both books reflect fairly well the broad span of research presented at BDAS 2019. This volume consists of 26 carefully selected papers that are assigned to seven thematic groups:

- Big Data and Cloud Computing
- Architectures, Structures, and Algorithms for Efficient Data Processing and Analysis
- Artificial Intelligence, Data Mining, and Knowledge Discovery
- Image Analysis and Multimedia Mining
- Bioinformatics and Biomedical Data Analysis
- Industrial Applications
- Networks and Security

The first group, containing four papers, is devoted to big data and cloud computing. Papers in this group discuss hot topics of diffused data processing and using big data technologies and scalable computing platforms in seismic signal analysis, monitoring people with IoT devices, generation of power in coal-fired power plants.

The second group contains four papers devoted to various database architectures and models, data structures, and algorithms used for efficient data processing. Papers in this group discuss data serialization and performance of various NoSQL data stores, integrity constraints, and schema decomposition in relational databases.

The third group contains six papers devoted to various methods used in data mining, knowledge discovery, and knowledge representation. Papers assembled in this group show a broad spectrum of applications of various exploration techniques, including decision rules, fuzzy and rough sets, clustering and classification algorithms, to solve many real problems.

The fourth group consists of three papers devoted to image analysis and processing. These papers discuss problems of face recognition, increasing image spatial resolution, and compression of medical images. The fifth group consists of three papers devoted to bioinformatics and biological data analysis. Papers in this group focus on protein structure comparison, clinical data annotation, and sports data preprocessing.

The sixth group consists of three papers proposing applications of data analysis to trading rules based on FOREX EUR/USD quotations, spatial data processing, and fuzzy modeling of the methane hazard rate. The last, seventh, group consists of three papers devoted to networks and security, including network protocol for synchronization of seismic data, SQLite-based e-mail servers, and building a security evaluation laboratory.

We hope that the broad scope of topics related to databases covered in this proceedings volume will help the reader to understand that databases have become an essential element of nearly every branch of computer science.

We would like to thank all Program Committee members and additional reviewers for their effort in reviewing the papers. Special thanks to Piotr Kuźniacki, builder and for 15 years administrator of our website bdas.polsl.pl, and Paweł Benecki and Mateusz Gołosz, new members of our team and copyeditors of this book.

The conference organization would not have been possible without the technical staff: Dorota Huget and Jacek Pietraszuk.

April 2019

Stanisław Kozielski
Dariusz Mrozek
Paweł Kasprowski
Bożena Małysiak-Mrozek
Daniel Kostrzewa

Organization

Program Committee

Chair

Stanisław Kozielski — Silesian University of Technology, Poland

Members

Klaus-Dieter Althoff	University of Hildesheim, Germany
Alla Anohina-Naumeca	Riga Technical University, Latvia
Desislava Atanasova	University of Ruse, Bulgaria
Sansanee Auephanwiriyakul	Chiang Mai University, Thailand
Sergii Babichev	J.E. Purkyně University in Ústí nad Labem, Czech Republic
Werner Backes	Sirrix AG Security Technologies, Bochum, Germany
Susmit Bagchi	Gyeongsang National University, Jinju, South Korea
Petr Berka	University of Economics, Prague, Czech Republic
Damir Blažević	Josip Juraj Strossmayer University of Osijek, Croatia
Victoria Bobicev	Technical University of Moldova, Chisinau, Republic of Moldova
Patrick Bours	Gjovik University College, Norway
Lars Braubach	University of Hamburg, Germany
Marija Brkić Bakarić	University of Rijeka, Croatia
Germanas Budnikas	Kaunas University of Technology, Lithuania
Peter Butka	Technical University of Košice, Slovakia
Rita Butkienė	Kaunas University of Technology, Lithuania
Sanja Čandrlić	University of Rijeka, Croatia
George D. C. Cavalcanti	Universidade Federal de Pernambuco, Brazil
Chantana Chantrapornchai	Kasetsart University, Bangkok, Thailand
Po-Yuan Chen	China Medical University, Taichung, Taiwan, University of British Columbia, BC, Canada
Yixiang Chen	East China Normal University, Shanghai, P.R. China
Andrzej Chydziński	Silesian University of Technology, Poland
Alba Como	University of Tirana, Albania
Armin B. Cremers	University of Bonn, Germany
Tadeusz Czachórski	IITiS, Polish Academy of Sciences, Poland
Sebastian Deorowicz	Silesian University of Technology, Poland
Alexiei Dingli	University of Malta, Malta
Jack Dongarra	University of Tennessee, Knoxville, USA
Andrzej Drygajlo	Ecole Polytechnique Federale de Lausanne, Switzerland
Denis Enachescu	University of Bucharest, Romania

Dušan Kolář	Brno University of Technology, Czech Republic
Zuzana Komínková Oplatková	Tomas Bata University in Zlín, Czech Republic
Daniel Kostrzewa	Silesian University of Technology, Poland
Aleksandar Kovačević	University of Novi Sad, Serbia
Stanislav Krajči	Pavol Jozef Šafárik University in Košice, Slovakia
Mojmír Křetínský	Masaryk University, Czech Republic
Genadijus Kulvietis	Vilnius Gediminas Technical University, Lithuania
Bora I Kumova	Izmir Institute of Technology, Turkey
Andrzej Kwiecień	Silesian University of Technology, Poland
Dirk Labudde	University of Applied Sciences, Mittweida, Germany
Jean-Charles Lamirel	LORIA, Nancy, France, University of Strasbourg, France
Dejan Lavbič	University of Ljubljana, Slovenia
Fotios Liarokapis	Masaryk University, Czech Republic
Sergio Lifschitz	Pontificia Universidade Catolica do Rio de Janeiro, Brazil
Antoni Ligęza	AGH University of Science and Technology, Poland
Ivica Lukić	Josip Jursj Strossmayer University of Osijek, Croatia
Ivan Luković	University of Novi Sad, Serbia
Eugenijus Mačikėnas	Kaunas University of Technology, Lithuania
Bozena Malysiak-Mrozek	Silesian University of Technology, Poland
Violeta Manevska	St. Clement of Ohrid University of Bitola, Republic of Macedonia
Yannis Manolopoulos	Open University of Cyprus, Cyprus
Saulius Maskeliūnas	Vilnius University, Lithuania
Marco Masseroli	Politecnico di Milano, Italy
Maja Matetić	University of Rijeka, Croatia
Zygmunt Mazur	Wroclaw University of Technology, Poland
Emir Mešković	University of Tuzla, Bosnia and Herzegovina
Biljana Mileva Boshkoska	Faculty of Information Studies, Novo Mesto, Slovenia
Mario Miličević	University of Dubrovnik, Croatia
Yasser F. O. Mohammad	Assiut University, Egypt
Tadeusz Morzy	Poznan University of Technology, Poland
Mikhail Moshkov	King Abdullah University of Science and Technology, Saudi Arabia
Dariusz Mrozek	Silesian University of Technology, Poland
Raghava Rao Mukkamala	Copenhagen Business School, Denmark
Mieczysław Muraszkiewicz	Warsaw University of Technology, Poland
Mariana Nagy	Aurel Vlaicu University of Arad, Romania
Sergio Nesmachnow	Universidad de la Republica, Uruguay
Laila Niedrīte	University of Latvia, Latvia
Mladen Nikolić	University of Belgrade, Serbia
Sven Nõmm	Tallinn University of Technology, Estonia
Tadeusz Pankowski	Poznan University of Technology, Poland
Martynas Patašius	Kaunas University of Technology, Lithuania

Witold Pedrycz	University of Alberta, Canada
Adam Pelikant	Lodz University of Technology, Poland
Ewa Piętka	Silesian University of Technology, Poland
Ewa Płuciennik	Silesian University of Technology, Poland
Bolesław Pochopień	Silesian University of Technology, Poland
Andrzej Polański	Silesian University of Technology, Poland
Horia F. Pop	Babeş-Bolyai University, Romania
Václav Přenosil	Masaryk University, Czech Republic
Hugo Proenca	University of Beira Interior, Portugal
Vytenis Punys	Kaunas University of Technology, Lithuania
Abdur Rakib	University of Nottingham, Semenyih, Selangor D.E, Malaysia
Zbigniew W. Ras	University of North Carolina, Charlotte, USA
Riccardo Rasconi	Italian National Research Council, Italy
Jan Rauch	University of Economics, Prague, Czech Republic
Marek Rejman-Greene	Centre for Applied Science and Technology in Home Office Science, UK
Jerzy Respondek	Silesian University of Technology, Poland
Blagoj Ristevski	St. Clement of Ohrid University of Bitola, Republic of Macedonia
Blaž Rodič	Faculty of Information Studies, Novo Mesto, Slovenia
Ewa Romuk	Medical University of Silesia, Poland
Corina Rotar	1 Decembrie 1918 University, Romania
Henryk Rybiński	Warsaw University of Technology, Poland
Christoph Schommer	University of Luxembourg, Luxembourg
Heiko Schuldt	University of Basel, Switzerland
Roman Šenkeřík	Tomas Bata University in Zlín, Czech Republic
Galina Setlak	Rzeszow University of Technology, Poland
Marek Sikora	Silesian University of Technology and EMAG, Poland
Hana Skalská	University of Hradec Králové, Czech Republic
Ivan Stanev	University of Sofia St. Kliment Ohridski, Bulgaria
Krzysztof Stencel	University of Warsaw, Poland
Borislaw Stoyanov	Konstantin Preslavsky University of Shumen, Bulgaria
Stanimir Stoyanov	Plovdiv University Paisii Hilendarski, Bulgaria
Przemysław Stpiczyński	Maria Curie-Skłodowska University, Poland
Dan Mircea Suciu	Babeş-Bolyai University, Romania
Snezhana Sulova	University of Economics Varna, Bulgaria
Dominik Ślęzak	University of Warsaw, Poland, Infobright Inc., Canada
Andrzej Świerniak	Silesian University of Technology, Poland
Todor Todorov	Veliko Tarnovo University, Bulgaria
Monika Tzaneva	University of National and World Economy, Bulgaria
Jüri Vain	Tallinn University of Technology, Estonia
Michal Valenta	Czech Technical University in Prague, Czech Republic
Irena Valova	Angel Kanchev University of Ruse, Bulgaria
Agnes Vathy-Fogarassy	University of Pannonia, Hungary
Karin Verspoor	University of Melbourne, Australia

Sirje Virkus	Tallinn University, Estonia
Daiva Vitkutė-Adžgauskienė	Vytautas Magnus University, Lithuania
Boris Vrdoljak	University of Zagreb, Croatia
Katarzyna Wac	University of Copenhagen, Denmark
Alicja Wakulicz-Deja	University of Silesia, Poland
Sylwester Warecki	Intel Corporation, San Diego, California, USA
Tadeusz Wieczorek	Silesian University of Technology, Poland
Jef Wijsen	University of Mons, Belgium
Piotr Wiśniewski	Nicolaus Copernicus University, Poland
Robert Wrembel	Poznan University of Technology, Poland
Stanisław Wrycza	University of Gdansk, Poland
Moawia Elfaki Yahia Eldow	King Faisal University, Saudi Arabia
Mirosław Zaborowski	IITiS, Polish Academy of Sciences, Poland
Grzegorz Zaręba	University of Arizona, Tucson, USA
Krzysztof Zieliński	AGH University of Science and Technology, Poland
Adam Ziębiński	Silesian University of Technology, Poland
Quan Zou	University of Electronic Science and Technology of China, P. R. China
Jānis Zuters	University of Latvia, Latvia

Organizing Committee

Bożena Małysiak-Mrozek
Dariusz Mrozek
Paweł Kasprowski
Daniel Kostrzewa
Paweł Benecki
Piotr Kuźniacki
Dorota Huget

Sponsoring Institutions

Technical co-sponsorship of the IEEE Poland Section

Contents

Image Analysis and Multimedia Mining

Bioinformatics and Biomedical Data Analysis

Industrial Applications

Networks and Security

Big Data and Cloud Computing

Nova: Diffused Database Processing Using Clouds of Components [Vision Paper]

Shahram Ghandeharizadeh[(⊠)], Haoyu Huang, and Hieu Nguyen

University of Southern California, Los Angeles, USA
{shahram,haoyuhua,hieun}@usc.edu

Abstract. Nova proposes a departure from today's complex monolithic database management systems (DBMSs) as a service using the cloud. It advocates a server-less alternative consisting of a cloud of simple components that communicate using high speed networks. Nova will monitor the workload of an application continuously, configuring the DBMS to use the appropriate implementation of a component most suitable for processing the workload. In response to load fluctuations, it will adjust the knobs of a component to scale it to meet the performance requirements of the application. The vision of Nova is compelling because it adjusts resource usage, preventing either over-provisioning of resources that sit idle or over-utilized resources that yield a low performance, optimizing total cost of ownership. In addition to introducing Nova, this vision paper presents key research challenges that must be addressed to realize Nova. We explore two challenges in detail.

1 Introduction

Database management systems (DBMSs) are a critical part of diverse applications in science, business, health-care, and entertainment. Today's DBMSs, both SQL and NoSQL [9], are complex and consist of numerous knobs to control their performance and scalability characteristics [23,29]. Numerous cloud providers offer one or more of these DBMSs as a service. For example, at the time of this writing, Amazon RDS offers MariaDB, MySQL, PostgreSQL, Oracle, Microsoft SQL Server, and Amazon Aurora as a DBMS in the cloud.

Nova proposes a departure from today's complex DBMSs, see Fig. 1. It envisions simple components that communicate using high speed networks such as those with *remote direct memory access* (RDMA) capabilities to realize a DBMS. A component may be a file system, a buffer pool manager, abstraction of data items as records, documents, and key-value pairs, indexing structures, data encryption and decryption techniques, compression techniques, a library of data operators such as the relational algebra operator select/project/join/etc., a query result cache, a concurrency control protocol, a crash recovery protocol, a query processing engine, a query language, a query optimizer, etc. These components may be services in a cloud and inter-operate to realize a DBMS dynamically. The application logic itself may run in Nova's infrastructure, external to

© Springer Nature Switzerland AG 2019
S. Kozielski et al. (Eds.): BDAS 2019, CCIS 1018, pp. 3–14, 2019.
https://doi.org/10.1007/978-3-030-19093-4_1

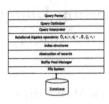

(a) A monolithic DBMS

(b) Nova's simple components will communicate using high speed networks

Fig. 1. Architecture of today's monolithic DBMS and the envisioned Nova

it on a device (e.g., a mobile client such as a smart-phone), or a hybrid where part of the application logic runs in the Nova cloud and other parts of it execute on devices.

The rest of this paper is organized as follows. Section 2 presents the vision of Nova and articulates its research challenges. Sections 3 and 4 focus on the first two research challenges (enumerated questions) of Sect. 2. Section 6 describes how Nova is related to several inspiring disciplines. Brief conclusions are provided in Sect. 7.

2   Overview and Research Challenges

When compared with today's service provider such as Amazon AWS, an application developer will use Nova without sizing a server. Instead, the developer will specify the desired performance (response time and throughput) and data availability (mean time to failure MTTF, mean time to data loss MTDL) characteristics of her application, leaving it to Nova to size its components to meet these specifications. Once the application is deployed, Nova will adjust the amount of resources allocated to each component to meet the application's performance goals in response to system load.

The Nova cloud may span hundreds of data centers operated by the same provider or different providers. It may host a large number of components for different DBMSs serving diverse applications. A Nova component may have variants suitable for different workloads. For example, a component that provides abstraction of data records may implement either a row store or a column store [27,35]. The row store is suitable for an online transaction processing (OLTP) workload while a column store is suitable for an online analytical processing (OLAP) workload.

Two or more components may inter-operate to provide a functionality. For example, a row store representation of records inter-operates with a storage component for storage and retrieval of data. Alternative variants of a storage components may be a block-store such as an Ext4 file system, an in-memory store such as RAMCloud [28], or a Log-Structured Merge-tree [26] (LSM-Tree) key-value store such as LevelDB [19] or RocksDB [13].

A component will have its own configuration knob [32,34] to finetune its performance and data availability characteristics. There will be intra-component communication for it to scale horizontally. Nova is responsible for adjusting the knobs of a component based on the characteristics of a workload, e.g., read to write ratio, rate of conflicts among concurrent accesses to data, pattern of access to data, etc.

Nova will be a transformative system because its simple components are well defined. It will monitor an application's workload and will configure each component to maximize its performance. A component will scale both vertically and horizontally, providing high availability and elasticity characteristics. Either an expert system or a machine learning algorithm may perform these tasks. These may be similar in spirit to [7,37,38] for fine-tuning performance of today's complex DBMSs. Over time, as the application workload evolves, Nova will identify components that are more suitable than the current DBMS components. If a replacement requires an expensive data translation and migration phase, it will notify a database administrator to confirm several candidate off-peak time slots for the component replacement. If a replacement is trivial, e.g., an optimistic concurrency control protocol instead of locking, it will put it in effect transparently and report both the change and the observed performance improvement.

The vision of Nova is made possible by recent advances in CPU processing, network communication, and storage technology. High speed networks such as RDMA and 5G are essential for implementing Nova's high performance components that inter-operate with one another. Several studies report RDMA bandwidths comparable to memory bandwidth, see [4] for an example. High density memory in the form of volatile DRAM and non-volatile SSD enable Nova to maintain state information and replicate this data for high availability. Inexpensive fast CPUs implement components that were traditionally software stacks of a monolithic system. Finally, operating systems such as LegoOS [33] are in synergy with Nova's vision, enabling Nova to provision the right amount of resources from abstraction of a device managed by LegoOS.

Nova's hypotheses are as follows. First, alternative protocols and specifications can be implemented as simple components with well defined functionality. This will enable Nova to bring together those that best satisfy the application's workload in a DBMS. Second, simple components are easier to manage and scale. Third, the data availability requirement of an application can be used to provide smarter execution paradigms. If replication of data across two or more servers satisfies the mean time to data loss (MTDL) of an application, for a component that requires durability property of transactions (e.g., LSM-Tree file system [26]), we propose to replicate its data and eliminate log records altogether. The idea is to provide a more efficient execution paradigm by using memory and network to either (a) eliminate state data such as log records or (b) perform expensive writes to Hard Disk Drive (HDD) and Solid State Disk (SSD) asynchronously. Section 4 presents this idea in greater detail.

Nova is broad in scope and raises many interesting research topics.

1. What is the overall software architecture of Nova? Nova's components must inter-operate to realize a DBMS. They may communicate using message passing or shared memory. The latter may be implemented using RDMA calls. Data encryption is a central component to secure data.
2. How does Nova implement the concept of a transaction and its atomicity, consistency, isolation, and durability (ACID) semantics? This includes an investigation of weaker forms of consistency.
3. How does Nova decide the number of each component for an application and across applications? System size impacts response time and throughput observed by an application. Load imposed by the workload also impacts these performance metrics and may be diurnal in nature. We envision Nova to use service level agreements required by an application to continuously monitor performance metrics and adjust the number of components dynamically in response to system load. The adjustment may be either detective or preventive.
4. How does Nova change one or more components of a deployed DBMS? Some of these may be driven by external factors. For example, an application developer may desire to switch from a relational model to a document model. Others may be driven internally. For example, a machine learning algorithm may identify a row store to be more appropriate than a column store, a lock-based protocol instead of an optimistic concurrency protocol, a log-structured file system instead of a regular Ext4 file system, and others. Some of these changes may require live data migration, translation, and re-formatting of data. Ideally, Nova should perform these operations with no disruption in processing application requests. They may impact system load. However, they can be processed during off-peak hours identified by Nova and confirmed by a system administrator.
5. How secure is Nova? This challenging research question examines the interplay between components selected for an application and where encryption is used to store and process data. Alternative Nova configurations may enhance security of data for an application. They may also require additional resources or slow down the application. It is important to provide an intuitive presentation of this tradeoff to an application developer to empower them to make an informed decision.
6. What would be an intelligent configuration advisor for Nova? Nova provides a rich spectrum of components with diverse tradeoffs. Ideally, a system architect should specify their workload and Nova's configuration advisor should suggest candidate components suitable for that workload. Similar to a query optimizer, the advisor must provide an explain feature for an architect to interrogate Nova's choice of components and how efficient they are in meeting the workload requirements.

The next two sections focus on the first two research questions.

3   Nova's Software Architecture

Nova's envisioned software architecture will be more than a classification of components with well defined interfaces. Components must inter-operate using high speed networks and RDMA. To illustrate the challenge, consider a traditional block-based buffer pool manager and a block-based file system. Nova will require these to inter-operate using RDMA. A key question is how will one scale the buffer pool manager to increase the amount of memory assigned to the buffer pool manager? One answer is to scale-up [10] as long as the server hosting the buffer pool manager has sufficient physical memory. However, once this memory is exhausted, the buffer pool manager must either migrate [21,25,36] to a server with a larger memory size or scale to run across multiple servers. Each has its own interesting research questions. With the first, how will Nova migrate the buffer pool manager without disrupting service? What happens to the RDMA connection of components that use the buffer pool manager? How will the buffer pool manager re-establish its RDMA connection with the file manager?

With the second approach, different instances of the buffer pool manager component assigned to different servers will be coordinated to act as one component. File system pages may be hash partitioned across these instances. The instances will be aware of one another and collaborate to route a request to the right instance. Research questions include: How will the instances communicate with one another, is it via RDMA or a regular message passing network? What is the trade off associated with separating the data fabric from the network fabric? Once scaled out, how will these instances communicate with the file system component using RDMA? Similarly, how will the software layer that uses the buffer pool manager communicate with the different buffer pool instances?

In addition to the above, Nova's architecture raises the following research questions:

1. Is it possible to scale each component independently? For example, for a workload with a low read-write ratio, is it possible to have tens of instances of the file system, 1 buffer pool manager with 1 lock manager and 1 log manager and a few instances of the indexing technique?
2. Is it possible to add components dynamically? For example, once the workload of an application evolves to exhibit a high read-write ratio (say 500 to 1 [5]), is it possible to insert a query result cache in its DBMS dynamically?
3. How to replace one or more components of a DBMS with another component to make it more suitable for a workload? For example, a LSM-Tree filesystems [26] such as LevelDB [19] and RocksDB [13] is suitable for workloads with a low read-write ratio. How would one use such a file system instead of an Ext4 file system? Given a workload with a low read-write ratio, how will Nova quantify the performance improvement observed with this change?
4. How to tolerate server failures without disruption of service? For example, if the server hosting the lock manager fails, is it important to recover its state at the time of failure? If so, how will Nova maintain the state of the lock manager and restore it to continue operation?
5. How will Nova monitor its components and control their placement across servers to resolve bottlenecks?

4 Nova Transactions

Nova's support for transactions with ACID semantics simplify design and implementation of applications [22,24]. They empower an application developer to reason about application behavior to debug and test it effectively. Transactions may be a feature of a component. For example, durability of writes as transactions may be provided by an implementation of LSM-Tree such as LevelDB or RocksDB. Several studies describe use of RDMA and data locality to scale a transaction processing workload such as TPC-C to a large number of nodes [6,40].

An accepted practice to provide atomicity and durability is to generate log records and flush them to persistent store prior to either a write (with LevelDB or RocksDB) or a transaction commit (with MySQL, Oracle, or MongoDB). Nova will offer alternatives to enhance performance. These alternatives are based on the key insight that today's racks of servers offer high mean time to failures, MTTF. Based on this, one may quantify mean time to data loss (MTDL) of replicating a data item two or more times. If this MTDL is lower than an application's tolerable MTDL then Nova may replicate data instead of generating log records to satisfy atomicity and durability properties of transactions. Should log records be required for auditing of transactions then these log records can be replicated across multiple servers. These log records are subsequently flushed to a HDD/SSD file system and deleted from memory asynchronously. This speeds up performance compared with today's state of the art that flushes log records to HDD/SSD synchronously.

To illustrate the superiority of the proposed approach, Fig. 2 shows an evaluation of MySQL using the TPC-C benchmark [31]. We show the transactions per minute of MySQL as a function of the number of concurrent threads generating the TPC-C workload. All results are gathered from a cluster of Emulab [39] nodes. Each node has two 2.4 GHz 64-bit 8-Core E5-2630 Haswell processors, 8.0 GT/s bus speed, 20 MB cache, 64 GB of RAM, and connects to the network using 10 Gigabits networking card. Each node runs Ubuntu OS version 16.04 (kernel 4.10.0). We adjust knobs of MySQL to maximize its performance, e.g., we increase its default memory size from 128 Megabytes to 48 Gigabytes.

We show two MySQL configurations. One using HDD and the other using SSD, labeled MySQL HDD and MySQL SSD, respectively. SSD is faster than HDD. Hence, tpmC with SSD is higher and levels off with 5 concurrent threads due to transactions waiting on one another in MySQL's lock manager.

When MySQL is extended with a write-back client-side cache [18] (top line), its performance is enhanced more than two folds. This is because all writes to the data store are performed by generating buffered writes across the network. Background threads apply the buffered writes to the data store asynchronously.

Write-back's tpmC also levels off beyond 5 threads due to contention for leases in the caching layer. Note that with the write-back cache, choice of SSD or HDD for MySQL does not impact[1] tpmC because almost all reads and writes are processed using the cache.

---

[1] The impact is on the amount of required memory because HDD slows down the rate at which buffered writes are applied and deleted.

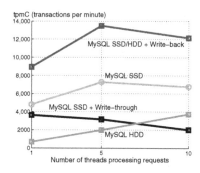

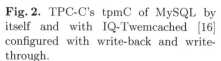

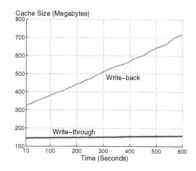

Fig. 2. TPC-C's tpmC of MySQL by itself and with IQ-Twemcached [16] configured with write-back and write-through.

Fig. 3. Memory size of IQ-Twemcached with write-back and write-through, 5 TPC-C threads, 1 warehouse.

In Fig. 2, the tpmC with the same cache configured using the write-through policy (black line) is lower than MySQL by itself. This is because 92% of TPC-C workload is writes. The overhead of updating the cache entries outweighs the benefits observed by reads that constitute 8% of the workload.

Figure 3 shows the size of memory required by write-back and write-through as TPC-C executes its workload. TPC-C generates new orders that increase both the database size and the number of entries in the caching layer. The difference between the write-back and write-through approximates the size of buffered writes generated by the write-back policy.

With write-back, we observe a 19% decrease in the reported tpmC when buffered writes are replicated thrice. We replicate buffered writes to satisfy the MTDL requirements of the target application. The buffered writes and their replicas are deleted asynchronously once they are applied to the data store.

Nova will improve performance of a component by replicating its transactional data in order to (a) make writes durable, (b) meet the MTDL requirements of an application, and (c) perform the transactional write to a mass storage device (HDD/SSD) asynchronously. We hypothesize the latter to enhance performance and scalability of the component. These replicas are deleted once the data is written to HDD/SSD. As demonstrated by the results of Figs. 2 and 3, these execution paradigms require memory. Nova will assume availability of servers with a large amount of memory to enable these execution paradigms.

There is a tradeoff between space, performance observed by the foreground load, and MTDL. Specifically, one may trade space for time and MTDL by maintaining a priority queue of read and write requests. Nova may give a low priority to processing log records to minimize impact on the foreground requests. This will cause replicas of the log records to stay across Nova's server for a longer time. However, it will enhance the performance observed by the foreground system load.

5   In-Memory Data Containers, MemDCs

To make the proposed ideas feasible, we present the concept of physical in-memory data containers (MemDCs). MemDC is envisioned as a component managed by Nova and created based on a policy. The policy may specify a MTDL requirement or the number of replicas for a data item. A MemDC may span servers across different racks of the same data center and potentially across data centers. Once data is assigned to a MemDC, the MemDC replicates it across its servers.

Design and implementation of MemDCs raise many interesting research questions. First, should a component that requires durable writes delegate persistent write of its data to MemDC? What is the tradeoff between the component performing the write asynchronously versus MemDC performing it? Does the former simplify component design? If the answer is affirmative then how does MemDC maintain the order in which data items should be written to the persistent store? The answer to the latter question is important when there are dependencies between log records and/or written data items.

Nova may consist of a large number (trillions) of MemDCs with different policies. How does Nova place these MemDCs across racks of servers in a data center and across data centers? How does it modify the initial placement in response to an evolving system load? A concept similar to MemDC is presented as a μshard by Facebook [3]. MemDC will be a super-set of a μshard in the form of memory for both application data and transient data such as log records.

In passing, we note that MTDL as a metric depends on mean time to repair (MTTR). A MemDC may detect loss of an in-memory replica due to a server failure and repair it by constructing a replica on a different server. This form of repair will enhance MTDL observed by an application.

Second, how does Nova implement non-idempotent writes? A non-idempotent write must be applied only once. In the presence of arbitrary failures, Nova may either (a) convert a non-idempotent write to an idempotent one or (b) maintain sufficient information to guarantee a non-idempotent write is applied only once. It is important to quantify the tradeoffs associated with these alternatives to develop a methodology for the components that utilize MemDCs.

Finally, based on our experiences with implementing a write-back policy using query result caches, there are a variety of race conditions that must be identified and addressed [17]. We anticipate Nova to use leases to address these race conditions. Leases [17,20] are different than locks in that they have a finite lifetime. This concept is useful when the server hosting the component that owns the lease fails. Once a lease expires, its data becomes available again.

6   Related Work

Vision of Nova is at the intersection of several inspiring disciplines, including extensible DBMSs, parallel DBMSs, web services (WS) and grid computing, federated DBMSs and polystores, and distributed systems. We describe these in turn.

Extensible DBMSs [12] advocate use of components that plug-n-play to provide customized data services to an application. Nova is different in two ways. First, its components will communicate using the network (instead of being part of a monolithic system). Second, we envision Nova as a framework that adjusts the components of a DBMS continuously to enhance efficiency without violating an application's service level agreements. It will do so by continuously monitoring the key metrics of an application's workload (e.g., its read to write ratio) and matching it with the capabilities of candidate components.

Nova is similar to parallel DBMSs [11] in that its components communicate with one another and the data may flow across multiple components, implementing both inter and intra component parallelism. It is different than parallel DBMSs in that the components are (a) shared among different applications in a cloud and (b) the components are not decided in advance. We envision Nova to adjust the number of components per required service level agreement (SLA) of an application. We also envision Nova to implement security policies across components shared among different applications.

Web services [1,2] and grid computing systems [15] consist of services that are assembled together. They may use security and encryption when exchanging data between services. These services communicate using IP network. Vision of Nova is different in that its micro-services are at a finer granularity, e.g., concurrency control protocols, buffer pools, etc. It envisions use of RDMA with servers accessing and sharing memory to implement these services efficiently, enabling them to scale horizontally.

Federated DBMSs such as Garlic [8] provide a single interface for different DBMSs. Their successor is a multi-store [30] or polystore [14] that supports multiple data models and databases, combining concepts from federated and parallel DBMSs. A polystore includes a middleware that provides an abstraction of data and implement a programming API for query planning, optimization, and execution using its databases. It may support multiple query languages for different applications. Similar to a parallel DBMS, a polystore may partition data to scale and use data flow execution paradigms for query processing. Nova is similar to polystores in that it envisions a DBMS of components that may support multiple data models and queries that execute across them. At the same time, it envisions switching from one data model to another if the workload justifies the switch. Nova is different than polystores in that it does not assume a DBMS as one of its components. It envisions a DBMS as a collection of components to be tailored together based on the application workload. A component of Nova may have a simple interface and functionality, e.g., get, put, append, etc. Moreover, Nova's components may share memory using fast networks such as RDMA. One may realize a polystore using multiple Nova DBMSs.

Finally, Nova is inspired by distributed systems such as LegoOS that implement services of an operating system as components. Nova is in synergy with such an operating system by implementing the corresponding DBMS service using the appropriate service of the operating system. Nova is different because it implements a query language such as SQL and ACID transactions.

7   Conclusions

This vision paper presents the vision of Nova as a DBMS realized using simple components. A cloud provider may deploy these components across one or more data centers. These components may run on devices and services of a modular operating system such as LegoOS [33], leveraging the scalability of its underlying hardware. The key advantage of simple components will be their simple knobs for scale up, scale out, high availability, and elasticity. Nova will select components of a DBMS with the objective to meet the performance requirements of an application workload. It will adjust the knobs of these components in response to the application workload and may change one or more components of a DBMS as the workload of the application evolves.

Acknowledgments. We gratefully acknowledge use of Utah Emulab network testbed [39] ("Emulab") for all experimental results presented in this paper. We thank anonymous BDAS 2019 reviewers for their valuable comments.

References

1. Alwagait, E., Ghandeharizadeh, S.: A comparison of alternative web service allocation and scheduling policies. In: 2004 IEEE International Conference on Services Computing, Shanghai, China, pp. 319–326, September 2004. https://doi.org/10.1109/SCC.2004.1358021
2. Alwagait, E., Ghandeharizadeh, S.: DeW: a dependable web services framework. In: RIDE-WS-ECEG 2004, March, Boston, MA, pp. 111–118 (2004). https://doi.org/10.1109/RIDE.2004.1281710
3. Annamalai, M., et al.: Sharding the shards: managing datastore locality at scale with akkio. In: OSDI, pp. 445–460. USENIX Association, Carlsbad (2018). https://www.usenix.org/conference/osdi18/presentation/annamalai
4. Binnig, C., Crotty, A., Galakatos, A., Kraska, T., Zamanian, E.: The end of slow networks: it's time for a redesign. Proc. VLDB Endow. **9**(7), 528–539 (2016). https://doi.org/10.14778/2904483.2904485
5. Bronson, N., Lento, T., Wiener, J.L.: Open data challenges at Facebook. In: 31st IEEE International Conference on Data Engineering, ICDE, Seoul, South Korea, 13–17 April 2015, pp. 1516–1519 (2015)
6. Cai, Q., et al.: Efficient distributed memory management with RDMA and caching. Proc. VLDB Endow. **11**(11), 1604–1617 (2018). https://doi.org/10.14778/3236187.3236209
7. Cao, Z., Tarasov, V., Tiwari, S., Zadok, E.: Towards better understanding of black-box auto-tuning: a comparative analysis for storage systems. In: USENIX Annual Technical Conference, Berkeley, CA, USA, pp. 893–907 (2018). http://dl.acm.org/citation.cfm?id=3277355.3277441
8. Carey, M.J., et al.: Towards heterogeneous multimedia information systems: the garlic approach. In: RIDE-DOM, pp. 124–131 (1995)
9. Cattell, R.: Scalable SQL and NoSQL data stores. SIGMOD Rec. **39**, 12–27 (2011)
10. Das, S., Li, F., Narasayya, V.R., König, A.C.: Automated demand-driven resource scaling in relational Database-as-a-Service. In: Proceedings of the 2016 International Conference on Management of Data, SIGMOD 2016, pp. 1923–1934. ACM, New York (2016). https://doi.org/10.1145/2882903.2903733

11. Dewitt, D.J., Ghandeharizadeh, S., Schneider, D.A., Bricker, A., Hsiao, H.-I., Rasmussen, R.: The Gamma database machine project. IEEE Trans. Knowl. Data Eng. **2**(1), 44–62 (1990). https://doi.org/10.1109/69.50905

12. Dittrich, K.R., Geppert, A. (eds.): Component Database Systems. Morgan Kaufmann Publishers Inc., San Francisco (2001)

13. Dong, S., Callaghan, M., Galanis, L., Borthakur, D., Savor, T., Strum, M.: Optimizing space amplification in RocksDB. In: CIDR 2017, 8th Biennial Conference on Innovative Data Systems Research, Chaminade, CA, USA, 8–11 January 2017, Online Proceedings (2017). http://cidrdb.org/cidr2017/papers/p82-dong-cidr17.pdf

14. Elmore, A., et al.: A demonstration of the BigDAWG polystore system. Proc. VLDB Endow. **8**(12), 1908–1911 (2015). https://doi.org/10.14778/2824032.2824098

15. Foster, I., Kesselman, C., Tuecke, S.: The anatomy of the grid: enabling scalable virtual organizations. Int. J. High Perform. Comput. Appl. **15**(3), 200–222 (2001)

16. Ghandeharizadeh, S., Yap, J., Nguyen, H.: IQ-Twemcached. http://dblab.usc.edu/users/iq/, https://github.com/scdblab/IQ-Twemcached

17. Ghandeharizadeh, S., Yap, J., Nguyen, H.: Strong consistency in cache augmented SQL systems. In: Middleware, December 2014

18. Ghandeharizadeh, S., Nguyen, H.: Design, implementation, and evaluation of writeback policy with cache augmented data stores. Technical report 2018-07, USC Database Lab (2018). Submitted to VLDB 2019: In Revision

19. Ghemawat, S., Dean, J.: LevelDB. https://github.com/google/leveldb. Accessed 15 Nov 2018

20. Gray, C., Cheriton, D.: Leases: an efficient fault-tolerant mechanism for distributed file cache consistency. SIGOPS Oper. Syst. Rev. **23**(5), 202–210 (1989). https://doi.org/10.1145/74851.74870

21. Kulkarni, C., Kesavan, A., Zhang, T., Ricci, R., Stutsman, R.: Rocksteady: fast migration for low-latency in-memory storage. In: SOSP, pp. 390–405. ACM, New York (2017). https://doi.org/10.1145/3132747.3132784

22. Lloyd, W., Freedman, M.J., Kaminsky, M., Andersen, D.G.: Don't settle for eventual: scalable causal consistency for wide-area storage with COPS. In: SOSP, October 2011

23. Ma, L., Aken, D.V., Hefny, A., Mezerhane, G., Pavlo, A., Gordon, G.J.: Query-based workload forecasting for self-driving database management systems. In: SIGMOD, Houston, TX, USA, 10–15 June 2018, pp. 631–645 (2018). https://doi.org/10.1145/3183713.3196908

24. Mehdi, S.A., Littley, C., Crooks, N., Alvisi, L., Bronson, N., Lloyd, W.: I can't believe it's not causal! Scalable causal consistency with no slowdown cascades. In: NSDI, Boston, MA, USA, 27–29 March 2017, pp. 453–468 (2017). https://www.usenix.org/conference/nsdi17/technical-sessions/presentation/mehdi

25. Memon, B.N., et al.: RaMP: a lightweight RDMA abstraction for loosely coupled applications. In: 10th USENIX Workshop on Hot Topics in Cloud Computing (HotCloud 2018). USENIX Association, Boston (2018). https://www.usenix.org/conference/hotcloud18/presentation/memon

26. O'Neil, P.E., Cheng, E., Gawlick, D., O'Neil, E.J.: The Log-Structured Merge-Tree (LSM-Tree). Acta Inf. **33**(4), 351–385 (1996). https://doi.org/10.1007/s002360050048

27. O'Neil, P.E., Quass, D.: Improved query performance with variant indexes. In: SIGMOD, Tucson, Arizona, USA, 13–15 May 1997, pp. 38–49 (1997). https://doi.org/10.1145/253260.253268

28. Ousterhout, J., et al.: The RAMCloud storage system. ACM Trans. Comput. Syst. **33**(3), 7:1–7:55 (2015). https://doi.org/10.1145/2806887
29. Pavlo, A., et al.: Self-driving database management systems. In: CIDR 2017, Chaminade, CA, USA, 8–11 January 2017, Online Proceedings (2017). http://cidrdb.org/cidr2017/papers/p42-pavlo-cidr17.pdf
30. Płuciennik, E., Zgorzałek, K.: The multi-model databases – a review. In: Kozielski, S., Mrozek, D., Kasprowski, P., Małysiak-Mrozek, B., Kostrzewa, D. (eds.) BDAS 2017. CCIS, vol. 716, pp. 141–152. Springer, Cham (2017). https://doi.org/10.1007/978-3-319-58274-0_12
31. Raab, F., Kohler, W., Shah, A.: Overview of the TPC-C benchmark. http://www.tpc.org/tpcc//detail.asp. Accessed 15 Nov 2018
32. Seltzer, M.: Beyond relational databases. Commun. ACM **51**(7), 52–58 (2008). https://doi.org/10.1145/1364782.1364797
33. Shan, Y., Huang, Y., Chen, Y., Zhang, Y.: LegoOS: a disseminated, distributed OS for hardware resource disaggregation. In: OSDI, pp. 69–87. USENIX Association, Carlsbad (2018). https://www.usenix.org/conference/osdi18/presentation/shan
34. Smolinski, M.: Impact of storage space configuration on transaction processing performance for relational database in PostgreSQL. In: Kozielski, S., Mrozek, D., Kasprowski, P., Małysiak-Mrozek, B., Kostrzewa, D. (eds.) BDAS 2018. CCIS, vol. 928, pp. 157–167. Springer, Cham (2018). https://doi.org/10.1007/978-3-319-99987-6_12
35. Stonebraker, M., et al.: C-Store: a column-oriented DBMS. In: VLDB, Trondheim, Norway, 30 August–2 September 2005, pp. 553–564 (2005). http://www.vldb.org/archives/website/2005/program/paper/thu/p553-stonebraker.pdf
36. Tsakalozos, K., Verroios, V., Roussopoulos, M., Delis, A.: Live VM migration under time-constraints in share-nothing IaaS-clouds. IEEE Trans. Parallel Distrib. Syst. **28**(8), 2285–2298 (2017). https://doi.org/10.1109/TPDS.2017.2658572
37. Van Aken, D., Pavlo, A., Gordon, G.J., Zhang, B.: Automatic database management system tuning through large-scale machine learning. In: SIGMOD, pp. 1009–1024. ACM, New York (2017). https://doi.org/10.1145/3035918.3064029
38. Wang, S., Li, C., Hoffmann, H., Lu, S., Sentosa, W., Kistijantoro, A.I.: Understanding and auto-adjusting performance-sensitive configurations. In: ASPLOS, pp. 154–168. ACM, New York (2018). https://doi.org/10.1145/3173162.3173206
39. White, B., et al.: An integrated experimental environment for distributed systems and networks. In: OSDI, December 2002
40. Zamanian, E., Binnig, C., Harris, T., Kraska, T.: The end of a myth: distributed transactions can scale. Proc. VLDB Endow. **10**(6), 685–696 (2017). https://doi.org/10.14778/3055330.3055335

Big Data in Power Generation

Marek Moleda[✉] and Dariusz Mrozek[✉]

Institute of Informatics, Silesian University of Technology,
ul. Akademicka 16, 44-100 Gliwice, Poland
marek.moleda@gmail.com, dariusz.mrozek@polsl.pl

Abstract. The coal-fired power plant regularly produces enormous amounts of data from its sensors, control and monitoring systems. The Volume of this data will be increasing due to widely available smart meters, Wi-Fi devices and rapidly developing IT systems. Big data technology gives the opportunity to use such types and volumes of data and could be an adequate solution in the areas, which have been untouched by information technology yet. This paper describes the possibility to use big data technology to improve internal processes on the example of a coal-fired power plant. Review of applying new technologies is made from an internal point of view, drawing from the professional experience of the authors. We are taking a closer look into the power generation process and trying to find areas to develop insights, hopefully enabling us to create more value for the industry.

Keywords: Big data · Power industry · Coal power plant · Predictive analytics

1 Introduction

Power plants are places of the coexistence of various types of IT and OT (operational technology) systems. A long lifetime of assets, non-hyper-competitive environment and regulation restrictions cause a substantial delay in the adoption of new technologies, for example, big data and cloud computing. Indisputably, especially in modern times data is an asset which can be utilized to create value and improve businesses. In this specific environment of centralized financial, ERP systems and distributed operation technology systems it is impossible to ignore outstanding opportunities for the application of big data analytic techniques. Desirable effects can be achieved by decreasing the cost of data storage as well as through still developing analytic tools. Many current activities, for example: making reports, spreadsheet calculations, intuitive decisions can be boosted, automatized or transformed to provide better performance and efficiency. There are also risks and problems we have to tackle as far as implementing this new technology is concerned in the IT environment of power plants.

S. Kozielski et al. (Eds.): BDAS 2019, CCIS 1018, pp. 15–29, 2019.
https://doi.org/10.1007/978-3-030-19093-4_2

2   Big Data Characterization

2.1   Big Data Definition

Big data refers to these types of data sets which processing causes a lot of problems or is simply impossible when traditional relational databases are employed. There are many definitions describing big data, and one of them says shortly "Big data is where parallel computing tools are needed to handle data". Big data not only refers to the size of data as the other features are well explained in, for example, the concept of three V's, which are: *volume, variety, velocity*. The occurrence of more than one of these features could mean we are dealing with big data. There are also two complementary features, which have come out recently and possess massive commercial use. These are *veracity* and *value* [20]. 5V's (Fig. 1) may be described as follows:

Volume: The quantity of generated and stored data. Most people define big data in tera or petabytes [22]. Machine-generated data is produced in much larger quantities than traditional relational data. For instance, a single jet engine can generate 10 TB of data in 30 min [7]. The number of variables and the frequency of data generation make this volume so big. For example, a steam boiler has about 20000 variables and produces 4–5 GB samples every month.

Variety: The type and nature of data. Big data draws from text, images, audio, video; and it completes missing pieces through data fusion [8]. Variety of sources use many different data types to analyze reasons for a variety of processes [13]. Data is categorized as: structured (e.g. relation-based databases), semi-structured (e.g., XML, JSON, RSS feed) and unstructured (e-mails, videos) [22].

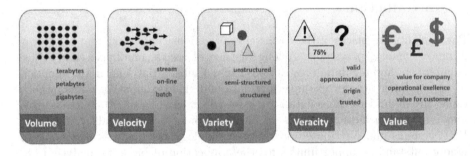

Fig. 1. The 5V model of Big data

Velocity: The frequency of data generation or the frequency of data delivery. For example, a continuous stream of data could be adjusted with a once-in-a-while event triggering data from a sensor. Although the majority of power system sensors are event-triggered, there are also sensors, for example, PMUs both at transmission and distribution level, which produce data streams at high rates [3].

Veracity: The quality of captured data. Means how much data is accurate (error free, raw or currently analyzed, integrated). Taking into consideration some information gathered from external sources, for instance, social media, we can observe that they cannot be fully trusted and its quality may be debatable.

Value: The worth derived from exploiting big data. Means internal value to the company. An important thing in DaaS (data as a service) and data monetization concepts [20].

2.2 Big Data Architecture and Components

Lambda Architecture. Described by Nathan Marz and James Warren lambda architecture is an elegant explanation of how big data works. It shows a way to achieve a scalable system with all requirements of big data system including low latency, high volume and error tolerance (listed in Table 1). Lambda architecture as it is shown in Fig. 2, contains three components: batch, speed, and serving layer [15].

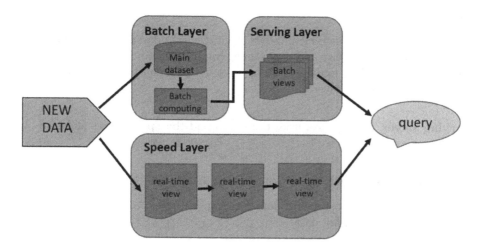

Fig. 2. Big data block schema

Batch Layer includes the main dataset called also "data lake". It is a repository of raw data in its native format. Data is stored in a distributed file system to provide greater scalability and capacity [14]. The repository recomputes all the data from the distributed file system to the batch views, which are accessible through fast queries. Recomputing is a slow process but the data is integral and accessible.

Table 1. Features of big data system

Low latency	Most applications require low delay regarding data access. It is not the most important thing concerning tasks similar to creating the annual report but in the case of real-time control or anomaly detection it is quite significant
Reliability and fault tolerance	Big data should be resistant to human errors and hardware breakdowns. High availability is provided by replication and distributed file system (e.g. Hadoop File System). Both failures and handmade errors can be fixed by renewed batch computing
Scalability	Ability to keep performance with fast grooving data storage and high load. Big data is vertically scalable by adding new clusters [18]
Extensibility	Big data is easy to develop. New functions and changes in the existing code do not need much effort in data migration and programming
Ad-hoc queries	Each and every query regarding the whole data set can be called at any time
Debug-ability	Big data allows to follow input and output (batch view) values

Speed Layer records incremental data from recent updates. It is a response to the requirement of velocity and it is a great complement to the batch layer. The speed layer stores data updates, taking place between consecutive recomputing processes, in the batch layer.

Serving Layer merges data from the Batch and the Speed layers and gives a real-time view of data.

2.3   Comparison with Traditional Data Analysis

In comparison with the standard data warehouse, there are some distinguished features characterizing big data.

From ETL to ELT. When we want to follow the traditional way of collecting and utilizing data, we must first design a database schema as well as collect and execute prepared operations. Taking that into consideration, we may observe a difference reflected in a simple fact that big data analysis is "closer" to data. It means that data is firstly collected in a simple form and then it is transformed and analyzed.

In-Database Analytics. Instead of copying data to other locations and processing it in a dedicated tool, it is possible now to do complex computing, machine learning operations or set operations within a database engine. This enables faster responses from databases, through which data could be fetched nearly in real time.

Distributed Computing. Data warehouses tend to be centralized systems. This kind of a system is vertically scaled, what means that to scale it up the system needs more RAM, a faster CPU or more CPUs. Distributed systems allow horizontal scaling by adding other machines. It makes big data solutions more cost-effective [17].

Unstructured Data. Decreasing the cost of data storage makes many types of unstructured data justified to store, even if its value remains unknown. By contrast, data warehouses operate only on relational databases and are designed for a specific purpose.

2.4   Electric Power Data

Following the big data definition, it is worth to check if we are really dealing with big data as defined by 5V's. Traditionally, operation technology systems work in distributed, firewalled environments to avoid noise and to provide adequate security.

The number of samples and control signals fulfill the criteria of velocity. Examples of some data charts from PGIM (Power Generation Information Manager) visualized in Grafana system are shown in Figs. 4, 5 and 6. However, designing big data in a way it could deal with island-like OT systems is a challenge. The growing number of devices with m2m (machine to machine) or Wi-Fi interfaces, could provide some additional information for analysis. Both volume and variety are characteristics of OPC servers (Open Platform Communications), which store historical data including process data, triggered events, and alarms.

The majority of data is either not logged, or it is overwritten very quickly. For example, in most protection relays and related sensors, the data collected is discarded shortly after internal use. If a pre-programmed event is not detected, then no data is automatically stored [3]. Accordingly, while many of the recently deployed or emerging power data measurement systems lie in the description of Big Data, the way that they are currently managed does not exactly match the spirit and purpose of Big Data. Once such hidden data is collected, managed, and analyzed, they will constitute the real Big Data in power systems [3].

There are also many external sources of data related to the energy sector, which can complete our big data set. For example, also shown in Fig. 3: weather conditions, GPS data, traffic information, social-media can support the real-time market or load management.

There is a large potential to use Big Data techniques in the analysis of power system data. Various data sources with smart analytics lead to intelligence while performing operational processes and tactical management.

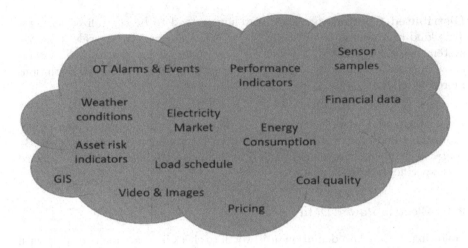

Fig. 3. Data types that could be used in Big Data analytic

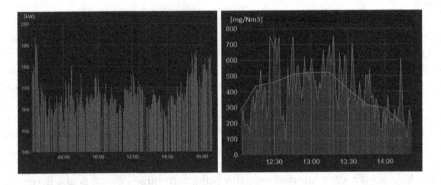

Fig. 4. Sample of process data: fan power and instant CO emission

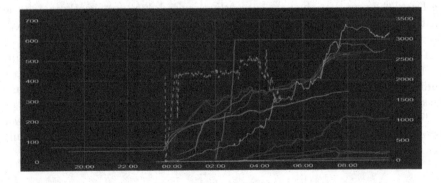

Fig. 5. Example graph of a power plant startup

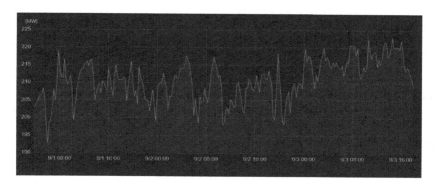

Fig. 6. Turbine active power

3   Big Data Analytics Approach

Big data analytics means using advanced analytic techniques operating on big data [22]. Hardware solutions and platforms like Hadoop [1] and Spark [2] allow for massive data collection and storage. These can well feed analytic tools to enhance its results [22]. The tools cover techniques, like data mining, machine learning (classification, regression, clustering), artificial intelligence (cognitive simulation, expert systems, perception, pattern recognition) statistical analysis, natural language processing, and advanced data visualization [3,12,23,24]. Data analytic tools can analyze huge amounts of data at speed impossible for humans without technology. However, any analytic method is useless, if no action is taken [5] (Table 2).

Table 2. Example way from data to wisdom [5]

From			To
Data	-3	Celsius	Information
Information	$-3\,^{\circ}$ C	At $3\,^{\circ}$ C it's cold out	Knowledge
Knowledge	$-3\,^{\circ}$ C	Need to dress warmly	Wisdom

Data analytics is categorized in some areas, depending on the purpose, scope, and techniques used:

Descriptive Analytics - interpretation of historical data. Helps to compare and understand data from the past, for example, annual reports comparison, assessment of generation unit capital project, key performance indicators as mean time between failures or month by month upkeep costs sets. It answers the question: "What happened?"

Diagnostic Analytics - a way to determine factors and causes of a particular event. Can be used for example in fraud detection or to understand failures from historical data. It answers the question: "Why did it happen?"

Predictive Analytics - this type of analytics uses statistics and modeling to determine future performance. It can help to predict financial trends, create reliability models or foresee asset management issues [5]. It is able to advise which feed pump should be running to give the best operational efficiency. It answers the question: "What can happen?"

Prescriptive Analytics - simulates possible paths and gives the best option according to predicted results. Can calculate the potential economic profits from steam-boiler renovation. It answers the question: "What should we do?"

Cognitive Analytics - uses advanced machine learning, cloud computing or artificial intelligence to give a real-time decision making aid. For example, can control in real time combustion process to achieve less nitrogen dioxide emissions.

4   Potential Benefits of Using Big Data in Power Generation

Big data and digitization process coming along with machine-generated and enterprise data can unlock new business opportunities. Not all of the areas are visible now, but many cases indicate some potentials to improve operational excellence and performance, as well as to reduce costs.

4.1   Fault Detection and Condition Monitoring

Early fault detection allows to improve system availability and to avoid additional downtime costs and regulatory fines. Data from control systems, diagnostic reports and videos can be analyzed in real time giving information on the current equipment condition; and therefore, potential failure occurrence. For example, based on data from SCADA (Supervisory Control And Data Acquisition) systems, it is possible to detect wind turbine failure by monitoring gearbox oil temperature, power output, and rotational speed [19]. Similarly, if used on other assets it could significantly reduce the operational and maintenance cost. Anomaly detection can also catch events not visible for control systems. For example, if a measurement tool has a predefined min-max range, the control system will not alarm us even though the reading is anomalous (as shown in Fig. 7).

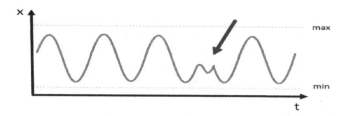

Fig. 7. Anomaly example

4.2   Operations-Planning Convergence

By integrating data from various sources and processes the possibility to convergence has been enhanced. The term operations-planning convergence refers to the ability of a utility enterprise to realize the future conditions of the power system with high probability and high accuracy. This is difficult to achieve without systematic data management and unified models [10]. Operational planning refers to preparation for weather, load, and generation conditions changes in the next minutes, hours, and days. There are various reasons for this convergence gap (e.g., diverse models, diverse data sources and data formats) and inefficient data management tools, which all can be overcome with the unified methods and systematic data management [3]. To simulate system behavior we can use analytic tools, e.g., predictive analysis, machine learning, stochastic analysis, etc. It is an interesting concept to create a digital copy of assets or the whole power plant.

Digital Twins is a digital power plant model allowing to simulate and visualize its performance and behavior in varied scenarios. Digital model would make it possible to determine how the real plant would respond to different conditions, supplies or even weather events [11].

4.3   Asset Management

Asset management is a systematic process of developing, operating, maintaining, upgrading, and disposing of assets cost-effectively. Big Data analytics may drive forward asset management maturity model to be integrated and automated.

Predictive Maintenance. Owing to equipment monitoring and fault detection it is possible to make a transition from preventive to predictive maintenance methods. Health-based maintenance of equipment provides better scheduled fixing plans minimizing planned and unplanned downtimes. It means:

- reducing unnecessary repairs of equipment in good condition,
- minimizing the probability of downtime by real-time health monitoring.

Predictive maintenance needs comprehensive information about the asset. Big data technology could fulfill this requirement in an inexpensive way of collecting data from existing systems.

Enhanced Planning. Complete assets data including current condition, repair costs, and risks, could be used for asset managers to diversify maintenance methods (including reactive or preventive maintenance where it is justified) [25]. Maintenance methods are described in more detail in Table 3. Implementing the right maintenance strategy can optimize costs and availability indicators. Critical equipment needs more predictive and preventive approach, while non-critical, cheap to repair devices sometimes could run to failure.

Table 3. Comparison of maintenance methods.

	Reactive	Preventive	Predictive
Description	"Fix on fail". Repair is done already when equipment is broken	"Time-based" maintenance. Equipment is serviced in regular time intervals based on the vendor's recommendations, MTBF (Mean time between failures) or other statistics	"Fix as required". Condition based maintenance. Service is done before failure is expected
Advantages	No service and inspection cost	Keeps equipment in high condition. Doesn't cause production halts	Less service costs. Better fault detection
Disadvantages	Less production availability. High overhead costs. Lost production costs	Requires planning. Vulnerable to random failures	High cost of monitoring and skills needed

Integrated Data. Another economical aspect refers to comparing asset performance before and after maintenance. Such analysis allows evaluating the profitability of service activities. Moreover, big data could also feed service crews with complete and accurate equipment information. The more adequate the information is, the faster the service task may be completed.

Procurement and Materials. Gaining knowledge about the current condition from big data could help in optimizing the procurement process, reducing its time and cost. It has also a positive influence on the inventory size and better supply planning.

4.4 Performance Optimization

Datasets coming from information technology systems and control systems used in the real-time analysis would constitute great feedback for operators controlling the production process. Understanding the power system as a black box, where parameters like steam pressure, steam temperature, and coal flow are inputs, and efficiency or generated power is an output, dramatically simplifies modeling process. Using data mining or machine learning techniques makes it possible to find the right way to optimize efficiency or emissions [6, 16].

Another example is to control additive chemicals applied to coal. Owing to advanced analysis, it is possible to adjust the number of additives to current load or other parameters. Such operation provides emissions compliance and prevents overfeeding chemicals [21].

Renewable power generation units and fluctuating market dictate new flexibility requirements to coal-based power plants. It means not only working in highly efficient constant conditions, but also quickly adjusting the load to the current demand. It needs start-up characteristics improvement by reducing starting and stopping time and optimization of load gradient.

4.5   Data Analysis and Visualization

Big data tools and techniques enable visualization and exploration of large volumes of data. It includes also enhancing the reporting ability of existing business intelligence systems and discovering new things for the enterprise. Unlike standard dashboards or charts, big data needs more advanced tools to present multidimensional, high-volume data. Visualization involves graphical representation of data structures and techniques making data more transparent, like aggregation and hierarchization. Data analysis is supported by advanced techniques and tools, like predictive analytics, data mining (IBM SPSS Modeler, KNIME, WEKA), statistical analysis (Matlab, RStudio, Python), complex SQL, data visualization (MS Power BI, Qlikview, Tableau), artificial intelligence (Keras, Tensorflow), or natural language processing (Natural Language Toolkit, Apache OpenNLP). Some spreadsheet-based reports could be substituted and automated, also the decision-making process could evolve from intuitive to data-based.

4.6   Demand Response

Electrical energy cannot be stored, so production is adjusted to temporary energy consumption. Demand prediction includes also forecasting energy consumption and power generation from renewable sources (wind turbines, solar plants). Gathering data from smart meters, social media, analyzing consumption patterns enables us to predict the expected energy consumption [9]. Moreover, we can forecast weather impact on predicted power generation from renewable sources (wind, solar). Complete information gives chances to use existing power supplies in an optimized way and aids in load planning (containing also energy collecting and power-to-heat strategies). Accurate energy forecasting allows to avoid imbalance costs and gives the possibility to gain more from the real-time market.

5   Challenges in Big Data Adoption

5.1   Siloed Data

Effective big data analytics needs comprehensive data access. One of the challenges is to face silo mentality, meaning the situation in which some departments do not wish to share their data with others. There are many factors fueling silo mentality such as poor communication or internal competition. Departments have different goals, priorities, and responsibilities, so they often do not collaborate with each other to achieve common business goals. A more important problem is to save data confidentiality in compliance with legal and security regulations. Sensitive data like customers personal data, financial or trading data still need to be restricted from unwanted access.

5.2   Cybersecurity

Power plants and energy utilities represent critical infrastructure where data and IT systems would be principally protected. More significance of IT and data makes it more vulnerable to cyberattacks. Known threats like data stealing, system disabling can cause huge financial losses. Moreover, cyberattacks more often could be aimed for energy utilities. An example could be an attack on the Ukrainian power grid in 2015 [26] when hackers using malware software and taking control on SCADA systems were able to cause outages in 30 substations. Considering data as a valuable asset needs to provide relevant activities and protection to assure its confidentiality, integrity, and backup.

5.3   Skills

Adapting big data in organization forces requires acquiring new skills covering both infrastructure architecture and data analytics. Data architects, data scientists or data engineers are some of the new professions created by big data. Besides, to leverage big data impact organizations need many business users with analytical skills. To attract good employees, companies will need to develop a distinct culture, career paths, and recruiting strategy for data and analytics talents. However, there are many talented engineers with analytical skills in the electricity industry, so many of analysts could be trained, not hired.

5.4   Leadership and Organization

Transforming the decision-making process in organizations into one based on data and analytics, besides technical skills, requires applying leadership, organizational structures and communication to make the expected revenue. Due to survey results most significant challenges are ensuring senior management involvement in data analysis activities and designing an appropriate organizational structure to support analytics activities [4]. Organizational structures should provide good communication between the analytics team and departments. Greater impact on both cost and revenue was achieved with hybrid structure meaning central analytics organization that coordinates with employees who are embedded in individual business units [4]. An important thing is supporting and involving analytics by CEOs to align activities close to their vision and strategy.

5.5   Architecture and Technology

There are many commercial and open sources of software, architecture, tools offering big data analytics. Chosen solution should fit in the organization needs and provide integration with existing systems, training and support. It is a big challenge to gather data from sensors and devices working in real-time systems, which was not intended for analysis in system building process because of high cost or lack of analytic tools in past times.

6   Conclusion

This article presents the characteristics of big data issues in the context of the power generation industry. Technology progress gives the opportunity to handle and analyze new streams of data, it also enhances current business intelligence based on data warehousing and gives the possibility to explore values from data in a new way. Big data analytics can be capitalized in many specific internal processes in power generation industry improving operating efficiency. It also helps to make better and faster decisions, using modeling and prediction to optimize maintenance and reliability. It is inevitable for big data to face some challenges; for example, to gain expected revenue organizations must create structures, acquire skills and change culture and mentality.

Acknowledgements. This work was supported by the Polish Ministry of Science and Higher Education as part of the Implementation Doctorate program at the Silesian University of Technology, Gliwice, Poland (contract No 0053/DW/2018), and partially, by the pro-quality grant for highly scored publications or issued patents of the Rector of the Silesian University of Technology, Gliwice, Poland (grant No 02/020/RGJ19/0167), and by Statutory Research funds of Institute of Informatics, Silesian University of Technology, Gliwice, Poland (grant No BK/204/ RAU2/2019).

References

1. Apache Hadoop homepage. https://hadoop.apache.org. Accessed 27 Oct 2018
2. Apache Spark homepage. https://spark.apache.org/. Accessed 27 Oct 2018
3. Akhavan-Hejazi, H., Mohsenian-Rad, H.: Power systems big data analytics: an assessment of paradigm shift barriers and prospects. Energy Rep. **4**, 91–100 (2018). https://doi.org/10.1016/j.egyr.2017.11.002. http://www.sciencedirect.com/science/article/pii/S2352484717300616

4. Brad Brown, J.G.: The need to lead in data and analytics (2016). https://www.mckinsey.com/business-functions/digital-mckinsey/our-insights/the-need-to-lead-in-data-and-analytics. Accessed 22 Oct 2018

5. Canadian Electricity Association: Data to wisdom. Big data and analytics in the Canadian electricity industry (2017). https://electricity.ca/library/data-to-wisdom/. Accessed 20 Oct 2018

6. Chongwatpol, J., Phurithititanapong, T.: Applying analytics in the energy industry: a case study of heat rate and opacity prediction in a coal-fired power plant. Energy **75**, 463–473 (2014)

7. Dijcks, J.P.: Big data for the enterprise (2014). https://www.oracle.com/assets/wp-bigdatawithoracle-1453236.pdf. Accessed 20 Sept 2018

8. Hilbert, M.: Big data for development: a review of promises and challenges. Dev. Policy Rev. **34**(1), 135–174 (2016)

9. Huang, Z., Luo, H., Skoda, D., Zhu, T., Gu, Y.: E-Sketch: gathering large-scale energy consumption data based on consumption patterns. In: 2014 IEEE International Conference on Big Data (Big Data), pp. 656–665. IEEE (2014)

10. Kezunovic, M., Xie, L., Grijalva, S.: The role of big data in improving power system operation and protection, pp. 1–9, August 2013. https://doi.org/10.1109/IREP.2013.6629368

11. Lawson, S.: Cloud-based 'digital twins' could make power plants more efficient (2015). https://www.networkworld.com/article/2987521/cloud-based-digital-twins-could-make-power-plants-more-efficient.html. Accessed 15 Dec 2018

12. Leskovec, J., Rajaraman, A., Ullman, J.D.: Mining of Massive Datasets. Cambridge University Press, Cambridge (2014)

13. Małysiak-Mrozek, B., Lipińska, A., Mrozek, D.: Fuzzy join for flexible combining big data lakes in cyber-physical systems. IEEE Access **6**, 69545–69558 (2018). https://doi.org/10.1109/ACCESS.2018.2879829

14. Małysiak-Mrozek, B., Stabla, M., Mrozek, D.: Soft and declarative fishing of information in big data lake. IEEE Trans. Fuzzy Syst. **26**(5), 2732–2747 (2018). https://doi.org/10.1109/TFUZZ.2018.2812157

15. Marz, N., Warren, J.: Big Data: Principles and Best Practices of Scalable Realtime Data Systems, 1st edn. Manning Publications, Shelter Island (2018)

16. Mohamed, O., Al-Duri, B., Wang, J.: Predictive control strategy for a supercritical power plant and study of influences of coal mills control on its dynamic responses. In: 2012 UKACC International Conference on Control (CONTROL), pp. 918–923. IEEE (2012)

17. Mrozek, D., Daniłowicz, P., Małysiak-Mrozek, B.: HDInsight4PSi: boosting performance of 3D protein structure similarity searching with HDInsight clusters in Microsoft Azure cloud. Inf. Sci. **349–350**, 77–101 (2016)

18. Mrozek, D.: Scalable Big Data Analytics for Protein Bioinformatics. CB, vol. 28. Springer, Cham (2018). https://doi.org/10.1007/978-3-319-98839-9

19. Qiu, Y., Chen, L., Feng, Y., Xu, Y.: An approach of quantifying gear fatigue life for wind turbine gearboxes using supervisory control and data acquisition data. Energies **10**(8), 1084 (2017)

20. Quitzau, A.: Transforming energy and utilities through big data and analytics (2014). https://www.slideshare.net/AndersQuitzauIbm/big-data-analyticsin-energy-utilities. Accessed 22 Oct 2018

21. Risse, M.: Using data analytics to improve operations and maintenance (2018). https://www.powermag.com/using-data-analytics-to-improve-operations-and-maintenance. Accessed 15 Sept 2018

22. Russom, P., et al.: Big data analytics. TDWI best practices report. Fourth Quart. **19**(4), 1–34 (2011)
23. Slavakis, K., Giannakis, G.B., Mateos, G.: Modeling and optimization for big data analytics: (statistical) learning tools for our era of data deluge. IEEE Signal Process. Mag. **31**(5), 18–31 (2014)
24. Tyagi, H., Kumar, R.: Optimization of a power plant by using data mining and its techniques. Int. J. Adv. Sci. Eng. Technol. **2**, 83–87 (2014)
25. Vesely, E.: Unsupervised machine learning: the path to industry 4.0 for the coal industry (2018). https://www.powermag.com/unsupervised-machine-learning-the-path-to-industry-4-0-for-the-coal-industry. Accessed 6 Oct 2018
26. Zetter, K.: The Ukrainian power grid was hacked again (2017). https://motherboard.vice.com/en_us/article/bmvkn4/ukrainian-power-station-hacking-december-2016-report. Accessed 20 Sept 2018

Using GPU to Accelerate Correlation on Seismic Signal

Dominika Pawłowska and Piotr Wiśniewski[✉]

Faculty of Mathematics and Computer Science,
Nicolaus Copernicus University, Toruń, Poland
{dpawlowska,pikonrad}@mat.umk.pl

Abstract. In analyzing the quality of seismic signal, the fundamental mathematical operation is the convolution of signal with basic signal. Analyses carried out in the field need solutions that can be executed by a single machine. Meanwhile the size of processed data from land seismic surveys is in order of tens of terabytes. In this article the efficient computation of convolution on GPU cores is proposed. We state that this approach if faster than even using parallel programming on CPU. It will be shown how big performance gain was achieved when using a graphic card that is several times less expensive than used CPU.

Keywords: Seismic signal · Convolution · CUDA

1 Introduction

Geophysical surveys used to comprehend the structure of Earth on different depths have applications in many respects. The best known is searching for deposits of natural resources, although nowadays it is more often used for making construction designs which leads to significant savings when building roads, high constructions, etc. One of the first stages of surveys is generating seismic wavelets and collecting the reflection wavelets - responses of different Earth strata. Wavelets are generated by using vibrator trucks or dynamite shots (that is why generating wavelet in seismic terminology is called shot). Signal of reflected wavelets (in seismic terminology called seismic traces) are gathered by geophones which are recording perpetually or only for specific time after the shot. In land surveys, to which we have access to, traces are recorded by about 1500 geophones (also called receivers). When receivers record continuously, important part of signal, which can be defined as data gathered for a few seconds after generating wavelets, has to be separated from any other data. This step is called arranging data, and is enormously vital for our study, which focuses mainly on the aforementioned data.

With signal of length 20 s and sample interval of 2 ms, we get single trace of size 40 kB. With survey using 1500 geophones for single shot we get 60 MB of data. In the survey few thousand shots were generated which gave us hundreds

S. Kozielski et al. (Eds.): BDAS 2019, CCIS 1018, pp. 30–39, 2019.
https://doi.org/10.1007/978-3-030-19093-4_3

of GB of input data. This amount of data inspires researchers to study and test the efficiency of seismic data processing systems. [1,2] The preliminary analysis of data gathered in field should be performed at the scene of the survey. Preliminary processing allows to estimate the quality of collected signal and indicate where the shots should be repeated. Repetition of the shots during the survey is incomparably cheaper than returning to the area afterwards.

The results of this paper were gathered during the creation of the software for collecting and arranging data gathered from geophones. One of the basic assumptions when creating software was the ability of running our program on workstation computer which will be a part of the seismic team equipment at the place of survey. Emerged software is used on machines with i9 7th generation processors. System is implemented in Java, which on one hand makes it platform independent, but on the other makes using low-level solutions difficult. Some GPU accelerated computing libraries exist, however they are still unsophisticated. Developed solution besides arranging gathered data was aimed to perform convolution of collected signal with basic signal. This allows to partially smooth out the noise with frequencies different than created by vibrating trucks. This approach is not used when wavelets are created by dynamite shot. The convolution theorem [3] allows to compute convolution using fast Fourier Transform but it still absorbs over 90% of CPU time. The size of data gained from seismic surveys is enormous. It is a classic example of big data convolution computation problem. Correlation of large seismic data is also great example of SIMD – single instruction multiple data. We have simple computations for very big amount of data. For this reason, convolution is a good candidate for processing on graphics processing units (GPU) with the use of CUDA. This was our inspiration to perform experiments presented in this paper.

CUDA is parallel computing platform developed by NVIDIA for general computing on graphical processing units. It provides drastic speed up of computing applications by harnessing the power of GPUs. GPU-accelerated libraries which provide highly-optimized functions for specific mathematical problems is its huge advantage. One of these libraries is cuFFT which provides GPU-accelerated FFT implementations which was another reason for testing how fast correlation will be performed with use of GPU. CUDA is based on C/C++ but wrappers to interact with CUDA from other programming languages exist. In the software we use jcuda library which also has the implementation of cuFFT.

The use of the capabilities of GPU in seismic field is shown for example in [4] where authors visualised large seismic data with CPU and GPU cooperation. Another example is [5] in which a cluster of workstations with high performance GPU is implemented and with advantage of the idleness of these computers and use of specific algorithms authors was able to significantly increase seismic data processing capacity. However our significant motivation were results of Karas and Svoboda in works [6,7], where authors used CUDA for 3D convolution on large seismic data. Karas and Svoboda suggest the solution based on dividing the problem into subproblems. It can be observed that mostly tested and implemented method for computation of convolution is dividing signal into

smaller parts. In this paper we propose the solution based on completely opposite idea. It uses the capabilities of cuFFT library and data profile to compute
convolution not on single signal or divided signal but gathers many signals and
perform convolution on batch of them. This paper is organized as follows: correlation details are described in Sect. 2, we present the idea behind it, mathematical details and properties useful for discussed problem. Section 3 is devoted the
implementation of solution based on CPU with parallel programming. Proposed
approach which uses GPU is discussed in Sect. 4. The last section presents the
conclusions, summarize the main contributions and discuss future work.

Contribution. The main contribution of presented paper is massive convolution
of large package of seisimic signals on GPU, what is much faster than convolution of such package on processor from HEDT (High-End Desktop Computers)
segment.

2   Correlation

For f, g: $\mathbb{R} \to \mathbb{R}$ we define the convolution as the function $y = f \star g$:

$$y(t) = \int_{-\infty}^{\infty} f(t - \tau) * g(\tau) d\tau$$

From the application standpoint the convolution is a function amplified in frequencies coexisting in both functions and attenuated otherwise. Figure 1 presents
the idea. In practical applications correlation of a signal with multiple other signals is called *correlation with this signal*. Correlation has applications in statistics, differential equations, digital image processing and also in signal processing,
where it is common to say that one function is a filter acting on the other function. In seismology it is used to bring out the features of one function given by
another function.

Computation of convolution based on definition is an ineffective method, as
naive algorithm has computational complexity of $O(n^2)$. Nonetheless, *convolution* theorem which states that convolution in time domain is a multiplication in
the frequency domain, allows to compute the convolution effectively. Pursuant
to this theorem, we can replace convolution in time domain with this formula in
frequency domain:

$$Y(\omega) = F(\omega) * G(\omega)$$

where $F(\omega)$ and $G(\omega)$ are Fourier Transforms of $f(t)$ oraz $g(t)$, respectively.

Seismic trace is a discrete signal which is why we compute the Fourier transform from the formula for discrete signal:

$$A_k = \sum_{n=0}^{N-1} a_n w_N^{-kn}$$

for $0 \leq k \leq N - 1$, where $w_N = e^{i\frac{2\pi}{N}}$ is a N degree root of 1.

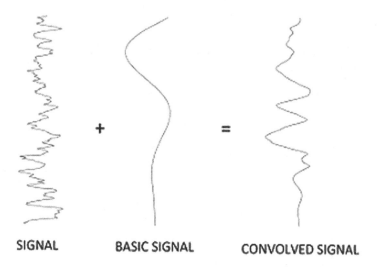

SIGNAL   **BASIC SIGNAL**   **CONVOLVED SIGNAL**

Fig. 1. Convolution example

In practical application we perform correlation on batch of signals for specific shot, which for considered application means around 1500 signals (traces) correlated with basic signal called pilot. The natural optimization is computing the Fourier transform for pilot firstly. The next stage is processing of seismic traces. It consists of following stages: changing the signal from time to frequency domain with Fourier transform, multiplication of trace and pilot in frequency domain and changing the result signal from frequency to time domain using inverse Fourier transform. Performing the Fourier transform is based on FFT algorithm [8]. The algorithm has computational complexity of $O(n \log(n))$, the multiplication of traces has complexity of$O(n)$, so using the convolution theorem we get the algorithm with computational complexity of $O(n \log(n))$.

3 Solution Based on CPU

As it was emphasized in the previous section, the main computational weight of correlation of batch of signals is changing the signal from time to frequency domain and back to time. These operations are performed based on Fast Fourier Transform algorithm. Signal is a function with discrete domain that why the discrete Fourier transform was used. There are different methods and approaches for DFT speed up, however for CPU convolution implementation case algorithm Colley-Tukey Radix-2 has been used. It is a "divide and conquer" algorithm: it recursively breaks down DFT of size n into two smaller parts of size $\frac{n}{2}$. The parts are interleaved - the first part consists of even-indexed elements, and the other of odd-indexed elements of base series. The Radix-2 algorithm first computes the DFTs for smaller parts and then combines the results into one DFT of the whole series. This algorithm can be performed recursively to reduce the time complexity to $O(n \log(n))$.

Function fft(A, N):

> if $N = 1$ then
> | return A
> end
>
> else
> > for $i \leftarrow 0$ to N do
> > | $Aodd[i] = A[i \cdot 2]$
> > | $Aeven[i] = A[i \cdot 2 + 1]$
> > end
> >
> > $Vodd \leftarrow fft(Aodd, N/2)$
> > $Veven \leftarrow fft(Aeven, N/2)$
> >
> > for $i \leftarrow 0$ to $N/2$ do
> > | $A[i] = Veven[i] + exp(-2 \cdot \pi \cdot k/N) \cdot Vodd[i]$
> > | $A[N/2 + i] = Veven[i] - exp(-2 \cdot \pi \cdot k/N) \cdot Vodd[i]$
> > end
> end
>
> return A

Algorithm 1. The Cooley-Tukey Radix-2 algorithm

In the software, the FFT Colley-Tukey Radix-2 algorithm is performed on the pair of signals (seismic traces) which accelerate the computation even more. Usually the Fourier transform algorithm input data is purely real and then it is changed to complex series with imaginary part which is equal to zero. In the system, the inputs of the algorithm are two signals: first is the real part and the second is imaginary part of complex series. Two traces are then treated as one signal on which we perform the convolution process.

In the case where processing correlation is performed on CPU first, the list of pairs of signals of specific batch is constructed. Then, with use of parallel stream mechanism, the correlation of pairs is performed based on above algorithms. This approach allows to effectively exploit logical processors (threads) of the CPU. Each thread is assigned its own memory for correlation computation on pairs, therefore problem of memory sharing is non-existent.

4   Solution Accelerate by GPU

Within CUDA platform, there is the cuFFT library for discrete Fourier transform computation which implements the FFT algorithm. It allows to effectively perform Fourier transform without the need of developing custom FFT CUDA implementation. The library provides optimized functions that perform

Function correlate(*traceOne, traceTwo, fftPilot, size*):

$commonTrace \leftarrow tracesToComplex(traceOne, traceTwo)$
$commonFFT \leftarrow FFT(commonTrace)$
$freqDomain \leftarrow ComplexArray[2][size]$

for $i \leftarrow 0$ **to** $size$ **do**
  $\quad freqDomain[0][i] =$
  $\quad fftPilot[i].conjugate().multiply(commonResult[0][i])$
  $\quad freqDomain[0][i] =$
  $\quad fftPilot[i].conjugate().multiply(commonResult[1][i])$
end

$timeDomain = IFFT(freqDomain[0], freqDomain[1]);$

return timeDomain

Algorithm 2. Corellation on pair of signals

the Fourier transform. One of the optimization features of the cuFFT is storing only nonredundant Fourier coefficients and, as the result for input data of size n, the $\frac{n}{2+1}$ complex coefficients are received. The library implements functions both for real-to-complex and complex-to-complex Fourier transform. In the software, we use only the real-to-complex transform as traces are purely real data, represented as arrays of floats. We omit the stage of conversion real to complex data. Because the heaviest part of the GPU solution is performed by CUDA functions it is difficult to estimate the complexity of this approach. In the first approach, correlation was implemented to perform on a single trace. Seismic trace through cuFFT was transformed to frequency domain, the convolution with pilot was performed and then, also using cuFFT, the result was transformed back to time domain. All correlation stages were performed using CUDA. It did not reveal any acceleration, so the approach was abandoned and the search for the better method began. While researching for different approach, the cuFFT functions that compute the parallel transform on batches were tested. In the system the batched execution means that we gather all traces for specific shot and perform correlation on all of them as a batch. Traces are then spread between GPU threads where the correlation computation is made. With that approach we get significant acceleration of correlation computing. The first stage of FFT on CUDA is creating a plan configuration for Fourier transform. cuFFT is provided with cufftPlanMany() function, where, among others, the size of a batch, that represents the number of traces on which the Fourier transform will be performed, and the size of a single trace is declared. After the plan is made the FFT algorithm can be performed. For real-to-complex FFT we use the cufftExecR2C() function. Then the result trace from the function is convolved with pilot on which the FFT was performed earlier, at the beginning of the processing. The last step is transforming the convolution result to the time domain using the inverse Fourier transform. For that cuFFT have function cufftExecC2R(), which needs its own configuration plan. The input of the system functions that perform FFT on many traces includes: the float array that is combined from all traces arrays, the number of traces (the size of batch) and the length of single trace.

Function `correlateMany`($traceList, batches, traceSize, fftPilotMany$)**:**

$size \leftarrow batches \cdot traceSize$
$traces \leftarrow newArray[size]$
for $trace \leftarrow 0$ **to** $batches$ **do**
$\quad$|$\quad$traces.add(traceList[trace])
end

$planFFT \leftarrow cuFFTplan(batches, traceSize, cufftExecC2R)$
$fftTraces \leftarrow executeManyFFT(planFFT, traces, batches, traseSize)$

$freqDomain \rightarrow newArray[size]$
for $i \leftarrow 0$ **to** $size$ **do**
$\quad$|$\quad$$freqDomain[i] \leftarrow fftPilot[i].conjugate().multiply(fftTraces[i])$
end

$planIFFT \leftarrow cuFFTplan(batches, traceSize, cufftExecR2C)$
$ifftTraces \leftarrow$
$executeManyIFFT(planIFFT, freqDomain, batches, traseSize)$

$timeDomain \leftarrow newArray[batches][traceSize]$
for $i \leftarrow 0$ **to** $batches$ **do**
$\quad$|$\quad$**for** $j \leftarrow 0$ **to** $traceSize$ **do**
$\quad$|$\quad$|$\quad$$timeDomain[i][j] \leftarrow ifftTaces[i]$
$\quad$|$\quad$**end**
end
return timeDomain

Algorithm 3. Correlation on batch of signals

In that approach significant acceleration was achieved, in comparison with the single trace GPU correlation, because of the device/host transfer of data on CUDA which is the most time-consuming part of the process. The first approach was slow because we were transferring small data from device to host, and host to device repeatedly. Now, we copy the big batch of data from device to host, perform the correlation and then transfer the results back from host to device.

In future work we would like to correlate traces gathered from 40 thousand of geophones. The main challenge will be allocating memory on host for that processing. For now, the graphic card with 8 GB frame buffer is more than enough, but in the future we might need more sophisticated approach that deals with limited GPU memory.

5   Performance Tests

Processing was tested on real seismic data which was gathered during 6 day-long seismic survey. The data was saved in 3 GB of files collected from around 4500 geophones, having the information about almost 3300 shots represented by around 5 million seismic signals.

The experiments where performed on different sets of data:

all data from 6 days All the packages of data.
one day data Shots from the first day of acquisition, around 550 packages.
three days data Random choosed 627 packages from three days.

Approximately, the single package consisted of around 1500 seismic traces, with each trace having around 14000 samples. The basic signal (pilot) had 9000 samples. Experiments where performed on machine with:

- Intel Core i9-7900X, 3.3 GHz, with 10 cores (20 threads)
- 64 GB RAM DDR4 2666 MHz
- Motherboard ASUS X299 WS PRO
- System Disk PCIe Samsung 960 pro 256 GB
- Storage Disk WD Gold 12 TB

and two graphic cards:

- NVIDIA GeForce GTX 1050 Ti, 4 GB, with 768 cores.
- NVIDIA GeForce GTX 1070 Ti, 8 GB, with 2432 cores (Table 1).

Table 1. Test results

Time	CPU time (ms)	GPU time (ms)	
		GeForce GTX 1050 Ti	GeForce GTX 1070 Ti
All data from 6 days	3038372	613530	605308
One day data	486483	95961	95194
Three days data	555280	117869	115023

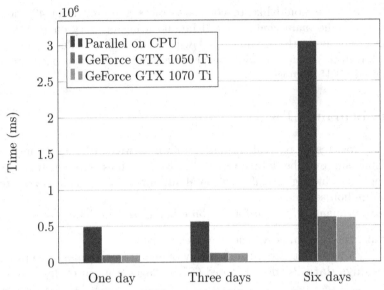

The results received from GPU are around five time faster then from CPU. On the other hand results of both GPUs are similar, but GTX 1070 Ti has more than 3 time more CUDA processors and 2 times more memory. This suggest two possibilities: First the problem could be in PCIe the same for both GPU and the most time is spent for transform data between main memory and GPU memory. The second possibility is that the data package size is still relatively small even for GTX 1050 Ti GPU.

6   Conclusions and Future Works

As we have shown it this paper, the use of CUDA threads substantially accelerate the preprocessing of seismic data as far as the convolution with basic signal is considered. Getting similar results on different graphic cards suggests that either communication with GPU is the bottleneck or, more likely, the GPU is not used to its full capabilities. There are three main goals for the further works. The first is the analysis of the acceleration quality when batching 40 000 signals and when signals in one batch have different sizes. Second is the implementation of the approach proposed by Karas in [7], where during the correlate computation of one batch the other batch will be transferred to GPU. This idea also will be extended to using multiple graphic cards. The third one is to optimize communication between main memory and GPU memory using lightweight compression proposed by Kaczmarski and Przymus in [9,10]. There are also other methods to be considered in order to optimize communication like Unified Virtual Addressing (UVA) or streams.

References

1. Piórkowski, A., Pieta, A., Kowal, A., Danek, T.: The performance of geothermal field modeling in distributed component environment. In: Sobh, T., Elleithy, K. (eds.) Innovations in Computing Sciences and Software Engineering, pp. 279–283. Springer, Dordrecht (2010). https://doi.org/10.1007/978-90-481-9112-3_47
2. Kowal, A., Piórkowski, A., Danek, T., Pieta, A.: Analysis of selected component technologies efficiency for parallel and distributed seismic wave field modeling. In: Sobh, T. (ed.) Innovations and Advances in Computer Sciences and Engineering, pp. 359–362. Springer, Dordrecht (2010). https://doi.org/10.1007/978-90-481-3658-2_62
3. Sacchi, M.D.: Statistical and transform methods in geophysical signal processing. University of Alberta, Edmonton, Canada (2012)
4. Xie, K., Wu, P., Yang, S.: GPU and CPU cooperation parallel visualisation for large seismic data. Electron. Lett. **46**, 1196–1197 (2010)
5. Souza, P., et al.: A cluster of workstations for seismic data processing using GPU. In: EAGE Workshop on High Performance Computing for Upstream (2014)
6. Pavel, K., David, S.: Algorithms for efficient computation of convolution. In: Ruiz, G., Michell, J.A., (eds.) Design and Architectures for Digital Signal Processing. IntechOpen, Rijeka (2013)
7. Karas, P., Svoboda, D., Zemčík, P.: GPU optimization of convolution for large 3-D real images. In: Blanc-Talon, J., Philips, W., Popescu, D., Scheunders, P., Zemčík, P. (eds.) ACIVS 2012. LNCS, vol. 7517, pp. 59–71. Springer, Heidelberg (2012). https://doi.org/10.1007/978-3-642-33140-4_6
8. Cooley, J.W., Lewis, P., Welch, P.: The Fast Fourier Transform algorithm and its applications. IBM Watson Research Center (1967)
9. Kaczmarski, K., Przymus, P.: Fixed length lightweight compression for GPU revised. J. Parallel Distrib. Comput. **107**, 19–36 (2017)
10. Przymus, P., Kaczmarski, K.: Compression planner for time series database with GPU support. Trans. Large Scale Data Knowl. Cent. Syst. **15**, 36–63 (2014)

Detection of Dangers in Human Health with IoT Devices in the Cloud and on the Edge

Mateusz Gołosz[ID] and Dariusz Mrozek[(✉)][ID]

Institute of Informatics, Silesian University of Technology,
ul. Akademicka 16, 44-100 Gliwice, Poland
{mateusz.golosz,dariusz.mrozek}@polsl.pl

Abstract. Smart bands are wearable devices that are frequently used in monitoring people's activity, fitness, and health state. They can be also used in early detection of possibly dangerous health-related problems. The increasing number of wearable devices frequently transmitting data to scalable monitoring centers located in the Cloud may raise the Big Data challenge and cause network congestion. In this paper, we focus on the storage space consumed while monitoring people with smart IoT devices and performing classification of their health state and detecting possibly dangerous situations with the use of machine learning models in the Cloud and on the Edge. We also test two different repositories for storing sensor data in the Cloud monitoring center – a relational Azure SQL Database and the Cosmos DB document store.

Keywords: Internet of Things · IoT · Data exploration ·
Cloud computing · Edge computing

1 Introduction

Along with the dynamic technological progress in recent years, the possibilities of using computer and electronic devices to revolutionize many spheres of human life grow. Internet of Things (IoT) is one of the most important technologies of today that have great potential to change the image of the world. As a network of connected, identifiable electronic devices that can communicate and exchange data, the IoT can be used in many areas of our life – starting from applications in intelligent construction, through supporting advanced processes in industry and manufacturing, to monitoring and informing us about many aspects of everyday activities and health state. Smart bands are wearable electronic IoT devices that allow monitoring the activities of people and some of their physiological parameters. For example, they can measure the number of steps taken by a monitored person while walking or jogging, and deliver information on the pulse, burned calories, and quality of sleep. Some of them are even able to monitor electrocardiography (ECG) signal. Smart bands are wirelessly connected to other units,

© Springer Nature Switzerland AG 2019
S. Kozielski et al. (Eds.): BDAS 2019, CCIS 1018, pp. 40–53, 2019.
https://doi.org/10.1007/978-3-030-19093-4_4

such as smartphones, tablets, laptops, personal computers or other dedicated devices that allow long-term storage of data or act as IoT gateways. The measurements obtained by the sensors located in the smart band are transmitted to a nearby IoT gateway with the use of a suitable, usually short-range and wireless communication protocol, like Bluetooth [3], ZigBee [3,7], ANT [3,8,10], Near Field Communications (NFC) [2,9], or WiFi. For smart bands, the most interesting is the Bluetooth Low Energy [12] (BLE) protocol thanks to its energy efficiency and simple implementation.

Smart bands can be used not only for personal monitoring as fitness trackers but also as important health indicators in remote healthcare monitoring of older people or people after some serious health-related incidents, like a heart attack or stroke. In such solutions data are not only collected on the nearby mobile device but usually transmitted to the monitoring center where they are continuously analyzed by the implemented software modules. In case of detection of a risk situation, the appropriate caregiver is notified and may take suitable action. Due to large scaling capabilities monitoring centers are eagerly located in Cloud platforms. Clouds provide almost unlimited resources for storing data and performing computations on the transmitted data. However, as the number of IoT devices available at the market and able to transmit the data to the Cloud grows rapidly, data processing (including filtering, aggregation, and combining) and data analysis (including the use of machine learning-based exploration models) can be also performed on the Edge. Edge computing is an alternative to the centralized data processing and analysis in the Cloud. It prevents network congestion and storage space overload, and in some situations, may eliminate unnecessary latency.

In this paper, we investigate the impact of the Cloud-based centralized and Edge-based distributed data analysis on the consumed Cloud storage space. We analyze two solutions that engage trained machine learning (ML) models for detection of anomalies in health on the basis of activity and physiological parameters of the monitored person – one that performs the detection in the Cloud, and one that performs it on the smartphone working as the IoT gateway.

2   Related Works

There are several related works that are devoted to sending and processing data gathered from IoT devices. Authors of [5] have given a general proposition of an architecture for a system that would exchange data between wearable devices and computing Cloud. However, their work has been focused mostly on the concept of actively supporting health services, diagnosis of disease in particular. Moreover, no real data gathered from the implementation of such a system has been presented. In [6], authors have proposed a solution to a problem that occurs in a different area - lack of coherency in both input and output interfaces. The implemented framework standardizes data regardless of its size, source device, format, and structure. Zhu et al. in [14] propose a model of a gateway for a sensor network, but it does not provide any details on how the given data is being processed. Instead, it presents a very general hardware implementation and a general overview of network packet construction, server architecture and overall flow of the transferred data without going further into processing the data once it has been sent. Yang et al. [11] proposed a wearable ECG monitoring system that utilizes the Cloud platform. The work covers the hardware implementation and data transportation model and investigates the risk of heart disease. In [4] Doukas and Maglogiannis show the usage of the IoT and cloud computing in pervasive healthcare, but instead of an ECG examination, they propose quite a unique implementation of its own wearable sensor system. The system is integrated into a sock and consists of multiple sensors measuring values such as heartbeat, motion, and temperature. However, none of the above works go into details when it comes to storing and processing gathered data. Chen et al. [1] also describe the process of transferring data from wearable devices to computing clouds, but with consideration for an improvement of the wearable devices themselves. The main emphasis has been put onto integrating multiple sensors which are available as separate modules into versatile smart clothing that would constantly monitor various health indicator as well as environmental parameters, such as air pollution. On the other hand, except introducing an architecture of a model being able to transfer data from IoT devices to the Cloud, Zhou et al. [13] focuses on an emerging problem with the privacy of data collected by such devices. They describe an efficient way of encrypting and anonymizing data in the process.

None of the works listed above concentrates on the amounts of data produced by wearable devices and on ways of reducing them to a minimum. One of the ways includes changing the point where most of the data are being processed, moving the processing from the cloud itself to another (Edge) device which takes a part in the earlier stage on the data flow. In the next section, we present and compare the Edge and Cloud-based standard architectures for data processing and analysis.

3 Classifying Data in the Cloud and on the Edge

For the purpose of the presented research, we created the health and activity monitoring system with the general architecture presented in Fig. 1. The architecture consists of:

- a wearable device with sensors measuring various parameters,
- a smartphone with the Android operating system,
- a data center located in the Cloud.

We decided to use the Xiaomi Mi Band 2 smart band as the wearable device. The smart band measures the following parameters:

- the number of steps made,
- pulse,
- the quality of sleep,

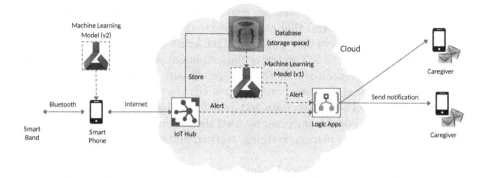

Fig. 1. General architecture of the health and activity monitoring system with the Machine Learning model implemented in the Cloud (v1) and on the Edge device (v2).

- the activity currently performed, identified on the basis of the steps taken,
- the time of measurement.

The Xiaomi Mi Band 2 device was selected based on the availability, popularity, economic issues and possibility to access raw data from the sensors. Most smart band manufacturers do not provide the possibility to extract raw data of sensor measurements, or such a feature is limited to one extraction per 24 h, which is too long period considering the requirement to check the current status of the monitored person. At the time of performed implementation of the monitoring system, there was no open-source wearable fitness tracker available on the market that would provide application programming interfaces to extract raw data. We extracted the sensor measurements from the Xiaomi Mi Band 2 in a reverse engineering process because there was no officially supported method of gathering raw data from the smart band. One of the best and most developed software for extracting such data is the open-source Gadgetbridge application for Android-based smartphones. The Gadgetbridge application is available on the GitHub platform, but its implementation is limited exclusively to Android OS. Extracting raw data from wearable devices on different operating systems, i.e., iOS would require reimplementation of software to extract data. Such efforts could eventually result in a failure due to more strict platform limitation in comparison to the Android, but further research is required in this field. Apart from the Xiaomi Mi Band 2, the Gadgetbridge supports several different devices, but does not provide unified interface for data extraction, which remains inconsistent among smart bands manufacturers. Supposedly, presented solution would work with every smart band, given the data gathered from other sensors was normalised, however, we haven't tested them in our solution. The Gadgetbridge application was installed on the smartphone, which served as the IoT gateway mediating data transfers to the monitoring data center.

The monitoring center in the developed system was established in the Microsoft Azure computing cloud. The Cloud was selected due to its high data security standards, global access to data with guaranteed bandwidth, capabilities to quickly build and add new resources, a wide range of tools, programming languages and different platforms that can be used to develop the IoT solutions, and relatively easy and intuitive user interface. The Azure cloud was used to gather data transmitted from the IoT gateway (smartphone) and to store the data in the database storage repository. As the storage repository, we tested Azure SQL Database – a relational database, and Cosmos DB – a document store. We also trained the ML model to classify data and detect possible dangers on the basis of raw sensor readings. For this purpose, we used the Machine Learning Studio – an Azure module that lets creation, training, testing, and manipulation of machine learning models.

The main goal of the developed solution was to determine whether a user of the wearable device (a monitored person) happened to be in a life-threatening situation. The process involves a binary classification, where one of the output defines that the user is safe, and the other indicates that there might be something wrong with the user's wellbeing. Multiple machine learning algorithms, like logistic regression, decision trees, support-vector machine were used, but since all of them produce the same binary output and all of the trained models use the same input data for classification, changing the ML algorithm did not affect the taken storage space in any way. Therefore, this work will not focus on describing certain algorithms and models used.

There are, however, two different approaches to locate the trained classification model and places where the dangers can be detected. They both influence the network traffic and the number of data sent to the Cloud. The first approach (Fig. 2) assumes that all data processing is done in the monitoring center, thus, all the data used for training the ML model are sent directly to the Cloud. The second approach (Fig. 3) limits the amount of data that needs to be sent by performing classification of the raw sensor readings on the Edge device before they are sent to the Cloud.

4   Impact on the Storage Space

In the architecture presented in Fig. 1 the classification of the health state of the monitored person is performed in the Cloud or on the Edge device (depending on the adopted approach, Figs. 2 and 3). Sensor readings constitute an input data set for the used classifier. The data set consists of the following variables:

Timestamp – a 10-digits integer value holding a time when data from the sensors were gathered

DeviceId – a unique identifier given for every unique device sending data to the Cloud; it holds subsequent integer values, starting from 1,

UserId – a unique identifier given for every unique user (monitored person), holds values from the same range as the DeviceId,

Raw intensity – a value, which describes the intensity of the action performed, it holds integers in a range from 0 to 99,

Steps – an integer value that holds the number of steps made by the monitored person,

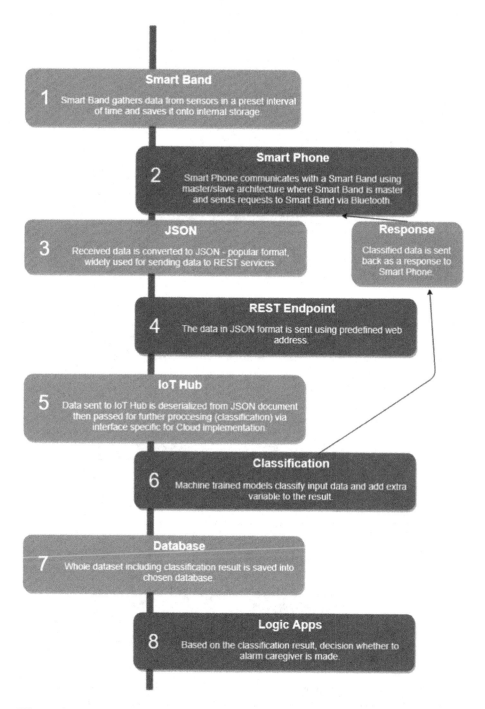

Fig. 2. Communication and flow of data in the system architecture with centralized data processing where all the data is transferred to the Cloud.

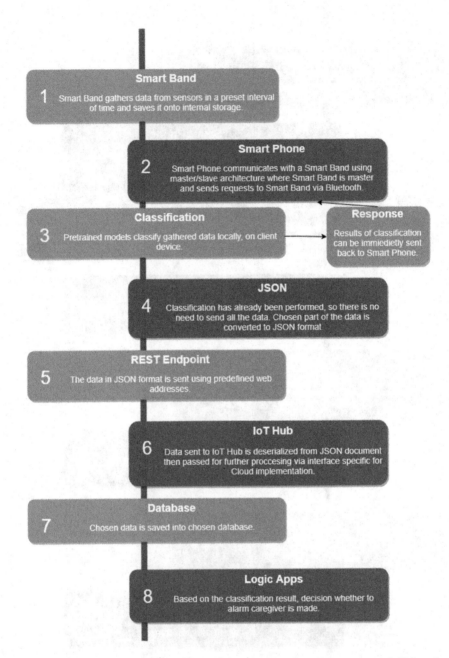

Fig. 3. Communication and flow of data in the system architecture with Edge data processing where classification occurs on the client's device (a smartphone).

Raw kind – an integer value holding values from a range of 0 and 99, which represents the activity conducted by the person; each value represents a unique action,

Heart rate – an integer value holding value of the measured heart rate.

The classifier labels the data by providing an additional attribute, called *healthy*, which is a binary value which holds either 0 (not healthy) or 1 (healthy).

4.1   Number of Transactions

The consumption of Cloud resources depends on several factors, including the time interval between successive sensor readings. Willing to provide real-time monitoring of a user, incoming readings are processed at once and depending on the architecture variant, whole or part of the data are immediately sent to the Cloud to be saved. For the purpose of our research, we assumed a time interval of 1 min, which defines how often data from the sensors is gathered and processed. This is a default, assumed value for every further analysis presented in this work. Assuming that every user is going to use a wearable device in a consistent manner, there will be 1,440 or less data transmission transactions performed per every 24 h. Pricing plans for the Azure Machine Learning Studio, a resource used for classifying data in the first, Cloud-based approach, have limits on both computing time and the number of transactions, whichever runs out first. In the approach which classifies data centrally in the Cloud, this might generate additional costs and limitations, presented in Table 1.

Table 1. Azure Machine Learning Studio - plan comparison

Plan	Max. users per day	Transaction limit	Cost (EUR)
Standard S1	69	100 000	84.44
Standard S2	1 388	2 000 000	843.36
Standard S3	34 722	50 000 000	8 432.99

4.2   Relational Database

The Azure cloud offers multiple options for storing data. For the purpose of the developed system, we tested two popular ways of storing data. The first one was the Azure SQL Database, which is a relational SQL database. The second one was Cosmos DB, which is a NoSQL document store that holds data in the JSON format. The relational database has been selected due to its high availability among all cloud solutions to data storage, thus given results could be potentially extended to other platforms. On the other hand, data gathered in the proposed solution is sent in chunks uniquely characterized only by timestamp and user, easily convertible to JSON format supported by other, efficient database - CosmosDB, that turned out to be the best NoSQL choice among available storage options on Azure Cloud. Performance of the Azure SQL relational database depends on the compute, storage, and IO resources used, referred to as *compute sizes*. Compute sizes are expressed in terms of Database Transaction Units (DTUs) for single databases, which constitute a currency in the DTU-based purchase model. Three different plans (also called as *tiers*) for relational databases presented in Table 2 define their capabilities, limitations, and costs.

Table 2. Single database DTU and storage limits

Plan	DTUs	Max. available storage (GB)	Min. cost per month (EUR)
Basic	up to 5	2	4.21
Standard	10–3 000	250	12.65
Premium	125–4 000	1 000	392.13

With the use of Azure monitoring tools, we noted the size of all data saved into the database during the period of one hour (covering 60 data transmission transactions between the IoT device and the Cloud), which was 30 kB. The minimum number of DTUs needed to perform the operation is equal to 0.02. In Fig. 4 we can observe how fast the size of data produced per hour increases with the growing number of active devices and monitored persons.

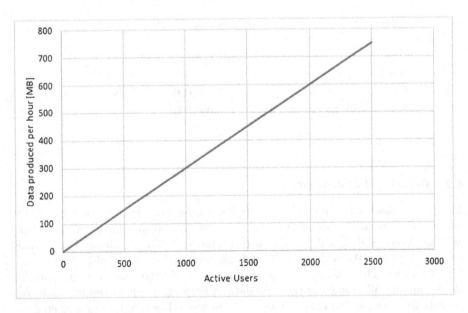

Fig. 4. The size of data produced per hour for the growing number of active devices and monitored persons

With the increasing amount of data produced per hour by registered IoT devices, the minimal number of DTUs increases as well, because more data must be saved during the same amount of time. The dependency between DTU needed for a database to operate correctly and the number of active users is presented in Fig. 5.

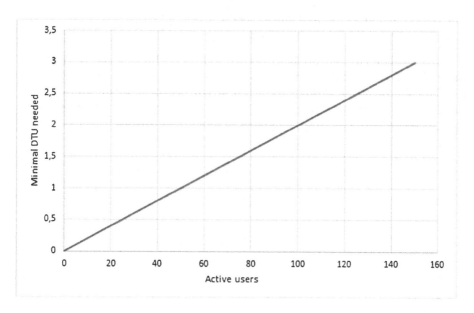

Fig. 5. Dependency between the minimum number of DTUs needed and the number of active IoT devices and monitored persons

4.3   Document Storage

The second database that we tested to store data was Cosmos DB that, instead of using relational tables, stores the data as JSON files. With the use of the Azure cloud monitoring tools, the data produced by a single user IoT device per one hour (comprising 60 transactions) and transferred as JSON files take 13 kB, which is less than half of the size of the storage space taken by relational data (30 kB). This should influence the costs of the storage. However, the comparison is not so straightforward. The Cosmos DB does not provide DTU-based model. Instead, it adapts to a number of transactions made, elastically increasing the price. Taking into account that both the Cosmos DB and the relational database are able to perform the same number of transactions, and that saving data as JSON documents takes 43% less space to store the same amount of data, we can compare pricing of both adopted storage approaches. The cost of storing data in the relational database depends on both the price of minimal number of DTUs needed for a database to operate and the storage space used. The pricing of the

Cosmos DB is different and is calculated (in EUR) with the use of the following formula:

$$Cost = g * 0.211 * req * h * 0.007 \qquad (1)$$

where:

g – storage space taken in GB,
req – the number of requests made per second,
h – the number of database active hours.

Figure 6 allows to compare both storage approaches by showing the minimum cost for both database solutions per one hour of database constant work. As can be seen the cost of the Cosmos DB when serving as the data storage space for the monitoring system for many IoT devices is much lower and the difference in costs increases with the growing number of monitored persons.

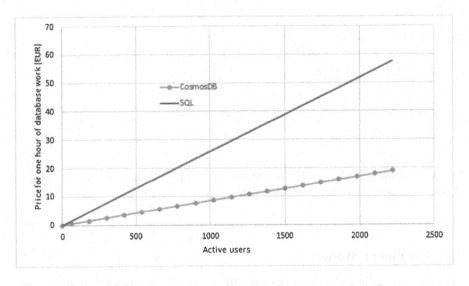

Fig. 6. Price for one hour of database work in relation to number of active users

4.4   Event-Based Cloud Connectivity and Data Transmission

In Sect. 4.1 we stated that the number of messages sent to the Cloud is constant in a given time interval. By implementing a mechanism that would reduce the number of messages sent to the Cloud would also decrease the overall minimum requirements for database efficiency, therefore reducing the costs. We have implemented two additional approaches in order to further reduce the amount of data sent to the Cloud. In the first approach, we assumed that we send the data from the smart band through the IoT device only when the activity performed by a monitored person changed since last measurement. The approach comes from an idea that state of the person whose life is endangered would be very likely to change, e.g., from standing to laying, from running to standing. The possible storage space savings would highly depend on the individual activity of the person during the day, but they would be seen at all times during the night, when most of monitored persons are most likely to perform the same action – sleeping. In this approach, the classification of the health state is still performed in the Cloud, but we reduce the number of data transmitted to the data center. The second approach assumes performing the classification of the health state on the IoT device (a smartphone) and determining if the user's life might be threatened. If the classification gives positive results (health state endangered), the data is sent to the Cloud. Figure 7 compares all approaches – (1) when all data are transmitted to the Cloud (with the classification in the Cloud), (2) when we transmit data only when the activity of the user changes (with the classification in the Cloud), and (3) when data are transmitted only when we detect possible dangerous situation (with the classification on the Edge IoT device working as the field gateway). As can be seen, the last approach allows reducing the amount of data transmitted to the Cloud the most effectively. However, data transmission when changing performed activity is also quite effective.

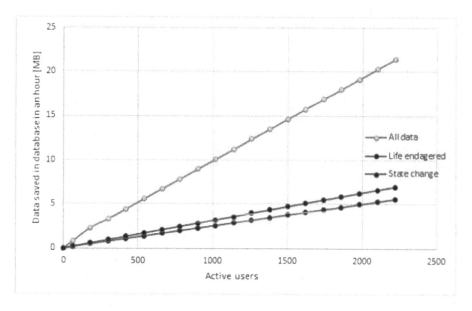

Fig. 7. Comparison between storage space consumed for three tested approaches – when all data are transmitted to the Cloud with detection of danger in the Cloud (All data), when data are transmitted only when the activity of the user changes with detection of danger in the Cloud (State change), and when data are transmitted only when we detect possible dangerous situation with detection of danger in the Edge IoT device (Life endangered).

5   Conclusions

The increasing popularity of IoT poses new challenges related to the amount of data that needs to be processed and stored. This work proposes a solution to the problem of choosing the most efficient and economical way of storing large amounts of data, by using a wearable smart band device and the Azure computing cloud. The stated problem is very common when it comes to IoT – a large number of devices produce relatively small amounts of data, but send the data over the network in short time intervals, which may lead to network congestion and raises the challenges of Big Data in monitoring centers. Considering the dynamic growth of the number of existing IoT devices and active users of smart bands, limiting the frequency of data transfers and using the computing power of client Edge devices seem to be an efficient way to decrease the storage space needed to perform tasks, like detection of health dangers through classification of data with machine learning models. Based on the Azure cloud pricing, it can be stated that most of the costs of the database storage space consumed in the scenario where a few thousands of IoT devices are connected to the Cloud at the same time would come not from storing the data, but more than 90% of the costs would flow from providing a sufficient number of concurrent transactions, which is multiple times more expensive than storage space itself.

The results of our experiments show that selection or classification of data on the Edge is an effective way of reducing the amount of data to be stored. As long as processing the data takes short enough for an end-user to comfortably use the product, the most efficient way would be processing all the data on the client's IoT device, periodically updating models used for classification.

Acknowledgments. This work was supported by Microsoft Research within Microsoft Azure for Research Award grant, pro-quality grant for highly scored publications or issued patents of the Rector of the Silesian University of Technology, Gliwice, Poland (grant No 02/020/RGJ19/0167), and partially, by Statutory Research funds of Institute of Informatics, Silesian University of Technology, Gliwice, Poland (grant No BK/204/RAU2/2019).

References

1. Chen, M., Ma, Y., Li, Y., Wu, D., Zhang, Y., Youn, C.H.: Wearable 2.0: enabling human-cloud integration in next generation healthcare systems. IEEE Commun. Mag. **55**(1), 54–61 (2017). https://doi.org/10.1109/mcom.2017.1600410cm
2. Coskun, V., Ozdenizci, B., Ok, K.: A survey on near field communication (NFC) technology. Wirel. Pers. Commun. **71**(3), 2259–2294 (2013). https://doi.org/10.1007/s11277-012-0935-5
3. Dementyev, A., Hodges, S., Taylor, S., Smith, J.: Power consumption analysis of Bluetooth Low Energy, ZigBee and ANT sensor nodes in a cyclic sleep scenario. In: 2013 IEEE International Wireless Symposium (IWS), pp. 1–4, April 2013. https://doi.org/10.1109/IEEE-IWS.2013.6616827

4. Doukas, C., Maglogiannis, I.: Bringing IoT and cloud computing towards perva- sive healthcare. In: 2012 Sixth International Conference on Innovative Mobile and Internet Services in Ubiquitous Computing. IEEE, July 2012. https://doi.org/10. 1109/imis.2012.26

5. Hassanalieragh, M., et al.: Health monitoring and management using Internet-of- Things (IoT) sensing with cloud-based processing: opportunities and challenges, June 2015. https://doi.org/10.1109/scc.2015.47

6. Jiang, L., Xu, L.D., Cai, H., Jiang, Z., Bu, F., Xu, B.: An IoT-oriented data storage framework in cloud computing platform. IEEE Trans. Ind. Inform. **10**(2), 1443– 1451 (2014). https://doi.org/10.1109/tii.2014.2306384

7. Malhi, K., Mukhopadhyay, S.C., Schnepper, J., Haefke, M., Ewald, H.: A zigbee- based wearable physiological parameters monitoring system. IEEE Sens. J. **12**(3), 423–430 (2012). https://doi.org/10.1109/JSEN.2010.2091719

8. Mehmood, N.Q., Culmone, R.: An ANT+ protocol based health care system. In: 2015 IEEE 29th International Conference on Advanced Information Networking and Applications Workshops, pp. 193–198, March 2015. https://doi.org/10.1109/ WAINA.2015.45

9. Pang, Z., Zheng, L., Tian, J., Kao-Walter, S., Dubrova, E., Chen, Q.: Design of a terminal solution for integration of in-home health care devices and services towards the Internet-of-Things. Enterp. Inf. Syst. **9**(1), 86–116 (2015). https:// doi.org/10.1080/17517575.2013.776118

10. Valchinov, E., Antoniou, A., Rotas, K., Pallikarakis, N.: Wearable ECG system for health and sports monitoring. In: 2014 4th International Conference on Wire- less Mobile Communication and Healthcare - Transforming Healthcare Through Innovations in Mobile and Wireless Technologies (MOBIHEALTH), pp. 63–66, November 2014. https://doi.org/10.1109/MOBIHEALTH.2014.7015910

11. Yang, Z., Zhou, Q., Lei, L., Zheng, K., Xiang, W.: An IoT-cloud based wearable ECG monitoring system for smart healthcare. J. Med. Syst. **40**(12) (2016). https:// doi.org/10.1007/s10916-016-0644-9

12. Zhang, T., Lu, J., Hu, F., Hao, Q.: Bluetooth low energy for wearable sensor-based healthcare systems. In: 2014 IEEE Healthcare Innovation Conference (HIC), pp. 251–254, October 2014. https://doi.org/10.1109/HIC.2014.7038922

13. Zhou, J., Cao, Z., Dong, X., Lin, X.: Security and privacy in cloud-assisted wireless wearable communications: challenges, solutions, and future directions. IEEE Wirel. Commun. **22**(2), 136–144 (2015). https://doi.org/10.1109/mwc.2015.7096296

14. Zhu, Q., Wang, R., Chen, Q., Liu, Y., Qin, W.: IOT gateway: BridgingWireless sensor networks into Internet of Things. In: 2010 IEEE/IFIP International Con- ference on Embedded and Ubiquitous Computing. IEEE, December 2010. https:// doi.org/10.1109/euc.2010.58

Architectures, Structures and Algorithms for Efficient Data Processing and Analysis

Serialization for Property Graphs

Dominik Tomaszuk[1]([✉])(iD), Renzo Angles[2,3](iD), Łukasz Szeremeta[1](iD),
Karol Litman[1], and Diego Cisterna[3]

[1] Institute of Informatics, University of Białystok,
Ciołkowskiego 1M, 15-245 Białystok, Poland
{d.tomaszuk, l.szeremeta}@uwb.edu.pl
[2] Department of Computer Science, Universidad de Talca, Curicó, Chile
rangles@utalca.cl
[3] Center for Semantic Web Research, Santiago, Chile
dcisterna@live.com

Abstract. Graph serialization is very important for the development
of graph-oriented applications. In particular, serialization methods are
fundamental in graph data management to support database exchange,
benchmarking of systems, and data visualization. This paper presents
YARS-PG, a data format for serializing property graphs. YARS-PG was
designed to be simple, extensible and platform independent, and to support all the features provided by the current database systems based on
the property graph data model.

Keywords: Serialization · Property graph · Graph database

1 Introduction

Data serialization is the process of converting data (obtained from a source system) into a format that can be stored (in the same system) or transmitted (to a
target system), and reconstructed later. Data serialization methods are applied
in several situations [6,21,27], in particular when an ETL process is required (i.e.
when the data need to be extracted, transformed and loaded). Data serialization implies the definition of a data format with specific syntax and semantics.
XML and JSON are two popular data formats today [17,33]. In the context
of database management, data serialization is very relevant for several reasons:
it is fundamental to support the interoperability of heterogenous databases; it
allows automatic data processing; it facilitates database benchmarking as the
same data can be shared among systems; it facilitates the translation to other
data formats; it results in a simple backup method; other manipulation and
visualization tools can read the data.

In the last years, the massive generation of large amounts of graph data has
motivated the development of graph-oriented database system, most of them
designed to support property graphs (i.e. labeled directed graphs where nodes
and edges can have label-value properties) [11,24,31]. Although these systems

---

The original version of this chapter was revised: The grant number was added
into the acknowledgments section. The correction to this chapter is available at
https://doi.org/10.1007/978-3-030-19093-4_27

© Springer Nature Switzerland AG 2019, corrected publication 2021
S. Kozielski et al. (Eds.): BDAS 2019, CCIS 1018, pp. 57–69, 2019.
https://doi.org/10.1007/978-3-030-19093-4_5

are very similar, they show variations in the implementation of the features presented by the property graph data model (as we will show in this paper).

The lack of a unique property graph data model directly influences the development of other components, including query languages and serialization formats. Although there are some graph data formats available (like GraphML or DotML), there is no a standard one, and none of them is able to cover all the features presented by the property graph data model.

In this paper we introduce YARS-PG, a data format to serialize property graphs. YARS-PG was designed to satisfy functional and non functional requirements. The functional requirements are related to the intrinsic features of the property graph data model. In this sense, YARS-PG is able to serialize property graphs containing multi-labeled nodes, multi-labeled edges, directed and undirected edges, mono-value and multi-value properties, and null values.

In terms of non functional requirements, we considered expressiveness, conciseness and readability. Expressiveness implies the types of objects and relationships that a serialization is able to express (i.e. data models). In this sense, YARS-PG allows to encode all the features presented by the property graph model. Conciseness is related to the number of extra syntactic elements used by the serialization. Note that such extra elements are required to parse the data in the right way. In this sense, YARS-PG provides a simple syntax with a reduced number of extra characters. Readability concerns the facilities to encode the structure of the data. In this case, YARS-PG is inspired on the syntax used by popular graph query languages (e.g. Cypher and Gremlin) to encode the structure of a property graph (i.e. nodes, edges and properties).

This article is organized as follow. First, we present a review of current graph database systems, selecting those oriented to support property graphs (Sect. 2). Next, we present a formal definition of the property graph data model, in such a way that it is general enough to cover all the data modeling features provided by a property graph (Subsect. 3.1). Such definition was used to compare current database systems (Subsect. 3.2). Next, we propose YARS-PG as a general and flexible format to serialize property graphs (Sect. 4). We present the syntax of the format and provide a comparison with other graph-oriented serialization formats (Sect. 5).

2   Review of Current Graph Database Systems

The current market of graph databases includes over 30 systems[1], most of which are designed to store and query property graphs. Table 1 shows a representative group of systems supporting property graphs. Some systems were not included for different reasons. We discard systems abandoned or no longer available, e.g. FlockDB, and GlobalsDB. We remove systems not supporting property graphs, e.g. HyperGraphDB, Graph Engine, Sqrrl, FaunaDB, and GRAKN.AI. We also discard systems focused on RDF (including Dgraph, GraphDB, Blazegraph, and Stardog). Other proposals are not strictly database systems, e.g. Giraph that is a graph processing framework, and HGraphDB that is an abstract layer.

---

[1] https://db-engines.com/en/ranking/graph+dbms.

Table 1 shows general information about the selected graph database systems. Specifically, we annotated the system's name, license types, the programming language that the system was implemented with, supported data models, and existence of a query language. We found that the systems allow four types of licenses: GNU GPL, Apache, GNU Affero General Public License (GNU AGPL)[2] and Commercial. Most systems provide a commercial version, and some of them provide an open source version. The most preferred programming language for system implementation is Java, followed by C++, C, Scala and C#. Although the selected systems are based on a graph-based data model, we found that other abstractions are also supported (i.e. some systems are multi-model). Almost every system supports a query language. The most commonly used is Gremlin, followed by openCypher.

Table 1. General information about graph databases

System	License				Prog. language			Data model						Query lang.
	GNU GPL	Apache	GNU AGPL	Commercial	C / C++	Java / Scala	C#	Graph	Relational	Object	Document	Column	Key-value	
Neo4j	•			•		•		•						•
Datastax				•		•		•				•		•
OrientDB		•		•		•		•	•	•		•		•
ArangoDB		•		•		•		•			•	•		•
JanusGraph		•		•		•		•						•
Neptune				•		•		•						•
TigerGraph				•	•	•		•						•
InfiniteGraph				•		•		•						
InfoGrid		•				•		•						
Sparksee				•	•			•						
Memgraph				•				•						•
VelocityDB				•			•	•		•				•
AgenGraph		•	•	•				•	•					•
TinkerGraph	•					•		•						•
HGraphDB	•					•		•						•

3   The Property Graph Data Model

In the most general sense, a property graph is a directed labelled multigraph with the special characteristic that each node or edge could maintain a set of

---

[2] GNU AGPL is a free license based on the GNU GPL and it is considered for any software that will commonly be run over a network.

property-value pairs. The primary components of a property graph are nodes, edges and properties. The secondary components are labels (for nodes, edges and properties) and data types for property values.

The notion of property graph was introduced by Rodriguez and Neubauer in [26]. It is possible to find variations in the basic definition [2,7,12,30], most of them related to the support for multiple labels for nodes and edges, or the occurrence of multivalue properties. In this section we provide a general definition which allows to exploit all the features of the property graph structure. Such definition is used to analyze the features covered by current graph databases systems.

3.1   Formal Definition of a Property Graph

Assume that L is an infinite set of labels (for nodes, edges and properties), and V is an infinite set of values (atomic or complex). Given a set S, we assume that $\mathcal{P}(S)$ is the power set of S, i.e. the set of all subsets of S, including the empty set $\emptyset$ and S itself.

Definition 1. *A property graph is a tuple $G = (N, E, P, \delta, \lambda, \rho, \sigma)$ where:*

1. *N is a finite set of nodes (also called vertices), E is a finite set of edges, and P is a finite set of properties, satisfying that $N \cap E \cap P = \emptyset$;*
2. *$\delta : E \to (N \times N)$ is a total function that associates each edge in E with a pair of nodes in N (i.e., δ is the usual incidence function in graph theory);*
3. *$\lambda : (N \cup E) \to \mathcal{P}(L)$ is a total function that associates a node/edge with a set of labels from L (i.e., λ is a labeling function for nodes and edges);*
4. *$\rho : P \to (L \times V)$ is a total function that associates each property with a pair label-value;*
5. *$\sigma : (N \cup E) \to \mathcal{P}(P)$ is a total function that associates each node or edge with a set of properties, satisfying that $\sigma(o_1) \cap \sigma(o_2) = \emptyset$ for each pair of distinct objects o_1, o_2 in the domain of σ.*

According to the above definition: N, E and P have no elements in common; given an edge e such that $\delta(e) = (n_1, n_2)$, we will say that n_1 and n_2 are the "source node" and the "target node" of e respectively, i.e. the edges are directed; nodes and edges could have zero or more labels; each property has a single label and a single value (although it could be complex); nodes and edges could have zero or many properties, and each property belongs to a unique node or edge.

Figure 1 shows a graphical representation of a property graph. Following our formal definition, the example property graph will be described as follows:

$N = \{n_1, n_2, n_3, n_4, n_5, n_6\}$

$E = \{e_1, e_2, e_3, e_4, e_5, e_6\}$

$P = \{p_1, p_2, p_3, p_4, p_5, p_6, p_7, p_8, p_9, p_{10}, p_{11}, p_{12},$
$\qquad p_{13}, p_{14}, p_{15}, p_{16}, p_{17}, p_{18}, p_{19}, p_{20}, p_{21}, p_{22}\}$

$\lambda(n_1) = \{\text{Author}\}, \ \sigma(n_1) = \{p_1, p_2\}, \ \rho(p_1) = (\text{fname}, \text{"John"}),$
$\rho(p_2) = (\text{lname}, \text{"Smith"})$

$\lambda(n_2) = \{\text{Author}\}, \ \sigma(n_2) = \{p_3, p_4\}, \ \rho(p_3) = (\text{fname}, \text{"Alice"}),$

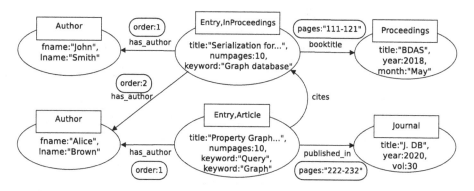

Fig. 1. Example of property graph representing bibliographic information

$\rho(p_4) = (\text{lname}, \text{``Brown''})$
$\lambda(n_3) = \{\text{Entry,InProceedings}\}, \sigma(n_3) = \{p_5, p_6, p_7\}, \rho(p_6) = (\text{numpages},10),$
$\rho(p_5) = (\text{title}, \text{``Serialization for...''}), \rho(p_7) = (\text{keyword}, \text{``Graph database''}),$
$\lambda(n_4) = \{\text{Entry, Article}\}, \sigma(n_4) = \{p_8, p_9, p_{10}, p_{11}\},$
$\rho(p_8) = (\text{title}, \text{``Property Graph...''}), \rho(p_9) = (\text{numpages},10),$
$\rho(p_{10}) = (\text{keyword}, \text{``Query''}), \rho(p_{11}) = (\text{keyword}, \text{``Graph''}),$
$\lambda(n_5) = \{\text{Proceedings}\}, \sigma(n_5) = \{p_{12}, p_{13}, p_{14}\}, \rho(p_{12}) = (\text{title}, \text{``BDAS''}),$
$\rho(p_{13}) = (\text{year},2018), \rho(p_{14}) = (\text{month}, \text{``May''})$
$\lambda(n_6) = \{\text{Journal}\}, \sigma(n_6) = \{p_{15}, p_{16}, p_{17}\}, \rho(p_{15}) = (\text{title}, \text{``J. DB''}),$
$\rho(p_{16}) = (\text{year},2020), \rho(p_{17}) = (\text{vol},30)$
$\delta(e_1) = (n_3, n_1), \lambda(e_1) = \{\text{has\_author}\}, \sigma(e_1) = \{p_{18}\}, \rho(p_{18}) = (\text{order},1)$
$\delta(e_2) = (n_3, n_2), \lambda(e_2) = \{\text{has\_author}\}, \sigma(e_2) = \{p_{19}\}, \rho(p_{19}) = (\text{order},2)$
$\delta(e_3) = (n_4, n_2), \lambda(e_3) = \{\text{has\_author}\}, \sigma(e_3) = \{p_{20}\}, \rho(p_{20}) = (\text{order},1)$
$\delta(e_4) = (n_4, n_3), \lambda(e_4) = \{\text{cites}\}$
$\delta(e_5) = (n_3, n_5), \lambda(e_5) = \{\text{booktitle}\}, \sigma(e_5) = \{p_{21}\},$
$\rho(p_{21}) = (\text{pages}, \text{``111--121''})$
$\delta(e_6) = (n_4, n_6), \lambda(e_6) = \{\text{published\_in}\}, \sigma(e_6) = \{p_{22}\},$
$\rho(p_{22}) = (\text{pages}, \text{``222--232''})$

3.2   Features of Current Property Graph Database Systems

Given the graph database systems presented in Sect. 2, we analyze their support for the features of the property graph data model presented above. Table 2 shows the results of our evaluation and are discussed below. We will use "some" to denote that a feature is covered by less than 50% of the systems, and "most" otherwise.

Node/Edge Labels. All the systems support labels for nodes and edges (zero, one or more). Some systems support nodes without labels, but unlabeled edges are not supported. Some systems support multiple labels for nodes, and just one system for allow many labels for edges.

Table 2. Property graph features supported by graph database systems.

System	Node labels			Edge labels			Edges				Properties			
	Zero	One	Many	Zero	One	Many	Directed	Undirected	Multiple	Duplicated	Mono-value	Multi-value	Null value	Duplicates
Neo4j	•		•		•		•		•	•	•			
Datastax		•			•		•		•	•	•			
OrientDB		•			•		•	•	•	•	•	•		
ArangoDB		•			•		•	•	•	•	•	•		
JanusGraph	•	•			•		•	•	•	•	•			
Amazon Neptune	•		•		•		•		•	•	•			
TigerGraph		•			•		•	•	•		•			
InfiniteGraph		•			•		•	•	•	•	•	•		
InfoGrid	•		•		•		•	•				•	•	
Sparksee		•			•		•	•	•	•	•		•	
Memgraph	•		•		•		•		•	•		•	•	
VelocityDB		•			•		•	•	•	•		•	•	
AgensGraph	•		•	•	•		•		•	•	•			
TinkerGraph		•			•		•		•	•	•			
HGraphDB		•			•		•		•	•	•		•	

Edges. All the systems support directed edges. More than a half of the systems allow undirected edges in an explicit way (recall that an undirected edge can be simulated with two directed edges, but the opposite is not possible). Practically all the systems allow multiple edges between a pair of nodes (i.e. they support multigraphs), and such edges could have the same label (i.e. the edges are independent of the labels).

Properties. A property is a pair $p = (l, v)$ where l is the property label (or property name) and v is the property value. A property have a single and unique label. Most systems support multivalue properties (e.g. emails for a person), a feature supported in two possible ways: properties with the same label, or properties with complex values (e.g. an array of strings). More than a half of the systems allow the *null* value to support the explicit description of an empty property[3]. The notion of duplicate property is not supported by current systems.

---

[3] This feature must not be confused with the *null* values allowed in the query language provided by the system.

4 YARS-PG Serialization

In this section we describe YARS-PG, a serialization for property graphs inspired in a serialization for RDF data called YARS [29] (we compare them in Subsect. 5.1). A YARS-PG serialization contains node declarations and relationship declaration (no order is required for them).

A *node declaration* begins with the object identifier (OID) of the node, followed by a list of node labels (inside squared brackets), a colon, and the properties of the node (inside braces). A *relationship declaration* contains the OID of the source node (inside parenthesis), a set of labels, a set of properties, and the OID of the target node. Relationships can be directed (->) or undirected (-). A relationship declaration is based on paths, following the syntax used in graph query languages like PGQL [25], Cypher [22] and G-CORE [3]. YARS-PG allows cyclic relationships and multiple relationships between the same pair of nodes.

A *property* is represented as a pair $p : v$, where p is the property label and v the property value. A property value could be atomic (e.g. string, integer, float, *null*, *true*, *false*) or complex (i.e. a list of atomic values).

The following example presents YARS-PG that is also showed in Fig. 1. The example presents a graphical representation of a property graph that contains bibliographic information. The node declarations are shown in lines 1–8. The relationship declarations are shown in lines 9–15.

```
1   Author01[Author]:{fname:"John",lname:"Smith"}
2   Author02[Author]:{fname:"Alice",lname:"Brown"}
3   EI01[Entry:InProc]:{title:"Serialization for...",
4                      numpages:10,keyword:"Graph database"}
5   EA01[Entry:Article]:{title:"Property Graph...",
6                      numpages:10,keyword:["Query", "Graph"]}
7   Proc01[Proceedings]:{title:"BDAS",year:2018,month:"May"}
8   Jour01[Journal]:{title:"J. DB",year:2020,vol:30}
9
10  (EI01)-[has_author {order:1}]->(Author01)
11  (EI01)-[has_author {order:2}]->(Author02)
12  (EA01)-[has_author {order:1}]->(Author02)
13  (EA01)-[cites]->(EI01)
14  (EI01)-[booktitle {pages:"111-121"}]->(Proc01)
15  (EA01)-[published_in {pages:"222-232"}]->(Jour01)
```

The main railroad diagram of a node definition is presented in Fig. 2. A node declaration begins with identifier (`ido`). The next part is a node label (`node_label`) nested in square brackets. Node properties (`prop`) are located in curly brackets. A parse tree of first two nodes are presented in Fig. 3. It has interior and leaf nodes. Interior nodes (e.g. `key`, `value`) are non-terminal symbols. Leaf nodes (e.g. `Author`, `fname`) are terminal symbols.

The main railroad diagram of a relationship declaration is presented in Fig. 4. A relationship declaration begins with first identifier (`ido`) nested in round brackets. The next part is a relationship label (`relationship_label`) with relationship

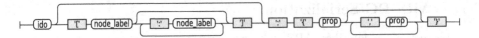

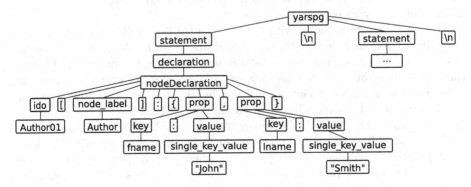

Fig. 2. Railroad diagram of a node declaration

Fig. 3. Parse tree fragment of first two lines in example

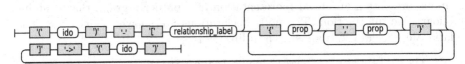

Fig. 4. Railroad diagram of directed relationship declaration

properties (`prop`). Properties begins and ends with curly brackets. The last element is the second identifier (`ido`) nested in round brackets.

The entire YARS-PG grammar in ANTLR 4 [23] and also in EBNF notation[4] has been made available in the GitHub repository [28]. We prepare three different parsers of YARS-PG in Java [19], Python [20], and C# [18].

5   Related Work

5.1   YARS-PG Versus YARS

YARS-PG is based on YARS [29], a concise RDF serialization proposed to facilitate data exchange between RDF and property graph databases.

The following example presents a YARS serialization.

```
1  :rdf: <http://www.w3.org/1999/02/22-rdf-syntax-ns#>
2  (a {value:<http://example.org/p#j>})
3  (b {value:<http://xmlns.com/foaf/0.1/Person>})
4  (a)-[:rdf:type]->(b)
```

---

[4] https://www.w3.org/TR/REC-xml/#sec-notation.

YARS-PG and YARS are textual and path-based. There are three main differences between YARS and YARS-PG. The first difference is the abandonment of support for RDF prefixes (line 1 and line 4) because this abbreviations are not useful for property graphs. The next one is the abandonment of support for URI datatypes between the < and > characters (line 2 and line 3) because property graphs do not have a special datatype for such references. The last change concerns handling of properties in edges, which were not provided by the mapping algorithm from RDF to YARS.

5.2   Graph Serialization Formats

In the property graphs field there are a few solutions for serializing graphs. It may be divided into four groups: formats that use XML, formats that use JSON, tabular-based serializations and text-based ones. Table 3 presents supporting details of those serializations, namely: key-value pair support, multi-values support, support of *null* as a special marker, node multi-labels support, unique label support, directed edges support, undirected edges support, multi-edges with the same label support, unstructured data support, and supported types of formats i.e. XML, JSON, textual, tabular.

The first group can be distinguished to GEXF [14], GraphML [9], DotML[5], DGML[6] and GXL [16,32]. Graph Exchange XML Format (GEXF) is syntax for describing complex networks structures, such as network nodes and edges, properties, hierarchies, and their associated data. It is dedicated for Gephi, which is network analysis and visualization software. Unfortunately, GEXF does not support multi-labels in nodes. Another serialization is GraphML. It supports properties for nodes and edges, hierarchical graphs, sub-graphs, and hyperedges. The advantage of this format is that it is widely adopted. However, the disadvantage, as in GEXF, is the lack of support for multi-labels and no grammar for a *null* value.

Another XML-based serialization is Dot Markup Language (DotML). This format is based on GraphViz DOT [10]. The disadvantage of this serialization is the lack of support for properties. Yet another syntax is Directed Graph Markup Language (DGML). This format supports cyclical and acyclic directed graphs. Unfortunately, DGML does not cope with most of the features considered in Table 3. The last XML-based format is Graph Exchange Language (GXL). It focuses on data interoperability between reverse engineering tools such as parsers, analyzers and visualizers. In its syntax, it is similar to GraphML and also has its disadvantages e.g. lack of support for multi-labels.

The second group are JSON-based serializations like GraphSON TinkerPop 2[7] and GraphSON TinkerPop 3[8]. GraphSON is a part of TinkerPop – the open source

---

[5] http://www.martin-loetzsch.de/DOTML/.

[6] https://docs.microsoft.com/en-us/visualstudio/modeling/directed-graph-markup-language-dgml-reference.

[7] https://github.com/tinkerpop/blueprints/wiki/GraphSON-Reader-and-Writer-Library.

[8] http://tinkerpop.apache.org/docs/current/reference/#graphson-reader-writer.

Table 3. Comparison of property graph serializations

System	Properties			Labels		Edges				Format				
	Pairs	Multiple	Null value	Multiple	Unique	Directed	Undirected	Multiple	Same label	Unstructured	XML	JSON	Textual	Tabular
GEXF	•					•	•	•	•	•	•			
GDF			○△			•		•	○▲					•
GML	•		○△			•	○▽	•					•	
GraphML	•	○			●▼	•	•	•		•	•			
Pajek NET				•		•	•	•						•
GraphViz DOT	•	○	○△	•		•	•		•	•			•	
UCINET DL	•			•		•	•	•	•				•	
Tulip TPL		•				•							•	
Netdraw VNA	•		○△		•	•		•	•					•
DotML				•		•		•	•	•	•			
S-Dot					•	•		•	•				•	
GraphSON TP2	•					•		•	•	•		•		
GraphSON TP3	•	○		•	•	•		•	•	•		•		
DGML	•					•		•		•	•			
GXL	•	○			●▼	•	•	•		•	•			
YARS-PG	•	•	•	•	○	•	•	•	•	•			•	

△ no grammar
▽ only global definition
▲ labels supported as properties
▼ only in the sense of identifiers

graph computing framework, which has its implementations for many databases. The third version of serialization, in contrast to GraphSON TinkerPop 2 supports multiple labels for nodes. However, these labels must be unique. GraphSON TinkerPop 3 also brings partial support for the possibility of defining several values for one key. Both versions support properties but do not support undirected edges. GraphSON TinkerPop 3 is not backward compatible.

There are also a few tabular-based formats including GUESS GDF [1], Pajek NET [5] and Netdraw VNA[9]. The first one is based on comma-separated values (CSV) [13] file format. GDF serialization is known from GUESS tool used to explore and visualize graphs. Blocks of declaration of vertices and edges are separated from each other. The format has rather basic capabilities and does not support properties or multiple values. Another tabular-based serialization is

---

[9] http://www.analytictech.com/Netdraw/NetdrawGuide.doc.

Pajek NET. Serialization allows multiple labels for nodes and undirected edges, but unfortunately, does not support properties. Netdraw VNA, unlike the previously discussed serializations from this group, allows for properties and has support for multiple edges with the same label. Additionally, labels in nodes must be unique. In this serialization, similarly to NET Pajek, the values in columns are separated by spaces. Unfortunately, serialization does not allow for several values for one key, as well as multiple labels for one node.

The last group is text-based syntaxes. This group includes GML [15], GraphViz DOT [10], UCINET DL [8], Tulip TLP [4], and S-Dot[10]. Graph Modelling Language (GML) is a simple structure based on nested key-values lists. The purpose of the structure was to provide flexibility as a universal format. Unfortunately, GML does not support multi-values. Another serialization is Graphviz DOT. This syntax is used in various fields. The format allows to collect data, but also to stylize the graph. The disadvantage of serialization is the lack of multigraph support. Yet another syntax is UCINET DL. This format based on matrixes and lists. The disadvantage of this serialization is the inability to use multi-value. The next syntax is Tulip TLP. This format has structure based on round brackets. The serialization allows to collect data, but also to stylize the graph. The last serialization belonging to textual group is S-Dot. This format is based on GraphViz DOT and on similar serialization DotML. Unfortunately, this format does not support properties.

Comparing the above serializations to YARS-PG, we can see that almost all features, listed in Table 3, are supported. The example in Sect. 4 shows key-value pairs in nodes (e.g. line 1) and in edges (e.g. line 11). This example also presents directed edges in lines 10–15, and multiple properties in line 6. Our proposal allows to use the same name of labels but the parser treats it as the same label.

6   Conclusions

This paper presents YARS-PG, a data serialization format for property graphs which is simple, extensible, and platform independent. YARS-PG supports all the features allowed by the current database systems based on the property graph data model, and can be adapted in the future to work with various database systems, visualization software and other graph-oriented tools.

The future work will focus on providing a binary and a compact version of this serialization, which will be faster and will make the serialization a good format for storing and sharing on the Web.

Acknowledgements. This work was supported by the National Science Center, Poland (NCN) under research grant Miniatura 2 (2018/02/X/ST6/00880) for Dominik Tomaszuk. This publication has received financial support from the Polish Ministry of Science and Higher Education under subsidy granted to the University of Bialystok for R&D and related tasks aimed at development of young scientists for Łukasz Szeremeta. Renzo Angles is funded by the Millennium Institute for Foundational Research on Data (Chile).

---

[10] http://martin-loetzsch.de/S-DOT/.

References

1. Adar, E.: GUESS: a language and interface for graph exploration. In: Proceedings of the SIGCHI Conference on Human Factors in Computing Systems, CHI 2006, pp. 791–800. ACM, New York (2006).https://doi.org/10.1145/1124772.1124889
2. Angles, R., Arenas, M., Barceló, M.A., Hogan, M.A., Reutter, M.A., Vrgoĉ, M.A.: Foundations of modern query languages for graph databases. CSUR **50**(5) (2017). https://doi.org/10.1145/3104031
3. Angles, R., et al.: A core for future graph query languages. In: Proceedings of the 2018 International Conference on Management of Data, SIGMOD 2018, pp. 1421–1432. ACM, New York (2018). https://doi.org/10.1145/3183713.3190654
4. Auber, D., et al.: TULIP 5 (2017)
5. Batagelj, V., Mrvar, A.: Pajek— analysis and visualization of large networks. In: Mutzel, P., Jünger, M., Leipert, S. (eds.) GD 2001. LNCS, vol. 2265, pp. 477–478. Springer, Heidelberg (2002). https://doi.org/10.1007/3-540-45848-4_54
6. Bhatti, N., Hassan, W., McClatchey, R., Martin, P., Kovacs, Z.: Object serialization and deserialization using XML. Advances in Data Management, vol. 1 (2000)
7. Bonifati, A., Fletcher, G., Voigt, H., Yakovets, N.: Querying graphs. In: Synthesis Lectures on Data Management. Morgan & Claypool Publishers (2018). https://doi.org/10.2200/S00873ED1V01Y201808DTM051
8. Borgatti, S.P., Everett, M.G., Freeman, L.C.: Ucinet for Windows: software for social network analysis (2002)
9. Brandes, U., Eiglsperger, M., Herman, I., Himsolt, M., Marshall, M.S.: GraphML progress report structural layer proposal. In: Mutzel, P., Jünger, M., Leipert, S. (eds.) GD 2001. LNCS, vol. 2265, pp. 501–512. Springer, Heidelberg (2002). https://doi.org/10.1007/3-540-45848-4_59
10. Ellson, J., Gansner, E.R., Koutsofios, E., North, S.C., Woodhull, G.: Graphviz and dynagraph – static and dynamic graph drawing tools. In: In: Jünger, M., Mutzel, P. (eds.) Graph Drawing Software. Mathematics and Visualization, pp. 127–148. Springer, Berlin (2004). https://doi.org/10.1007/978-3-642-18638-7_6
11. Guminska, E., Zawadzka, T.: EvOLAP graph – evolution and OLAP-aware graph data model. In: Kozielski, S., Mrozek, D., Kasprowski, P., Małysiak-Mrozek, B., Kostrzewa, D. (eds.) BDAS 2018. CCIS, vol. 928, pp. 75–89. Springer, Cham (2018). https://doi.org/10.1007/978-3-319-99987-6_6
12. Hartig, O.: Reconciliation of RDF* and Property Graphs. Technical reports. http://arxiv.org/abs/1409.3288 (2014)
13. Hausenblas, M., Wilde, E., Tennison, J.: URI Fragment Identifiers for the text/csv Media Type. RFC 7111, RFC Editor, January 2014. http://www.rfc-editor.org/rfc/rfc7111.txt
14. Heymann, S.: Gephi. In: Alhajj, R., Rokne, J. (eds.) Encyclopedia of Social Network Analysis and Mining, pp. 612–625. Springer, New York (2014). https://doi.org/10.1007/978-1-4614-6170-8_299
15. Himsolt, M.: GML: a portable graph file format (1997). http://www.uni-passau.de/fileadmin/files/lehrstuhl/brandenburg/projekte/gml/gml-technical-report.pdf
16. Holt, R.C., Winter, A., Schürr, A.: GXL: toward a standard exchange format. In: Proceedings of the Seventh Working Conference on Reverse Engineering, pp. 162–171, November 2000. https://doi.org/10.1109/WCRE.2000.891463
17. Kangasharju, J., Tarkoma, S.: Benefits of alternate xml serialization formats in scientific computing. In: Proceedings of the Workshop on Service-Oriented Computing Performance: Aspects, Issues, and Approaches, pp. 23–30. ACM, New York (2007). https://doi.org/10.1145/1272457.1272461

18. Litman, K.: YARSpg Parser C Sharp 0.3 (GitHub), December 2018. https://doi.org/10.5281/zenodo.2285046

19. Litman, K.: YARSpg-Parser-Java 0.3 (GitHub), December 2018. https://doi.org/10.5281/zenodo.2284679

20. Litman, K.: YARSpg Parser Python 0.4 (GitHub), December 2018. https://doi.org/10.5281/zenodo.2285247

21. Maeda, K.: Comparative survey of object serialization techniques and the programming supports. Int. J. Comput. Inf. Eng. 5(12) (2011)

22. Marton, J., Szárnyas, G., Varró, D.: Formalising openCypher graph queries in relational algebra. In: Kirikova, M., Nørvåg, K., Papadopoulos, G.A. (eds.) ADBIS 2017. LNCS, vol. 10509, pp. 182–196. Springer, Cham (2017). https://doi.org/10.1007/978-3-319-66917-5_13

23. Parr, T.: The Definitive ANTLR 4 Reference. Pragmatic Bookshelf (2013)

24. Płuciennik, E., Zgorzałek, K.: The multi-model databases – a review. In: Kozielski, S., Mrozek, D., Kasprowski, P., Małysiak-Mrozek, B., Kostrzewa, D. (eds.) BDAS 2017. CCIS, vol. 716, pp. 141–152. Springer, Cham (2017). https://doi.org/10.1007/978-3-319-58274-0_12

25. van Rest, O., Hong, S., Kim, J., Meng, X., Chafi, H.: PGQL: a property graph query language. In: Proceedings of the Fourth International Workshop on Graph Data Management Experiences and Systems, GRADES 2016, pp. 1–6. ACM, New York (2016). https://doi.org/10.1145/2960414.2960421

26. Rodriguez, M.A., Neubauer, P.: Constructions from dots and lines. Bull. Am. Soc. Inf. Sci. Tech. 36(6), 35–41 (2010)

27. Sumaray, A., Makki, S.K.: A comparison of data serialization formats for optimal efficiency on a mobile platform. In: Proceedings of the 6th International Conference on Ubiquitous Information Management and Communication, pp. 1–6. ACM (2012). https://doi.org/10.1145/2184751.2184810

28. Szeremeta, Ł.: YARS-PG ANTLR4 grammar (GitHub), February 2019. https://doi.org/10.5281/zenodo.2555898

29. Tomaszuk, D.: RDF data in property graph model. In: Garoufallou, E., Subirats Coll, I., Stellato, A., Greenberg, J. (eds.) MTSR 2016. CCIS, vol. 672, pp. 104–115. Springer, Cham (2016). https://doi.org/10.1007/978-3-319-49157-8_9

30. Tomaszuk, D., Pak, K.: Reducing vertices in property graphs. PLoS ONE 13(2), 1–25 (2018)

31. Warchał, Ł.: Using Neo4j graph database in social network analysis. Stud. Informatica 33(2A), 271–279 (2012). https://doi.org/10.21936/si2012_v33.n2A.147

32. Winter, A., Kullbach, B., Riediger, V.: An overview of the GXL graph exchange language. In: Diehl, S. (ed.) Software Visualization. LNCS, vol. 2269, pp. 324–336. Springer, Heidelberg (2002). https://doi.org/10.1007/3-540-45875-1_25

33. Yusof, K., Man, M.: Efficiency of JSON for data retrieval in big data. Ind. J. Electr. Eng. Comput. Sci. 7, 250–262 (2017)

Evaluation of Key-Value Stores
for Distributed Locking Purposes

Piotr Grzesik$^{(\boxtimes)}$ and Dariusz Mrozek

Institute of Informatics, Silesian University of Technology, ul. Akademicka 16,
44-100 Gliwice, Poland
pj.grzesik@gmail.com, dariusz.mrozek@polsl.pl

Abstract. This paper presents the evaluation of key-value stores and corresponding algorithms with regard to the implementation of distributed locking mechanisms. Research focuses on the comparison between four types of key-value stores, etcd, Consul, Zookeeper, and Redis. For each selected store, the underlying implementation of locking mechanisms was described and evaluated with regard to satisfying safety, deadlock-free, and fault tolerance properties. For the purposes of performance testing, a small application supporting all of the key-value stores was developed. The application uses all of the selected solutions to perform computation while ensuring that a particular resource is locked during that operation. The aim of the conducted experiments was to evaluate selected solutions based on performance and properties that they hold, in the context of using them as a base for building a distributed locking system.

Keywords: Redis · Etcd · Consul · Zookeeper · Raft · Paxos · Zab ·
Redlock · Distributed computing · Cloud computing ·
Amazon Web Services · Python · Distributed locks

1 Introduction

The requirement of mutual exclusion in concurrent processing was identified over 50 years ago in paper [9] by Edsger Dijkstra. At present, most of the operating systems implement primitives that can be used to satisfy that requirement on a single machine, however, the dynamic growth of distributed computer systems in areas like artificial intelligence, data warehousing, and processing requires us to solve the locking problem in distributed systems in a reliable and fault-tolerant manner. Algorithms, like Paxos [22], Raft [25] or Zab [15] and their corresponding safety and liveness properties enabled and inspired implementation of consistent key-value stores like etcd [10], Consul [6] and Zookeeper [14], which can be used as a base for building distributed locking solutions; all of them offering primitives that make building such systems easier. Figure 1 shows the interaction between distributed locking service and two separate clients. Firstly, Client 1 acquires the lock and during a period of time when that lock is held, requests from Client

© Springer Nature Switzerland AG 2019
S. Kozielski et al. (Eds.): BDAS 2019, CCIS 1018, pp. 70–81, 2019.
https://doi.org/10.1007/978-3-030-19093-4_6

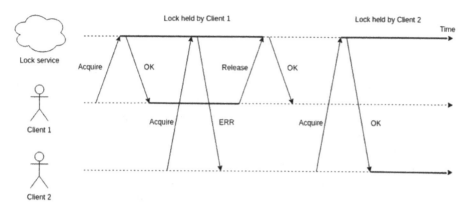

Fig. 1. Interaction with lock service by two clients

2 to acquire lock will fail. After the lock is released by Client 1, Client 2 can successfully acquire distributed lock.

When evaluating distributed locking mechanisms, depending on a use case and correctness requirements, it is very important to determine, if the mechanism satisfies some or all of the following properties:

- Safety – property that satisfies the requirement of mutual exclusion, ensuring that only one client can acquire and hold the lock at any given moment.
- Deadlock-free – property that ensures that, eventually, it will be possible to acquire and hold a lock, even if the client currently holding the lock becomes unavailable.
- Fault tolerance – property ensuring that as long as the majority of the nodes of the underlying distributed locking mechanism are available, it is possible to acquire, hold, and release locks.

Deadlock-free and fault tolerance can be also summed up as one, liveness property.

The aim of the research presented in the paper is to evaluate and compare key-value stores with corresponding algorithms in regard to available mechanisms that can be used for performing distributed locking. For the purposes of testing, four solutions were selected, etcd, Redis [26] with Redlock [28], Consul and Zookeeper. Selection of the key value stores was made based on the properties of their underlying algorithms like Raft and Zab as well as their popularity and widespread adoption. All proposed services were evaluated based on performance, their safety, deadlock-free and fault tolerance properties as well as on ease-of-use both in client applications and in setting up infrastructure needed for them to run in a fault tolerant manner. As a base for research, a small Python 3.7 application was developed, that is able to perform distributed locking with all of the mentioned solutions.

2   Related Works

In the literature, there is few research concerning the comparison of key-value stores performance. In the research [23], Gyu-Ho Lee is comparing key write performance of Zookeeper, etcd, and Consul working in three-node clusters. The author is measuring performance with regards to disk bandwidth, network traffic, CPU utilization as well as memory usage. The author concludes the research with claims that etcd offers better throughput and average latency while using less memory than other solutions. However, it is also noted that Zookeeper offers the lowest minimal latency, at the cost of potential higher average latency.

Patrick Hunt in his analysis [12] evaluates Zookeeper server latency under varying loads and configurations, which included operations like creating permanent znodes[1], setting, getting and deleting znodes, as well as creating ephemeral znodes and performing znodes watches. He observed that for a standalone server, additional cores (testing was performed with 1, 2 and 4 CPUs) did not provide significant performance gains. He also observed that in general, Zookeeper is able to handle more operations per second with a higher number of clients (around 4 times more operations per second between 1 and 10 clients).

Redis documentation [27] offers a detailed performance tests. It also highlights factors that are critical for Redis performance, such as significantly worse performance while running on a virtual machine compared to running without virtualization on the same set of hardware. It also notes that due to single-threaded nature of Redis, it performs much better on CPUs with larger caches and fewer cores, such as Intel Sandy Bridge CPUs, which perform up to 50% better in comparison to corresponding AMD Opteron CPUs.

The liveness and safety properties of Raft were presented in paper [25] by Diego Ongaro, where the same properties of Zab, the algorithm underpinning Zookeeper, were presented [15] by Junqueira, Reed and Serafini. Redlock was developed and described by Salvatore Sanfilippo, however safety property was later disputed by Martin Kleppmann, which presented in his article [19], that under certain conditions, the safety property of Redlock might not hold, suggesting that Redlock might not be the most optimal solution where correctness is the main objective.

In his paper [4], Mike Burrows from Google described an architecture for Chubby, lock service for distributed systems. One of the important decisions made by the team developing Chubby was to create a separate service, instead of a library, motivating that decision by the easier implementation for clients (in comparison to integrating consensus protocol into the applications). Chubby uses Paxos as an underlying consensus mechanism, ensuring safety property.

Kyle Kingsbury in his works [17,18] evaluated etcd, Consul and Zookeeper with Jepsen [13], the framework for distributed system verification. During verification, it was revealed that Zookeeper is able to preserve linearizability in the presence of network partition and leader election. The same did not hold true for

---

[1] Zookeeper uses a hierarchical namespace, where every node is called znode.

Consul and etcd, which experienced stale reads[2]. This research prompted main-
tainers of both etcd and Consul to provide mechanisms that enforce consistent
reads.

While there are a few research works focusing on certain aspects like performance or correctness of the selected solutions, none of them provides a detailed
comparison of them in the context of using them as a building block of a distributed locking system.

3   Key-Value Stores

Key-value stores, also commonly called key-value databases, are certain type of
NoSQL [29] databases, which employ, unlike established SQL databases, schema-
less data model. As the name suggests, stored data is represented in form of
key-value pairs, where values can be arbitrary binary objects, which makes it
the most flexible data store from application perspective. Thanks to the simple
structure, key-value stores like Redis can offer very high performance in comparison to traditional SQL databases as well as other types of NoSQL databases
[16]. For purposes of this research, the most interesting key-value stores are those
that offer strict data consistency and high availability (falling into CP of CAP
theorem presented by Brewer [3]) which in combination with performance, can
serve as a solid base for building fast and responsive distributed locking mechanisms. Out of the existing stores, four were selected: etcd, Consul, Zookeeper, and
Redis, based on their wide-spread usage (in Hadoop ecosystem [11], Kubernetes
[20], Nomad [24]), properties of underlying consensus algorithms and ease-of-use.

3.1   Etcd

Etcd is an open source, distributed key-value store written in Go, currently
developed under Cloud Native Computing Foundation [5]. It uses the Raft consensus algorithm for management of highly available replicated log, being able
to tolerate node failures, including the failure of the leader. In addition, it offers
dynamic cluster membership reconfiguration. Etcd enables distributed coordination by implementing primitives for distributed locking, write barriers and leader
election. It uses persistent, multi-version, concurrency-control data model. Client
libraries in languages like Go, Python, Java, and others are available, as well as
command-line client "etcdctl".

3.2   Consul

Consul is an open source, distributed key-value store written in Go, developed
by Hashicorp. In addition to being a consistent key-value store, Consul can be
used for health checking, service discovery or as a source of TLS certificates for

---

[2] Stale read is a read operation that fetches result which does not reflect all updates
to the given value.

providing secure connections between services in a system. It also has support for multiple data centers, with a separate Consul cluster in each data center. Similarly to etcd, it uses the Raft consensus algorithm for managing replicated log and also can be used for distributed coordination with sessions mechanism enabling distributed locks. In addition, Consul offers several consistency modes for reads, depending on the application needs. Consul nodes can be either servers or clients with only servers taking part in Raft consensus protocol. Clients use the Gossip protocol to communicate with each other.

3.3 Zookeeper

Zookeeper is an open source, highly available coordination system, written in Java. Initially developed at Yahoo, currently is a project maintained by Apache Software Foundation [2]. Zookeeper uses the Zab atomic broadcast protocol. Nodes store data in a form of a hierarchical namespace, resembling file system, where each node in Zookeeper's tree is called a znode. Zookeeper's main use cases are leader election, group membership, a configuration store, distributed locking and priority queues. While not being strictly a key-value store, it can be also used and qualified as such for purposes of this research. Zookeeper's Java client library, Apache Curator, offers high-level API with implemented recipes for elections, locks, barriers, counters, caches, and queues. Similar recipes are also available as a part of the Python client library, kazoo.

3.4 Redis

Redis is an open source, in-memory key-value database, that offers optional durability. It was developed by Salvatore Sanfilippo and is written in ANSI C. Redis offers support for data structures such as lists, sets, sorted sets, maps, strings, hyperloglogs, bitmaps, and streams. Redis exhibits very high performance in comparison to other database offerings [16]. According to the DB-Engines ranking [8], it is the most popular key-value database. While mostly used for caching, queuing and Pub/Sub, it can also be used, in combination with the Redlock, as a base for building distributed locking mechanism.

4   Environment and Implementation

For the purposes of testing, a small Python 3.7 application was developed, that can use each of the stores presented in the previous chapter to acquire distributed lock, simulate a short computation that requires the lock to be held, and release the lock afterwards. In addition, it measures the time taken to acquire a distributed lock. Consul, etcd, and Zookeeper were tested as 3-node clusters, using configuration presented on Fig. 2. The whole needed infrastructure was deployed in Amazon Web Services cloud offering as EC2 instances, each of which has been provisioned in different availability zone in the same geographical region, eu-central-1, to ensure fault tolerance of a single availability zone., while maintaining latency between cluster instances in sub-milliseconds range. All instances (including the client node) are of type "t2.medium", which have the following specification:

- OS - Amazon Linux 2 [1]
- AMI[3] ID - ami-0cfbf4f6db41068ac
- CPU - 2 vCPUs of 3.3 GHz Intel Scalable Processor
- 4 GiB of RAM
- 8 GiB of EBS[4] SSD storage

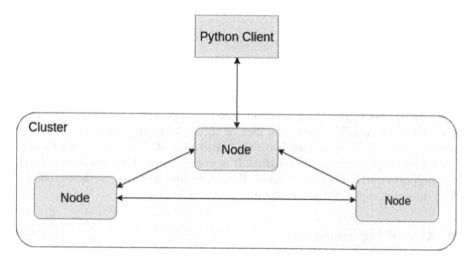

Fig. 2. Environment configuration diagram for etcd, Consul and Zookeeper

Redis has been deployed in a similar environment, however the usage of the Redlock locking algorithm required to use 3 standalone Redis nodes instead of having 3-nodes cluster. Configuration is presented on Fig. 3.

4.1   Etcd Implementation and Configuration

To build an etcd cluster, the etcd in version 3.3.9 was used, compiled with Go language with version 1.10.3. The cluster was configured to ensure sequential consistency model [21], which satisfies the safety property of the solution. To communicate with the cluster, the client application used the python-etcd3 library, which implements locks using the atomic compare-and-swap mechanism,

---

[3] Amazon Machine Images, image that is used to create virtual machines using Amazon Elastic Compute Cloud.
[4] Elastic Block Storage, persistent block storage offering from Amazon Web Services.

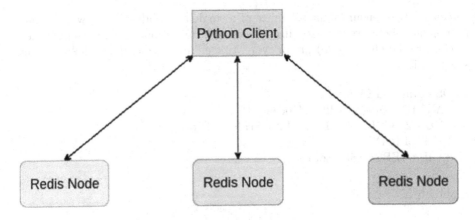

Fig. 3. Environment configuration diagram for Redis

that checks if the given key is already set and if not, atomically sets it to a given value, thus acquiring the lock. If the key is already set, that means the lock is already acquired. To ensure deadlock-free property, the TTL (Time-To-Live) for a lock is set, which releases the lock, if the lock-holding client crashes or is network partitioned away. Thanks to the underlying Raft consensus algorithm, fault tolerance property is also satisfied. It is worth noting that etcd from version 3.2.0 supports native Lock and Unlock RPC methods, however, python-etcd3 as of yet does not support those methods.

4.2 Consul Implementation

Consul cluster was assembled using version 1.4.0, compiled with Go 1.11.1 and configured with "consistent" consistency mode [7], ensuring the safety of the underlying Raft protocol. Usage of the Raft consensus protocol also ensures that fault tolerance property is satisfied. Application acquires locks by using the consul-lock Python library, which implements locks using sessions mechanism with the check-and-set operation. According to the Consul documentation [6], depending on the selected health checking mechanism, either safety or liveness property may be sacrificed. The selected implementation uses only session TTL health check, which guarantees that both properties hold, however, it's worth noting that TTL is applied at the session level, not on specific lock and session timeout either deletes or releases all locks related to the session.

4.3 Zookeeper Implementation

Zookeeper cluster was build using version 3.4.13, running on Java 1.7, with Open-JDK Runtime Environment. Thanks to the Zab protocol properties, safety and fault tolerance requirements are satisfied. To implement application interacting with the Zookeeper cluster, the Python kazoo library with lock recipe was used.

This implementation takes advantage of Zookeeper's ephemeral and sequential znodes. On lock acquire, a client creates znode with ephemeral and sequential flag and checks, if sequence number of that znode is the lowest, and if that is true, the lock is acquired. In opposite situation, client can watch the znode and be notified when the lock is released so it can once again try to acquire the lock. Thanks to usage of ephemeral znodes, we can achieve deadlock-free property, because if the client holding lock becomes unavailable, the ephemeral node will be destroyed resulting in release of the lock.

4.4 Redis with Redlock Implementation

The last solution is build on top of 3 standalone Redis nodes with version 5.0.3. The application implementing locking mechanisms is using the Redlock algorithm, provided by the Python redlock library. The Redlock algorithm works by sequentially trying to acquire lock on all independent Redis instances and the lock is acquired when it was successfully acquired on majority of the nodes. Specific details of Redlock algorithm are presented in [28]. Thanks to the requirement to lock on the majority of the nodes before considering the lock to be acquired, safety property is satisfied. Similarly to the other solutions, the deadlock-free property is satisfied based on lock timeouts. Fault tolerance is satisfied as well, by tolerating up to $(N/2) - 1$ failures of the independent Redis nodes. It is worth noting, that under specific circumstances, described by Martin Kleppmann in his article [19], it is possible for the Redlock to lose the safety property, making it not suitable for applications in which correctness cannot be sacrificed under any circumstances.

4.5 Implementations Summary

In this chapter, we described the implementation and configuration details related to all selected solutions. In cases of Zookeeper, etcd and Consul, the safety, deadlock-free and fault tolerance properties are satisfied thanks to selected cluster configurations, their mechanisms like sessions for Consul and ephemeral nodes for Zookeeper, as well as their underlying algorithms like Raft and Zab. In the case of Redis, the deadlock-free property is satisfied thanks to the selected Redlock algorithm and the size of the cluster used for experiments allows for failure of one node, hence satisfying the fault tolerance property. The safety property for Redlock is not always satisfied, which was described by Martin Kleppmann in his analysis [19]. In the next chapter, we focused on the performance evaluation of all selected solutions.

5 Performance Experiments

To evaluate the performance of the selected solutions, we performed several experiments. Firstly, we conducted the experiment to evaluate the behavior of all solutions for a workload where multiple processes have to acquire distributed locks on various resources, but where we do not experience the clients competing for a particular lock. In this step, for each service, we simulated concurrent traffic from 1, 3, and 5 different processes that are trying to acquire locks on different lock keys, with each simulation lasting 2 min. Secondly, we ran a simulation to evaluate the behavior of selected solutions for a workload, where we experience high contention over a particular lock key, with multiple processes trying to acquire the same lock simultaneously. In this step, once again, we simulated concurrent traffic from 1, 3, and 5 different client applications, but this time, all of them tried to acquire the same lock key. Based on the results of the simulations, we examined metrics related to lock acquire time for each configuration and evaluated all of the solutions with regard to expected workload patterns and requirements.

5.1 Acquiring Different Lock Keys

In the first simulation, we evaluated the locking of different lock keys. Obtained results are presented in Table 1. For the etcd, we observe that with an increasing number of client processes, the average lock acquire time is also increasing, which is similar for Consul and Zookeeper as well. For Redis, we see very small changes in the average lock time, with the average time for 5 processes being even lower than for 3 processes. Redis also has the lowest average lock acquire time from all of the tested solutions, with 1.3 ms for 5 processes, Zookeeper is next with 5.9 ms after that is etcd with 7.68 ms and Consul at the end with 16.52 ms. Maximum lock acquires time, the 90th percentile of lock acquire time, and the 99th percentile of lock acquire time were also tested, to evaluate the stability of expected lock acquire time. In all tested solutions with 5 processes, the 90th percentile was about 20% higher than average time, however, while for etcd, Consul and Zookeeper 99th percentile was about 2 times higher than the average, for Redis we observed 4 times higher value of the 99th percentile in comparison to the average time. However, even with that change, the 99th percentile of lock acquire time of Redis was still almost 2 times smaller than those of other tested solutions.

Table 1. Summary of results for different keys simulations

	etcd			Consul			Zookeeper			Redis		
No. of proc.	1	3	5	1	3	5	1	3	5	1	3	5
Avg (ms)	4.91	6.8	7.68	11.7	13.6	16.52	3.64	4.53	5.9	1.2	1.36	1.3
Median (ms)	4.66	6.49	7.33	11.5	13.12	16.01	3.41	4.4	5.74	1.12	1.26	1.25
Min (ms)	3.99	4.43	4.19	10.5	10.42	10.38	2.86	2.79	2.76	0.95	0.97	0.98
Max (ms)	20.97	21.75	28.43	28.3	45.38	119.71	36.64	15.35	19.98	12.82	14.13	9.42
90p (ms)	5.51	8.32	9.31	12.42	16.09	20.29	4.3	5.97	7.4	1.32	1.56	1.49
99p (ms)	9.38	13.41	16.1	15.09	19.74	25.07	7.49	7.65	9.97	2.93	3.37	5.07

5.2   Contesting over the Same Lock Keys

In the second simulation, we tested the behavior of locking applications that were contesting over the same lock key. Results gathered during the simulation are presented in Table 2. While inspecting the average lock acquire time, we observed a general trend of increasing time with more client processes for all tested solutions. It is, however, worth noting, that while for etcd, Consul and Zookeper changes in average lock time between 1, 3 and 5 processes are relatively small, for Redis we observe 3 times increase while changing from 1 to 3 processes and almost 5 times increase for 5 processes. Even with that behavior, Redis still has the lowest average lock time among all tested solutions, but only 0.24 ms lower (for 5 concurrent processes) than Zookeeper. Results of the maximum, the 90th percentile and the 99th percentile of the lock acquire time reveal that Redis-based solution is prone to having instances of very high lock acquire times. While the 90th percentile is very small (1.32 ms), the 99th percentile is over 200 ms with maximum lock acquire time being 411.2 ms. All other solutions do not exhibit such anomalies, all having the 99th percentile for 5 processes roughly 2 times higher than average lock acquire time. The best performing solution based on that metric is Zookeeper, with 7.49 ms and 10.17 ms of the 99th percentile for 3 and 5 processes, respectively.

Table 2. Summary of results for the same keys simulations

No. of proc.	etcd			Consul			Zookeeper			Redis		
	1	3	5	1	3	5	1	3	5	1	3	5
Avg (ms)	4.91	6.98	7.6	11.7	13.84	16.36	3.64	4.82	5.94	1.2	3.78	5.7
Median (ms)	4.66	6.48	7.22	11.5	13.49	15.89	3.41	4.65	5.79	1.12	1.15	1.1
Min (ms)	3.99	4.23	4.16	10.5	10.62	10.18	2.86	2.84	3.07	0.95	0.96	0.93
Max (ms)	20.97	38.7	47.13	28.3	151.96	95.05	36.64	20.93	19.98	12.82	414.52	411.2
90p (ms)	5.51	8.5	9.16	12.42	16	20.85	4.3	6.3	7.4	1.32	1.3	1.3
99p (ms)	9.38	14.14	16.79	15.09	20.1	26.25	7.49	8.9	10.17	2.93	5.07	203.87

5.3   Results Summary

Comparing results from both simulations, we can observe that for the etcd, Zookeeper and Consul there is little difference in performance between situations where client processes are contesting over the same key or are concurrently acquiring different lock keys. The same does not hold for Redis, which exhibits stable and the best performance of all tested solutions in case of clients accessing different lock keys, while being unstable with workload involving multiple clients contesting over the same key, having the 99th percentile for 5 processes as high as 203.87 ms, over 10 times more than other tested solutions. Considering both workloads, the most performing solution is Zookeeper, which in worst case scenario (5 processes, different lock keys) offers average lock time of 5.94 ms, with the maximum of 19.98 ms, the 90th percentile of 7.4 and the 99th percentile of 10.17 ms. It is worth noting that if upper lock time bound is not essential for the given application, then the best performing solution is Redis, which for the same workload offers the 90th percentile lock acquire time of 1.3 ms.

6 Concluding Remarks

The aim of this research was to implement and compare distributed locking applications built on top of selected key-values stores. To evaluate the performance of selected solutions, an application supporting distributed locking was developed. We believe that experiments and analysis of the results presented in this paper complement performance comparisons presented in [23], as well as analysis of safety and correctness by Kyle Kingsbury [17,18] and Martin Kleppmann [19] to serve as a comprehensive guide to selecting the base for building distributed locking systems.

All solutions were also evaluated based on ease-of-use both from the infrastructure and implementation perspective. When it comes to setting up infrastructure, Redis with Redlock is the easiest to setup, because all nodes are independent and can be easily replaced. For Zookeeper, Consul and etcd, it is slightly more complicated, because all nodes have to form a cluster communicating with each other. Zookeeper, prior to version 3.5.0 also does not offer cluster membership changes without restarting nodes. All solutions have easy to use Python client libraries, however, Zookeeper's kazoo with corresponding recipes is the most mature and robust client library.

The second axis on which the comparison was performed is related to safety, deadlock-free and fault tolerance properties. Zookeeper, etcd and Consul with correct configurations satisfy all considered properties, making them suitable for applications for which those properties are a strict requirement. Redis with Redlock satisfy both deadlock-free and fault tolerance properties, however, under some circumstances, safety property might be sacrificed, which makes it a questionable choice for cases where safety and correctness is a top priority.

The third and final comparison of all solutions was based on metrics related to lock acquire time. It was observed that for workload patterns where we do not observe high contention over the same lock key, Redis is the best performing solution, however with the growing number of concurrent processes trying to acquire the same lock key, Redis performance deteriorates and we observe cases in which worst-case lock acquire time is 10 to 20 times higher than other solutions. For workloads with such patterns and for which worst-case scenario performance is critical, Zookeeper emerged as the most suitable solution.

References

1. Amazon Linux 2. https://aws.amazon.com/amazon-linux-2/. Accessed 9 Jan 2019
2. Apache Software Foundation. https://www.apache.org/. Accessed 11 Jan 2019
3. Brewer, E.: CAP twelve years later: how the "Rules" have changed. Computer **45**, 23–29 (2012). https://doi.org/10.1109/MC.2012.37

4. Burrows, M.: The Chubby lock service for loosely-coupled distributed systems. In: 7th USENIX Symposium on Operating Systems Design and Implementation (OSDI) (2006)
5. Cloud Native Computing Foundation. https://www.cncf.io/. Accessed 11 Jan 2019
6. Consul Documentation. https://www.consul.io/docs/index.html. Accessed 11 Jan 2019
7. Consul consensus protocol. https://www.consul.io/docs/internals/consensus.html. Accessed 11 Jan 2019
8. Corporation, N.: DB engines ranking. https://db-engines.com/en/ranking
9. Dijkstra, E.: Solution of a problem in concurrent programming control. Commun. ACM **8**(9), 569 (1965)
10. etcd Documentation. https://etcd.readthedocs.io/en/latest/. Accessed 10 Jan 2019
11. Hadoop Documentation. https://hadoop.apache.org/. Accessed 11 Jan 2019
12. Hunt, P.: Zookeeper service latencies under various loads and configurations. https://cwiki.apache.org/confluence/display/ZOOKEEPER/ServiceLatency Overview. Accessed 9 Jan 2019
13. Jepsen - Distributed Systems Safety Research. https://jepsen.io/. Accessed 9 Jan 2019
14. Junqueira, F., Reed, B., Hunt, P., Konar, M.: ZooKeeper: wait-free coordination for Internet-scale systems. In: Proceedings of the 2010 USENIX conference on USENIX Annual Technical Conference, June 2010
15. Junqueira, F., Reed, B., Serafini, M.: Zab: high-performance broadcast for primary-backup systems. In: IEEE/IFIP 41st International Conference on Dependable Systems and Networks (DSN), pp. 245–256, June 2011
16. Kabakus, A.T., Kara, R.: A performance evaluation of in-memory databases. J. King Saud Univ. Comput. Inf. Sci. **29**(4), 520–525 (2016)
17. Kingsbury, K.: Jepsen: etcd and Consul. https://aphyr.com/posts/316-call-me-maybe-etcd-and-consul. Accessed 9 Jan 2019
18. Kingsbury, K.: Jepsen: Zookeeper. https://aphyr.com/posts/291-call-me-maybe-zookeeper. Accessed 9 Jan 2019
19. Kleppmann, M.: How to do distributed locking. http://martin.kleppmann.com/2016/02/08/how-to-do-distributed-locking.html. Accessed 9 Jan 2019
20. Kubernetes Documentation. https://kubernetes.io/. Accessed 11 Jan 2019
21. KV API Guarantees. https://coreos.com/etcd/docs/latest/learning/api_guarantees.html. Accessed 11 Jan 2019
22. Lamport, L.: The part-time parliament. ACM Trans. Comput. Syst. **16**(2), 133–169 (1998)
23. Lee, G.H.: Exploring performance of etcd, zookeeper and consul consistent key-value datastores. https://coreos.com/blog/performance-of-etcd.html. Accessed 9 Jan 2019
24. Nomad Documentation. https://www.nomadproject.io/. Accessed 11 Jan 2019
25. Ongaro, D., Ousterhout, J.: In search of an understandable consensus algorithm. In: Proceedings of the 2014 USENIX Conference on USENIX Annual Technical Conference, pp. 305–320, June 2014
26. Redis Documentation. https://redis.io/documentation. Accessed 11 Jan 2019
27. How fast is Redis? https://redis.io/topics/benchmarks. Accessed 9 Jan 2019
28. Distributed locks with Redis. https://redis.io/topics/distlock. Accessed 11 Jan 2019
29. Sullivan, D.: NoSQL for Mere Mortals, 1st edn. Addison-Wesley Professional, Boston (2015)

On Repairing Referential Integrity Constraints in Relational Databases

Raji Ghawi[⊠][iD]

Bavarian School of Public Policy, Technical University of Munich,
Richard-Wagner-Straße. 1, 80333 Munich, Germany
raji.ghawi@tum.de

Abstract. Integrity constraints (ICs) are semantic conditions that a database should satisfy in order to be in a consistent state. Typically, ICs are declared with the database schema and enforced by the database management system (DBMS). However, in practice, ICs may not be specified to the DBMS along with the schema, this is considered a bad database design and may lead to many problems such as inconsistency and anomalies. In this paper, we present a method to identify and repair missing referential integrity constraints (foreign keys). Our method comprises three steps of verification of candidate foreign keys: data-based, model-based, and brute-force.

Keywords: Relational databases · Integrity constraints · SQL · Validation · Verification

1 Introduction

Integrity constraints capture an important aspect of every database application. They are semantic conditions that a database should satisfy in order to be an appropriate model of reality [6]. There are several types of integrity constraints including: primary keys, functional dependencies, and referential integrity constraints (known as foreign keys). These constraints are derived from the semantics of the data and of the miniworld it represents. It is the responsibility of the database designers to identify integrity constraints during database design. Some constraints can be specified to the DBMS and automatically enforced [13]. A common assumption in data management is that databases can be kept consistent, that is, satisfying certain desirable integrity constraints. In practice, and for many reasons[1], a database may not satisfy those integrity constraints, and for that reason it is said to be *inconsistent*.

Inconsistency is an undesirable property for a database. Therefore, as a database is subject to updates, it should be kept consistent. This goal can be achieved in several ways. One of them consists in declaring the ICs together with the schema, thus, the DBMS will take care of keeping the database consistent,

---

[1] For example, when merging data from different sources.

© Springer Nature Switzerland AG 2019
S. Kozielski et al. (Eds.): BDAS 2019, CCIS 1018, pp. 82–96, 2019.
https://doi.org/10.1007/978-3-030-19093-4_7

by rejecting transactions that may lead to a violation of the ICs [6]. Another possibility is the use of triggers or active rules that are created by the user and stored in the database. They react to updates of the database by notifying a violation of an IC, rejecting a violating update, or compensating the update with additional updates that restore consistency. Another common alternative consists of keeping the ICs satisfied through the application programs that access and modify the database, i.e., from the transactional side. However, the correctness of triggers or application programs with respect to ensuring database consistency is not guaranteed by the DBMS.

In practice, for many reasons, integrity constraints *are not specified* to the DBMS along with the database schema. In particular, foreign keys could be missing because of (1) lack of support for checking foreign key constraints in the host system, (2) fear that checking such constraints would impede database performance, or (3) lack of database knowledge within the development team. Absence of integrity constraints specification could be considered a bad database design and may lead to many problems such as inconsistency, and anomalies. Anomalies may cause redundancy (during insertion or modification), accidental loss of information (during deletion), waste of storage space, and generation of invalid and spurious data during joins on base relations with matched attributes that may not represent a proper (foreign key, primary key) relationship.

In this paper, we address the problem of *absence of referential integrity constraints* specification, that is, when integrity constraints are not specified to the DBMS. We present a method to identify and repair missing foreign keys in a relational database. In this method, we first identify candidate foreign keys, then we conduct a thorough validation process of those candidates. The validation process comprises three types of verification, namely: data-based, model-based and brute-force verification. The objective of this process is to find valid foreign keys such that they are then specified to, and enforced by the DBMS.

The paper is organized as follows. Section 2 reviews related works. We present some preliminaries and the problem definition in Sect. 3. Then, we give an overview of the solution in Sect. 4 where we discuss identification and validation of candidate foreign keys. Then, we present the three verification steps: data-based (Sect. 5), brute-force (Sect. 6), and model-based (Sect. 7).

2   Related Work

A considerable amount of research has been done in the area of repairing inconsistent databases. Some works focused on *data cleaning* techniques to cleanse the database from data that participates in the violation of the ICs (see [20] for an overview). Other works have addressed the problem of *Consistent Query Answering* (CQA), that is, computing consistent answers over inconsistent database (see [2,5,9,10,22]). Such works rely on the notion of *database repair*, which is a new database instance that is consistent with respect to the ICs, and minimally differs from the inconsistent database at hand. Some of these works are implemented in prototypes systems, such as Hippo [12], and ConQuer [14]. However, these works assume that the integrity constraints are correctly specified; and the problem is in the data. In contrast, in our work, we assume that the integrity constraints are not correctly specified to the DBMS. We focus on identifying and repairing missing referential integrity constraints.

Another related area is the discovery of inclusion dependencies in a given database. Many papers have been addressed approximate and exact discovery of inclusion dependencies [15, 21, 23], and different discovery strategies have been proposed, based on inverted indices [17], sort-merge joins [4], and distributed data aggregation [16]. Research has also devised algorithms for exact discovery of n-ary inclusion dependencies, such as Mind [17] and Binder [19], and for approximate discovery, such as Faida [15]. These works assume data to be complete or consistent, hence proposed discovery methods are mainly based on data.

What distinguish our present work is that we address a two-fold problem. First, foreign keys are missing, therefore their discovery is needed. Second, data itself is not assumed to be consistent or complete, therefore repairing the database instance is also needed. Thus, our proposed solution combines repairing the database schema (specifying valid foreign keys), and repairing the database instance (removing dangling values, when necessary).

In literature, various kinds of repair semantics have been proposed, based on database operations used, and the type of constraints/dependencies. For inclusion dependencies, three types of repairs are possible in general:

1. Tuple-insertion-based repairs [8] – New tuples are inserted in order to satisfy violated constraints. This repair semantics is applied when the database at hand is considered to be incomplete and is then completed via additional tuple insertions. Repairing inclusion dependencies with tuple-insertion requires that values have to be invented for them. This leads to possibly infinitely many repairs. Moreover, value inventions are in general non-deterministic, and complex to handle [6]; and they can lead to the undecidability of consistent query answering [8].

2. Tuple-deletion-based repairs [11] – Tuples that violate constraints are deleted. This class of repairs assume that the database instance at hand is closed, and no insertions of new tuples are accepted [6], therefore integrity-restoration actions are limited to tuple deletions. A good reason for adopting this kind of repair semantics is that, when we insert tuples to enforce inclusion dependencies, we may have to invent data values for the inserted tuples.

3. Null-insertion based repairs [7, 18] – Under this repair semantics, inclusion dependencies are repaired by insertions of null values to restore consistency. Null values can also be used for value invention as required by tuple-insertion-based repairs of referential ICs.

In this present paper, we adopt tuple-deletion-based repairs in the first place, but we also consider null-insertion-based repairs are possible and valid in our case study. However, we do not address tuple-insertion repairs as we assume the database instance is closed, in the sense that no insertions of new tuples are allowed, because they require value invention.

3   Preliminaries and Problem Definition

In relational databases, referential integrity is a property of data stating that references within it are valid. It requires every value of one attribute of a relation to exist as a value of another attribute in a different (or the same) relation. Formally, referential integrity constraints are expressed in terms of inclusion dependencies [1]. Let R be a relation schema and $X = A_1, \cdots, A_n$ a sequence of attributes (possibly with repeats) from R. For an instance I of R, the projection of I onto the sequence X, denoted $I[X]$, is the n-ary relation $I[X] = \{\langle t(A_1), \cdots, t(A_n) \rangle \mid t \in I\}$.

Let $\boldsymbol{R}$ be a relational schema. An inclusion dependency (IND) over $\boldsymbol{R}$ is an expression of the form $\sigma \doteq R[A_1, ..., A_m] \subseteq S[B_1, ..., B_m]$ where: R, and S are (possibly identical) relation names in $\boldsymbol{R}$, $A_1, ..., A_m$ is a sequence of distinct attributes of R, and $B_1, ..., B_m$ is a sequence of distinct attributes of S. An instance $\boldsymbol{I}$ of $\boldsymbol{R}$ satisfies σ, denoted $I \models \sigma$, if $\boldsymbol{I}(R)[A_1, ..., A_m] \subseteq \boldsymbol{I}(S)[B_1, ..., B_m]$. The left-hand side of an IND is referred to as *dependent* attribute(s) and the right-hand side as *referenced* attribute(s) [19]. Both of these attribute sequences must be of the same size m. An IND is said to be unary if m = 1, otherwise it is n-ary. Notice that we do not consider any semantics for null values as they do not contribute to INDs: we simply ignore them.

In this study, we have a relational database that has a large number of tables. This database is poorly designed and suffers from several design problems. However, it is running and a complex application operates on top of it, thus a redesign from scratch is not possible. Our task is to repair the database and get rid of the design issues. Given a relational database that is accessed by an application program, and has a large number of tables; assume the following: (1) all tables have primary keys (declared with the schema), (2) most of tables do not have declared foreign keys, (3) some tables have data records while others do not, (4) all data records are identified using *universally unique identifiers* (UUID); the problem is to identify missing foreign keys and declare them in the database schema such that the integrity constraints are enforced by the DBMS.

Actually, in our case-study, the reasons of inconsistency were mainly the lack of database knowledge within the development team, and inappropriate use of ORM (Object-Relational Mapping) technique [3]. As consequence, integrity constraints were enforced by the application code only, without DBMS support; hence when some records are deleted via the application, their referencing records are not removed, leaving the database in an inconsistent state.

4   Solution Outlines

Identification of Foreign Keys. Our legacy database consists of a large number of tables. All the tables have primary keys (PKs). However, only a subset of the foreign keys are defined at the database level, while the majority of the relationships among the tables are not defined in terms of foreign keys. Thus, our major objective is to identify the foreign keys and define them at the database level in order to enforce the referential integrity constraints. Our method goes as follows. First, we conduct an exhaustive identification of candidate foreign keys. Then, we validate candidate foreign keys using both the data and the application code. Finally, we define valid foreign keys explicitly in the database, and correct invalid foreign keys.

The first step is to obtain an exhaustive list of candidate foreign keys. The result is a table of four columns: table, column, reference table, and reference columns. Table 1 shows a sample of such candidate foreign keys. The identification is performed manually by first investigating the columns names. According to the naming convention used in our database, if a column name contains a name of primary key of another table, it is probably a foreign key that references this primary key. Also, if the column name contains 'parent' this indicates that the column is probably a foreign key that references (the primary key of) its own table. When necessary, the application code is also examined to confirm the candidacy of each foreign key. By the end of this process, we obtain a list of candidate foreign keys.

Table 1. Examples of candidate foreign keys

Table	Column	Reference table	Reference column
department	department_branch_id	branch	branch_id
department	department_parent	department	department_id
department_section	department_section_department_id	department	department_id

Validation of Candidate Foreign Keys. Identified candidate foreign keys need to be validated. Such a validation is conducted over several steps. First, if the owner table of a foreign key has data records ($I(R) \neq \emptyset$), then we use *data-based verification* where data records are used to verify the validity of the foreign key. That is, we check whether the current instance satisfy the constraint or not. Candidate foreign keys that fail in the data-based verification step are subject to another verification step called *brute-force verification*, to find their potential reference tables and/or their known values that have no reference (dangling values.) However, if the owner table is empty ($I(R) = \emptyset$), we use a *model-based verification* where the application program associated with database is used to validate the foreign key (Sect. 7). In any case, candidate foreign keys that successfully pass the data-based verification or model-based verification are considered valid, and thus are declared explicitly in the database schema. Foreign keys that are verified using brute-force step are corrected manually, and dangling values are removed as we will see in Sect. 6. Figure 1 illustrates the overall process of validation of candidate foreign keys.

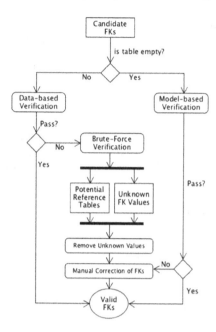

Fig. 1. Process of validation of candidate foreign keys

5   Data-Based Verification

The idea behind data-based verification is the following: a candidate foreign key is valid if the database instance satisfies this foreign key constraint. Formally, let $\boldsymbol{R}$ be a database schema with an instance $\boldsymbol{I}$, and let $\sigma \doteq R[A] \subseteq S[B]$ be a candidate foreign key over $\boldsymbol{R}$, this candidate foreign key σ is considered valid if the current instance $\boldsymbol{I}$ satisfies it ($\boldsymbol{I} \models \sigma$.) In order to verify whether a candidate foreign key $\sigma \doteq R[A] \subseteq S[B]$ is satisfied, the sets of distinct values of the referencing column $R[A]$ and the referenced column $S[B]$ (denoted $\boldsymbol{I}(R)[A]$ and $\boldsymbol{I}(S)[B]$, respectively) are extracted from the database instance $\boldsymbol{I}$. Based on the definition of inclusion dependency, if $\boldsymbol{I}(R)[A] \subseteq \boldsymbol{I}(S)[B]$, then the referential integrity constraint holds, and $\sigma \doteq R[A] \subseteq S[B]$ is indeed a valid foreign key.

The successfully validated foreign keys, are then defined explicitly in the database in order to enforce the referential integrity constraint. This could be done using the following SQL script:

```
ALTER TABLE R ADD FOREIGN KEY (A) REFERENCES S(B);
```

Once a foreign key is defined with the schema, it will be enforced by the DBMS; that is, if a user attempts to insert a tuple that violates the referential IC (e.g., a tuple with a value $t[A] \notin \boldsymbol{I}(S)[B]$), the DBMS will reject this insertion, keeping the constraint satisfied and the database consistent.

Example 1. Consider a database schema with three relations $P_1(A, B)$, $P_2(C, D)$ and $P_3(E)$, and two candidate foreign keys $\sigma_1 \doteq P_1[B] \subseteq P_2[C]$ and $\sigma_2 \doteq$

$P_2[D] \subseteq P_3[E]$. Consider an instance I' such that $I'(P_1) = \{(a,b),(c,d)\}$, $I'(P_2) = \{(b,e),(d,f),(g,h)\}$ and $I'(P_3) = \{(e),(h),(i)\}$. Then we have:

$$\{b,d\} = I'(P_1)[B] \subseteq I'(P_2)[C] = \{b,d,g\}$$
$$\{e,f,h\} = I'(P_2)[D] \nsubseteq I'(P_3)[E] = \{e,h,i\}$$

This means that the instance I' *satisfies* the constraint $\sigma_1 \doteq P_1[B] \subseteq P_2[C]$, but not the constraint $\sigma_2 \doteq P_2[D] \subseteq P_3[E]$. In this example, the first candidate foreign key is *valid* and therefore is declared with the schema, whereas the second candidate is *invalid* and should be verified again with brute-force verification.

6   Brute-Force Verification

Candidate foreign keys that do not successfully pass the data-based verification (i.e., do not satisfy the condition $I(R)[A] \subseteq I(S)[B]$) are subject to another inspection called brute-force verification. The idea behind this test is the following: the values of a candidate foreign key are compared with the values of the primary keys of all tables in the database. When those values *match* (possibly partially) the values of a primary key of a table, this table is considered a *potential reference table* for the candidate foreign key. If there is no match with any table, then such unmatched values are considered unknown and should be removed to keep the consistency of the database. In this section, we present brute-force verification as an algorithm which takes a candidate foreign key as input, and returns a set of potential referenced tables and a set of unknown values as output. According to the outcomes of the algorithm, we present possible solutions for different cases.

When the condition $I(R)[A] \subseteq I(S)[B]$ is not satisfied, this means that there is at least an element $\hat{a} \in I(R)[A]$ such that $\hat{a} \notin I(S)[B]$. Let us denote the set of such elements $\hat{A}$, then we have:

$$\hat{A} = \{a \mid a \in I(R)[A] \land a \notin I(S)[B]\} = I(R)[A] - I(S)[B]$$

In some sense, this set contains incorrect tuples of $I(R)[A]$, i.e., tuples that violate the integrity constraint; thus we want to find their correct references if any. However, besides those incorrect tuples, $I(R)[A]$ could contain other correct tuples that reference $S[B]$. Let us denote the set of correct tuples as A^*, then:

$$A^* = \{a \mid a \in I(R)[A] \land a \in I(S)[B]\} = I(R)[A] \cap I(S)[B]$$

Note that $I(R)[A] = \hat{A} \cup A^*$. The set A^* could be empty, and in this case there is no data-based evidence that S is a potential referenced table for the candidate foreign key. However, if this set A^* is not empty, then the table S remains a potential reference table for $R[A]$.

Let T be a relation schema, and let K_T be the primary key of T, the set of values of this primary key in the database instance is: $I(T)[K_T]$.

The brute-force verification is depicted in Algorithm 1, that takes an input a candidate foreign key $\sigma \doteq R[A] \subseteq S[B]$, and returns a set $\widehat{A}$ of dangling values of $R[A]$, and a set Z of potential referenced tables.

First, the set Z is initially empty, whereas the set A^* equals the intersection $\boldsymbol{I}(R)[A] \cap \boldsymbol{I}(S)[B]$. If this set is not empty, then S is indeed a potential reference table, thus it is added to Z. Then, the set of dangling values $\widehat{A}$ initially equals the difference $\boldsymbol{I}(R)[A] - \boldsymbol{I}(S)[B]$. For every table T in the database $\boldsymbol{R}$, we extract the set of values of T's primary key: $\boldsymbol{I}(T)[K_T]$, and find Q the intersection of this set with $\widehat{A}$. If this intersection Q is empty, the loop continues. But if it is not empty, this means that we found a potential referenced table T for the candidate foreign key $R[A]$, therefore, we append this table T to the set of potential referenced tables Z, and we remove the set Q from $\widehat{A}$, because this subset is not considered dangling any more, since we found its originating table T. Finally, the algorithm returns the set of potential referenced tables Z that have been found, and the set of remaining dangling values $\widehat{A}$.

Algorithm 1. Brute-Force Verification

Input: a candidate foreign key $(\sigma \doteq R[A] \subseteq S[B])$
Output: a set of potential referenced tables Z and a set of dangling values $\widehat{A}$

1: $Z \leftarrow \emptyset$
2: $A^* \leftarrow \boldsymbol{I}(R)[A] \cap \boldsymbol{I}(S)[B]$ ▷ correctly referenced values
3: **if** $A^* \neq \emptyset$ **then**
4:    $Z \leftarrow Z \cup \{S\}$
5: $\widehat{A} \leftarrow \boldsymbol{I}(R)[A] - \boldsymbol{I}(S)[B]$ ▷ dangling values
6: **for** $T \in \boldsymbol{R}$ **do**
7:    **find** $\boldsymbol{I}(T)[K_T]$ ▷ extract the values of T's pk
8:    $Q \leftarrow \widehat{A} \cap \boldsymbol{I}(T)[K_T]$
9:    **if** $Q \neq \emptyset$ **then**
10:       $\widehat{A} \leftarrow \widehat{A} - Q$
11:       $Z \leftarrow Z \cup \{T\}$
12: **return** $Z, \widehat{A}$

Algorithm 1 is executed for every candidate foreign key $\sigma \doteq R[A] \subseteq S[B]$, that has failed in the data-based verification. Based on the outcomes of the algorithm, we distinguish three cases:

- **Case 1.** $Z = \emptyset$, the FK has no potential referenced tables at all.
- **Case 2.** $|Z| = 1$, the FK has one potential referenced table.
- **Case 3.** $|Z| > 1$, the FK has more than one potential referenced tables.

6.1   Case 1. No Potential Referenced Tables

The first case is the easiest one to solve. It means that we are unable to find an alternative table that could be referenced by $R[A]$. This proves that the originally candidate referenced table S should be actually the correct referenced table, even if there is no data-based evidence. Anyway, the problem in this case is only with dangling values $\widehat{A}$. Therefore, the solution is simply to remove the dangling values $\widehat{A}$, and declare the foreign key with the schema. The solution can be expressed in SQL as shown in Listing 1.1.

Listing 1.1. Solution of Case 1

```
--- Remove dangling values
DELETE FROM R WHERE A IN A;

--- Define the foreign key
ALTER TABLE R
ADD FOREIGN KEY (A) REFERENCES S(B);
```

Example 2. Consider a schema with three relations $P_1(A, B)$, $P_2(C)$ and $P_3(D)$ and a candidate foreign key $\sigma \doteq P_1[B] \subseteq P_2[C]$. Consider $I'(P_1) = \{(a, b)\}$, $I'(P_2) = \{(c)\}$ and $I'(P_3) = \{(d)\}$. Clearly, the instance I' does not satisfy σ and the candidate foreign key fails the data-based verification. With brute-force verification, we have $Z = \emptyset$ and $\widehat{B} = \{b\}$. Therefore, the solution is to admit the candidate foreign key σ as valid and declare it with the schema, and to remove the dangling value b, either by removing the entire tuple (a, b), or setting b to null, $I'(P_1) = \{(a, NULL)\}$.

6.2   Case 2. One Potential Referenced Table

The second case is also easy to solve, because the only potential referenced table found in Z must be the correct referenced table that we are looking for. This found table could be the same candidate referenced table S, or another table. Anyway, let us denote it U. The solution is simply to correct the foreign key to be: $\sigma' \doteq R[A] \subseteq U[K_U]$ where K_U is the primary key of U. This implies the removal of the dangling values $\widehat{A}$, and defining the correct foreign key explicitly:

Listing 1.2. Solution of Case 2

```
--- Remove dangling values
DELETE FROM R WHERE A IN A;

--- Define the correct foreign key
ALTER TABLE R
ADD FOREIGN KEY (A) REFERENCES U(K_U);
```

Example 3. Consider a schema with three relations $P_1(A, B)$, $P_2(C)$ and $P_3(D)$ and a candidate foreign key $\sigma \doteq P_1[B] \subseteq P_2[C]$. Consider $I'(P_1) = \{(a, b), (c, d)\}$, $I'(P_2) = \{(e), (f)\}$ and $I'(P_3) = \{(b), (d)\}$. Clearly, $I' \not\models \sigma$ and the candidate foreign key fails the data-based verification. With brute-force verification, we have $Z = \{P_3\}$ and $\widehat{B} = \emptyset$. The solution here is to correct the candidate foreign key to be: $\sigma' \doteq P_1[B] \subseteq P_3[D]$.

6.3   Case 3. Many Potential Referenced Tables

The third case is tricky, as there are many different potential referenced tables $Z = \{U_1, U_2, \cdots, U_n\}$ that are referenced by $R[A]$.

Example 4. Consider a schema with three relations $P_1(A, B)$, $P_2(C)$ and $P_3(D)$ and a candidate foreign key $\sigma \doteq P_1[B] \subseteq P_2[C]$. Consider $I'(P_1) = \{(a, b), (c, d)\}$, $I'(P_2) = \{(b), (e)\}$ and $I'(P_3) = \{(d), (f)\}$ (Fig. 2). Clearly, $I' \not\models \sigma$ and the candidate foreign key fails the data-based verification. With brute-force verification, we have $Z = \{P_2, P_3\}$ and $\widehat{B} = \emptyset$.

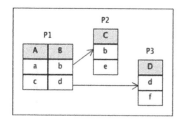

Fig. 2. Database schema of Example 4

This case requires a careful inspection of the database schema and the application code by the database designer in order to know exactly the reason of this situation, and to figure out an appropriate solution. For instance, we could distinguish two cases:

Case 3-1. For instance, it could be the case that only one of these tables is the correct one, and the others are not. Here, the solution is similar to the solution of Case 2 above, but with an additional removal of erroneous values. That is, let us consider that U_1 is the correct table, and the other tables $U_2, \cdots, U_n$ should not be referenced by $R[A]$.

In this case, the only valid foreign key is: $\sigma_1 \doteq R[A] \subseteq U_1[K_{U_1}]$ which should be declared in the schema. In addition to removing dangling values $\widehat{A}$ from $R[A]$, we need also to remove the erroneous values that reference any of $U_2, \cdots, U_n$ tables. This solution can be expressed in SQL as shown in Listing 1.3.

Example 5. In the previous example (Example 4), we found two potential referenced tables $Z = \{P_2, P_3\}$. If we consider P_2 is the correct reference table, then the valid foreign key is $\sigma \doteq P_1[B] \subseteq P_2[C]$ (Fig. 3-a), and the erroneous data record is (c, d). But if we consider P_3 is the correct reference table, then the valid foreign key is $\sigma' \doteq P_1[B] \subseteq P_3[D]$ (Fig. 3-b), and the erroneous data record is (a, b).

Listing 1.3. Solution of Case 3-1

```
--- Remove dangling values
DELETE FROM R WHERE A IN Â;

-- Remove erroneous values
DELETE FROM R
WHERE A IN (SELECT K_U2 FROM U2);
...
DELETE FROM R
WHERE A IN (SELECT K_Un FROM Un);

--- Define the correct foreign key
ALTER TABLE R
ADD FOREIGN KEY (A) REFERENCES U1(K_U1);
```

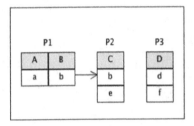

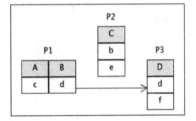

(a) P_2 is the correct reference table   (b) P_3 is the correct reference table

Fig. 3. Solutions for Case 3-1 (Example 5)

Case 3-2. Another possibility, is that all the potential tables $Z = \{U_1, U_2, \cdots, U_n\}$ are considered correct and should be indeed referenced by $R[A]$. This case contradicts with the basic principles of database design, and must be solved radically. Based on the semantics of the relations, two solutions are possible:

Case 3-2, Solution 1. Replace the column A in R, by n columns $A_1, A_2, \cdots, A_n$ that reference $U_1, U_2, \cdots, U_n$, respectively. This means, we replace the candidate foreign key $\sigma \doteq R[A] \subseteq S[B]$ by the following foreign keys:

$$\sigma_1 \doteq R[A_1] \subseteq U_1[K_{U_1}]$$
$$\sigma_2 \doteq R[A_2] \subseteq U_2[K_{U_2}]$$
$$\cdots$$
$$\sigma_n \doteq R[A_n] \subseteq U_n[K_{U_n}]$$

This solution implies that we first delete dangling values, then we add new columns $A_1, A_2, \cdots, A_n$ to R. Now, we should update R such that each column A_i contains the values of A that reference $U_i[K_{U_i}]$, for $i = 1, \cdots, n$. Finally, we drop the column A, and define the foreign keys. This solution can be expressed using SQL as shown in Listing 1.4.

Example 6. In Example 4, we found two potential referenced tables $Z = \{P_2, P_3\}$. If both tables are considered correct and should be referenced by $P_1[B]$, then the solution presented above is to replace the column B by two columns B_1 and B_2, such that we have two foreign keys instead of one, namely: $P_1[B_1] \subseteq P_2[C]$, and $P_1[B_2] \subseteq P_3[D]$ (Fig. 4).

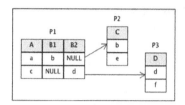

Fig. 4. Solution 1 of Case 3-2 (Example 6)

Listing 1.4. Solution 1 of Case 3-2

```
--- Remove dangling values
DELETE FROM R WHERE A IN Â;

--- Add new columns to R
ALTER TABLE R
    ADD COLUMN A₁ ... ,
    ...
    ADD COLUMN Aₙ ...;

--- Migrate data values from A to the new columns
UPDATE R SET A₁ = A
WHERE A IN (SELECT K_U₁ FROM U₁);
...
UPDATE R SET Aₙ = A
WHERE A IN (SELECT K_Uₙ FROM Uₙ);

--- Drop column A
ALTER TABLE R DROP COLUMN A;

--- Finally, define the correct foreign keys
ALTER TABLE R
    ADD FOREIGN KEY (A₁) REFERENCES U₁(K_U₁),
    ...
    ADD FOREIGN KEY (Aₙ) REFERENCES Uₙ(K_Uₙ);
```

Case 3-2, Solution 2. Replace the whole table R by n new tables, $R_1, R_2, \cdots, R_n$, each of which will be referencing one of the referenced tables $U_1, U_2, \cdots, U_n$. In this case, each one of those tables R_i will contain a version of the column A_i that references $U_i[K_{U_i}]$. This means that we also replace the candidate foreign key $\sigma = R[A] \subseteq S[B]$ by the following foreign keys:

$$\sigma_1 \doteq R_1[A_1] \subseteq U_1[K_{U_1}]$$
$$\sigma_2 \doteq R_2[A_2] \subseteq U_2[K_{U_2}]$$
$$\cdots$$
$$\sigma_n \doteq R_n[A_n] \subseteq U_n[K_{U_n}]$$

This solution implies that we first delete dangling values, then we create new tables $R_1, R_2, \cdots, R_n$, including the definition of foreign keys. Then, data records should be migrated from R to those new tables such that each new table R_i contains the records that reference his corresponding reference table U_i for $i = 1, \cdots, n$. Finally, table R can be dropped safely. This solution can be expressed using SQL code as shown in Listing 1.5.

Listing 1.5. Solution 2 of Case 3-2

```
-- Remove dangling values
DELETE FROM R WHERE A IN Â;

-- Create new tables, including defining correct foreign keys
CREATE TABLE R₁ (
   ..., A₁, ...
   FOREIGN KEY A₁ REFERENCES U₁(K_U₁)
);
...
CREATE TABLE Rₙ (
   ..., Aₙ, ...
   FOREIGN KEY Aₙ REFERENCES Uₙ(K_Uₙ)
);

-- Migrate data from R to the new tables
INSERT INTO R₁ (..., A₁, ...)
SELECT ..., A, ... FROM R
WHERE A IN (SELECT K_U₁ FROM U₁);
...
INSERT INTO Rₙ (..., Aₙ, ...)
SELECT ..., A, ... FROM R
WHERE A IN (SELECT K_Uₙ FROM Uₙ);

-- Finally, drop table R
DROP TABLE R;
```

Example 7. In Example 4, we found two potential referenced tables $Z = \{P_2, P_3\}$. If both tables are considered correct and should be referenced by $P_1[B]$, then the solution presented above is to replace the table $P_1(A, B)$ by two tables $P_{1,1}(A_1, B_1)$ and $P_{1,2}(A_2, B_2)$, such that we have two foreign keys: $P_{1,1}[B_1] \subseteq P_2[C]$, and $P_{1,2}[B_2] \subseteq P_3[D]$, as shown in Fig. 5.

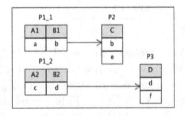

Fig. 5. Solution 2 of Case 3-2 (Example 7)

7   Model-Based Verification

This type of verification is applied for candidate foreign keys whose tables do not have data records, and it is based on the application program. In our case study, each database table has an associated model (Java bean). Model-based verification is based on the following idea: we inspect the model associated with owner table of the candidate foreign key. If this model has a reference to the model associated with referenced table of the foreign key, then we consider that the foreign key is valid. Formally, given a candidate foreign key $\sigma = R[A] \subseteq S[B]$, we inspect the models M_R and M_S associated with the tables R and S, respectively. If the model M_R contains a field a of type M_S, this means that the table R has a column C_a which is probably a foreign key references the table S associated with the model M_S.

8   Conclusion

In this paper, we have presented a method for repairing referential integrity constraints in relational databases. This method is applied when the constraints are not correctly specified in terms of foreign keys in the database schema. The method starts with identifying candidate foreign keys. Then, these candidates are verified using three types of verification: model-based, data-based, and brute-force verification. Valid foreign keys are declared with the database schema and, thus, are enforced by the DBMS to keep the database consistent.

Our method has been applied on a real-world database composed of 167 tables. We have identified 393 candidate foreign keys. Among them, there are 246 foreign keys whose tables have records, hence they are subject to data-based validation; whereas the remaining 147 foreign keys have empty tables, thus they are subject to model-based verification. Among the 246 foreign keys that have been verified using data-based validation, there are 232 that passed, while 14 foreign keys only have failed, thus they are subject to brute-force validation. As a result of brute-force validation, 4 FKs have no potential reference tables (case 1), while 9 FKs have one potential reference table (case 2), and only one FK has three potential reference tables (case 3). For all the 14 foreign keys there were dangling records in the range of 1 to 7 records per table (in total: 31 records). Among the 147 foreign keys that have been verified using model-based validation, only 4 FKs have failed the test and hence have been manually corrected.

References

1. Abiteboul, S., Hull, R., Vianu, V.: Foundations of Databases. Addison-Wesley, Boston (1995). http://webdam.inria.fr/Alice/
2. Arenas, M., Bertossi, L., Chomicki, J.: Consistent query answers in inconsistent databases. In: Proceedings of the 18th Symposium on Principles of Database Systems, PODS 1999, pp. 68–79. ACM, New York (1999)
3. Barry, D., Stanienda, T.: Solving the Java object storage problem. Computer **31**(11), 33–40 (1998)

4. Bauckmann, J., Leser, U., Naumann, F., Tietz, V.: Efficiently detecting inclusion dependencies. In: ICDE, pp. 1448–1450. IEEE Computer Society (2007)
5. Bertossi, L.: Consistent query answering in databases. SIGMOD Rec. **35**(2), 68–76 (2006)
6. Bertossi, L.: Database Repairing and Consistent Query Answering. Morgan & Claypool Publishers (2011)
7. Bravo, L., Bertossi, L.: Semantically correct query answers in the presence of null values. In: Grust, T., et al. (eds.) EDBT 2006. LNCS, vol. 4254, pp. 336–357. Springer, Heidelberg (2006). https://doi.org/10.1007/11896548_27
8. Calì, A., Lembo, D., Rosati, R.: On the decidability and complexity of query answering over inconsistent and incomplete databases. In: Proceedings of the 22nd Symposium on Principles of Database Systems, PODS 2003, pp. 260–271. ACM, New York (2003)
9. Chomicki, J.: Consistent query answering: opportunities and limitations. In: 17th International Workshop on Database and Expert Systems Applications (DEXA 2006), 4–8 September 2006, Krakow, Poland, pp. 527–531 (2006)
10. Chomicki, J.: Consistent query answering: five easy pieces. In: Schwentick, T., Suciu, D. (eds.) ICDT 2007. LNCS, vol. 4353, pp. 1–17. Springer, Heidelberg (2006). https://doi.org/10.1007/11965893_1
11. Chomicki, J., Marcinkowski, J.: Minimal-change integrity maintenance using tuple deletions. Inf. Comput. **197**(1–2), 90–121 (2005)
12. Chomicki, J., Marcinkowski, J., Staworko, S.: Hippo: a system for computing consistent answers to a class of SQL queries. In: Bertino, E., et al. (eds.) EDBT 2004. LNCS, vol. 2992, pp. 841–844. Springer, Heidelberg (2004). https://doi.org/10.1007/978-3-540-24741-8_53
13. Elmasri, R., Navathe, S.: Fundamentals of Database Systems, 6th edn. Addison-Wesley Publishing Company, Boston (2010)
14. Fuxman, A., Fazli, E., Miller, R.J.: ConQuer: efficient management of inconsistent databases. In: Proceedings of the 2005 ACM SIGMOD International Conference on Management of Data, pp. 155–166. ACM, New York (2005)
15. Kruse, S., et al.: Fast approximate discovery of inclusion dependencies. In: Datenbanksysteme für Business, Technologie und Web (BTW 2017) (2017)
16. Kruse, S., Papenbrock, T., Naumann, F.: Scaling out the discovery of inclusion dependencies. In: Datenbanksysteme für Business. Technologie und Web (BTW 2015), pp. 445–454. Gesellschaft für Informatik e.V, Bonn (2015)
17. Marchi, F.D., Lopes, S., Petit, J.M.: Unary and N-ary inclusion dependency discovery in relational databases. J. Intell. Inf. Syst. **32**(1), 53–73 (2009)
18. Molinaro, C., Greco, S.: Polynomial time queries over inconsistent databases with functional dependencies and foreign keys. Data Knowl. Eng. **69**(7), 709–722 (2010)
19. Papenbrock, T., Kruse, S., Quiané-Ruiz, J.A., Naumann, F.: Divide & conquer-based inclusion dependency discovery. VLDB Endow. **8**(7), 774–785 (2015)
20. Rahm, E., Do, H.H.: Data cleaning: problems and current approaches. IEEE Data Eng. Bull. **23**, 2000 (2000)
21. Rostin, A., Albrecht, O., Bauckmann, J., Naumann, F., Leser, U.: A machine learning approach to foreign key discovery. In: 12th International Workshop on the Web and Databases, WebDB 2009, Rhode Island, USA (2009)
22. Wijsen, J.: Consistent query answering under primary keys: a characterization of tractable queries. In: Database Theory - ICDT 2009, 12th International Conference, St. Petersburg, Russia, 23–25 March 2009, Proceedings, pp. 42–52 (2009)
23. Zhang, M., Hadjieleftheriou, M., Ooi, B.C., Procopiuc, C.M., Srivastava, D.: On multi-column foreign key discovery. VLDB Endow. **3**(1–2), 805–814 (2010)

Interactive Decomposition of Relational Database Schemes Using Recommendations

Raji Ghawi$^{(\boxtimes)}$ (iD)

Bavarian School of Public Policy, Technical University of Munich,
Richard-Wagner-Straße. 1, 80333 Munich, Germany
raji.ghawi@tum.de

Abstract. Schema decomposition is a well known method for logical database design. Decomposition mainly aims at redundancy reduction and elimination of anomalies. A good decomposition should preserve dependencies and maintain recoverability of information. We propose a semi-automatic method for decomposing a relational schema in an interactive way. A database designer can build the subschemes step-by-step, guided by quantitative measures of decomposition "goodness". At each step, a ranked set of recommendations are provided to the designer to guide him to the next possible actions that lead to a better design.

Keywords: Relational databases · Functional dependencies · Schema decomposition · Normal forms · Recommendation

1 Introduction

Logical database design is a wide field that has been very well studied in the past decades. Schema decomposition is a well known method for logical database design which mainly aims at eliminating anomalies and reducing redundancy. Decomposing a relation involves splitting its attributes to make the schemes of new relations.

In fact, careless selection of a relational database schema can lead to redundancy and related anomalies. It also introduces the potential for several kinds of errors. Therefore, one of the motivations for performing a decomposition is that it may eliminate anomalies and reduce redundancy. Normal forms have long been studied as a means of reducing redundancies caused by data dependencies in the process of schema design.

Besides anomaly elimination, literature suggests two other properties that are desired in a decomposition [1,16,27]: information recoverability and dependency preservation. Recoverability of information means that the original relation can be recovered by joining the relations in the decomposition. Dependency preservation means that the set of dependencies of the original schema are preserved within the subschemes of the decomposition. Among all the data dependencies

© Springer Nature Switzerland AG 2019
S. Kozielski et al. (Eds.): BDAS 2019, CCIS 1018, pp. 97–108, 2019.
https://doi.org/10.1007/978-3-030-19093-4_8

which have been proposed in the literature, functional dependencies (FDs) are the most common kind of constraints in a relational database.

In this paper, we propose a semi-automatic method for decomposing a relational schema in an interactive way. A database designer can build the sub-schemes step-by-step guided with quantitative measures of the decomposition "goodness". Our goal is to provide richer insights that help a database designer have a better understanding of the consequences of different decomposition choices and make design decisions accordingly. Therefore, at each step, a ranked set of recommendations are provided to the designer that guide him to the possible next actions that lead to a better design. In order to do so, quantitative "goodness" measures of the decomposition need to be defined.

The paper is organized as follows. Section 2 provides some preliminaries, and Sect. 3 discusses the properties of good decomposition. Quantitative measures of decomposition goodness are defined in Sect. 4. Then, in Sect. 5 we present our proposed method for interactive decomposition using recommendations. Section 6 concludes the paper.

2 Preliminaries

Let $R(A_1, \cdots, A_n)$ be a relation scheme, and X and Y be subsets of $\{A_1, \cdots, A_n\}$. We say X functionally determines Y, denoted $X \rightarrow Y$ if, in whatever instance r of R, any two tuples that agree on X values they also agree on Y values. We say a relation instance r satisfies functional dependency $X \rightarrow Y$ if for every two tuples t_1 and t_2 in r such that $t_1[X] = t_2[X]$, it is also true that $t_1[Y] = t_2[Y]$.

Let F be a set of FD's for relation scheme R, and let $X \rightarrow Y$ be a FD. We say F logically implies $X \rightarrow Y$, if every instance r for R that satisfies the dependencies in F also satisfies $X \rightarrow Y$. The closure of F, denoted F^+, is the set of functional dependencies that are logically implied by F.

Armstrong [3] defined a complete and sound set of inference rules for functional dependencies:

- *Reflexivity.* If $Y \subseteq X$, then $X \rightarrow Y$ holds (called trivial FD).
- *Augmentation.* If $X \rightarrow Y$ holds, then $XZ \rightarrow YZ$ holds.
- *Transitivity.* If $X \rightarrow Y$ and $Y \rightarrow Z$ hold, then $X \rightarrow Z$ holds.

Let F be a set of FD's on set of attributes U, and let X be a subset of U, then, the closure of X (with respect to F), denoted X^+, is the set of attributes A such that $X \rightarrow A$ can be deduced from F by Armstrong's axioms. The FD $X \rightarrow Y$ follows from a given set of dependencies F using Armstrong's axioms if and only if $Y \subseteq X^+$; where the closure of X is taken with respect to F.

If R is a relation scheme with attributes $A_1, \cdots, A_n$ and functional dependencies F, and X is a subset of $A_1, \cdots, A_n$, we say X is a *superkey* of R if $X \rightarrow A_1, \cdots, A_n$ is in F^+. That is, X functionally determines all attributes of R. We say X is a *key* of R if X is a superkey and no proper subset $Y \subset X$ is a superkey of R. Clearly, a relation can have more than one key. An attribute A

is called *prime* if it belongs to any key. Thus, an attribute that does not belong to any key is called *non-prime*.

Let F and E be two sets of FD's, we say that F and E are equivalent, denoted $F \equiv E$, if and only if $F^+ = E^+$. That is, we can deduce all FD's of E from F, and vice versa. Moreover, we say that E is a minimal cover of F if and only if $E \equiv F$ and there is no proper subset $E' \subset E$ such that $E' \equiv F$.

If R is a relation scheme with attributes $A_1, \cdots, A_n$ and functional dependencies F, and S is a subscheme of R with attributes $A_{i_1}, \cdots, A_{i_k} \subseteq A_1, \cdots, A_n$. The projection of F over S, denoted F_S is the set of functional dependencies $X \to Y$ in F^+ such that $(X \cup Y) \subseteq A_{i_1}, \cdots, A_{i_k}$.

Let R be a relation schema with FDs F.

- We say that R is in *First Normal Form* (1NF) if all the attributes are atomic (single-valued).
- We say that R is in *Second Normal Form* (2NF) [8] if it is in 1NF and for a non-prime attribute A, whenever $X \to A$ is in F^+, then X is not a proper subset of any key.
- We say that R is in *Third Normal Form* (3NF) if whenever $X \to A$ is in F^+, then either X is a superkey of R or A is a prime attribute.
- We say that R is in *Boyce-Codd Normal Form* (BCNF) if whenever $X \to A$ is in F^+, then X is a superkey of R.

Clearly, if R is in 3NF then it is also in 2NF; and if it is in BCNF then it is also in 3NF.

Other normal forms are formulated in the literature, such as Fourth Normal Form (4NF) [17], Fifth Normal Form (5NF) [11,18] (also known as Project-Join Normal Form, PJNF), Sixth Normal Form (6NF) [10,12], and Essential Tuple Normal Form (ETNF) [9]. However, these normal forms are based on other types of dependencies: multivalued-dependencies (MVD's) and join-dependencies (JD's). Therefore, they are beyond the scope of this paper.

A decomposition of a relation scheme $R(A_1, \cdots, A_n)$ is its replacement by a collection $\delta = \{R_1, \cdots R_k\}$ of subsets of R, called *subschemes*, such that $R = R_1 \cup \cdots \cup R_k$. The subsets R_i's are not required to be disjoint. If r is an instance of R then the sub-instance of a subscheme R_i is the projection of r on R_i, that is, $r_i = \pi_{R_i}(r)$.

One of the motivations for performing a decomposition is that it may eliminate (insert, delete and update) anomalies and reduce redundancy.

3   What Is a Good Decomposition?

A decomposition of a relation scheme R is its replacement by a collection $\delta = \{R_1, \cdots R_k\}$ of subsets of R, called *subschemes*, such that $R = R_1 \cup \cdots \cup R_k$. The subsets R_i's are not required to be disjoint. If r is an instance of R then the sub-instance of a subscheme R_i is the projection of r on R_i, that is, $r_i = \pi_{R_i}(r)$.

Literature shows three main properties that a decomposition is desired to have: [4,21,22]: Elimination of Anomalies, Recoverability of Information, and Preservation of Dependencies.

1. *Elimination of Anomalies* can be described in terms of normal forms. The literature shows that certain undesirable anomalies can be avoided when the database scheme is in a normal form w.r.t the given dependencies [22].
2. *Recoverability of Information* (or lossless-join) means that the original relation can be recovered by taking the natural join of the relations in the decomposition. In fact, any decomposition gives back at least the tuples with which we start, but a carelessly chosen decomposition can give tuples in the join that were not in the original relation [19]. Formally, a decomposition $\delta = \{R_1, \cdots, R_n\}$ of a schema R is recoverable iff for whatever instance r of R: $\pi_{R_1}(r) \bowtie \cdots \bowtie \pi_{R_n}(r) = r$. Lossless-join property can be checked using *chase* test [2].
3. *Preservation of Dependencies* means that all the functional dependencies that hold in the original relation can be deduced from the FD's in the decomposed relations. Formally, a decomposition $\delta = \{R_1, \cdots, R_n\}$ of a schema R with a set of FD's F is dependency preserving iff: $\bigcup_{i=1}^{n} F_{R_i} \equiv F$ where F_{R_i} is the projection of F over R_i. An algorithm to test the dependency preservation property is given in [5].

The literature shows several well known works on decomposition methods that achieve some of the above desired properties. Bernstein [6] proposed a "synthetic" approach for dependency-preserving decomposition into 3NF. Tsou and Fischer [26] proposed a lossless-join decomposition into BCNF. Biskup et al. [7] give a 3NF decomposition with a lossless join and dependency preservation. Demba [13] propose an algorithmic approach for database normalization up to third normal form.

However, it has been shown that there is no decomposition that guarantees all the three properties at once. That is, sometimes, decomposition into BCNF can lose the dependency-preservation property, while decomposition into 3NF does not guarantee to eliminate all redundancy due to FD's [19]. Moreover, it is important to remember that not every lossless-join decomposition step is beneficial, and some can be harmful [27]. Codd has argued that we should not insist that a relation schema be in a given normal form. Rather, the database designer should be aware of the issues and have a warning flag that if the relation schema is not in a given normal form, then certain problems may arise.

A database designer may wish to thoroughly investigate many decompositions in order to choose a good one for his design. However, the number of possible decompositions of a schema is exponential; thus, examining the whole the search space of decompositions is almost unfeasible. In order to overcome this drawback, we suggest to navigate step-wisely through this space. That is, from some 'current' decomposition, the designer can investigate the neighbors of that decomposition that are one step away; where a step means an action of adding or removing of a subscheme or an attribute.

In this paper, we propose an interactive decomposition method, that supports this idea of navigational search for good decomposition. Our goal is to guide the database designer throughout the decomposition process with quantitative measures that assesses the "goodness" of the decomposition.

In the literature, there are many works that have proposed tools and solutions to support database decomposition, such as Mirco [15], RDBNorma [14], and JMathNorm [28]. Some of such tools have educational purposes only, including: [20,23–25]. However, to the best of our knowledge, there is yet no tool that support interactive decomposition in the manner we do in this paper, where recommendations (equipped with quantitative measures) for next steps are provided to guide the user during the decomposition process.

In the next section, we define decomposition "goodness" measures.

4   Decomposition Goodness Measures

Let R be a relation schema with FD's F. Let D_R be the set of all possible decompositions over R, then a **goodness measure** is a function: $\vartheta : D_R \longmapsto [0,1]$. Let $\delta \in D_R$ be a decomposition of k subschemes: $\delta = \{R_1, \cdots, R_k\}$, we define four goodness measures:

4.1   Join-Lossless Measure ϑ_J

This is a strict measure that takes value 1 when δ is join-lossless, and 0 otherwise:

$$\vartheta_J(\delta) = \begin{cases} 1 & \text{if } \delta \text{ is join lossless} \\ 0 & \text{otherwise} \end{cases}$$

4.2   Dependency-Preservation Measure ϑ_P

This measure can also be defined in a strict fashion: $\vartheta_P(\delta) = 1$ when δ is dependency-preserving, 0 otherwise.

$$\vartheta_P(\delta) = \begin{cases} 1 & \text{if } \delta \text{ is dependency-preserving} \\ 0 & \text{otherwise} \end{cases}$$

Alternatively, it can be defined in a relaxed fashion:

$$\vartheta_P(\delta) = \frac{|F_{pr}|}{|F|}$$

where $F_{pr} \subseteq F$ is the set of FD's preserved by δ.

4.3   Normal-Forms Measure ϑ_N

Let R_i be a subscheme in δ, then the normal-forms measure of R_i is defined as:

$$\vartheta_N(R_i) = \begin{cases} 3 & \text{if } R_i \text{ is in BCNF} \\ 2 & \text{if } R_i \text{ is not in BCNF, but in 3NF} \\ 1 & \text{if } R_i \text{ is not in 3NF, but in 2NF} \\ 0 & \text{if } R_i \text{ is not in 2NF} \end{cases}$$

The normal-forms measure ϑ_N of a decomposition δ is then the normalized average of the normal-forms measure of its components:

$$\vartheta_N(\delta) = \frac{1}{3k} \sum_{i=1}^{k} \vartheta_N(R_i)$$

Clearly, ϑ_N would be 0 when none of the subschemes is in 2NF, and reaches 1 when all the subschemes are in BCNF.

4.4   Structural Issues Measure ϑ_I

When a decomposition is built in an interactive way by adding or removing subschemes and/or attributes, structural *issues* may arise. These issues can be classified in different types as shown in Table 1.

Let Ω_δ be the set of all the structural issues occurring in a decomposition δ, the Structural-Issues Measure ϑ_I is defined as:

$$\vartheta_I(\delta) = \frac{1}{1 + |\Omega_\delta|}$$

Clearly, ϑ_I can not be 0; it would be 1 when the decomposition δ does not suffer any structural issues ($\Omega_\delta = \emptyset$).

Table 1. Types of structural issues

Issue	Formulation
A subsheme has no attributes	$\exists R_i \in \delta : R_i = \emptyset$
A subscheme has one attribute only	$\exists R_i \in \delta : |R_i| = 1$
A subscheme is the same as the original schema	$\exists R_i \in \delta : R_i = R$
An attribute is not mentioned in any relation	$\exists a \in R, \forall R_i \in \delta : a \notin R_i$
Two subschemes have exactly the same attributes	$\exists R_i, R_j \in \delta : R_i = R_j$
A subscheme is a proper subset of another	$\exists R_i, R_j \in \delta : R_i \subset R_j$
A subscheme has no shared attributes with others	$\exists R_i \in \delta, \forall R_j \in \delta/R_i : R_i \cap R_j = \emptyset$

The list shown in Table 1 contains the most common and frequent types of structural issues. These types are what we consider in the current version of our tool. However, we do not claim that this list of issues is exhaustive.

Moreover, the formula of Structural-Issues Measure can be modified to obtain a smoother decay of score when the number of issues increases, for example, using $\vartheta_I(\delta) = \frac{3}{3+|\Omega_\delta|}$.

Total Score θ

To summarize the above goodness measures, a total score θ is defined as their weighted average. Let w_J, w_P, w_N, and w_I be non-negative weights that sum up to 1. These weights signify the degree of importance of the goodness measures, $\vartheta_J, \vartheta_P, \vartheta_N$, and ϑ_I, respectively. The total score θ is computed as:

$$\theta(\delta) = w_J \vartheta_J(\delta) + w_P \vartheta_P(\delta) + w_N \vartheta_N(\delta) + w_I \vartheta_I(\delta)$$

Example 1. Let $R(A, B, C, D, E)$ be a schema with FD's $F = \{A \rightarrow B, E \rightarrow G, B \rightarrow DE\}$, let $\delta = \{R_1(B, D, E), R_2(C, E, G)\}$ be a decomposition of R. Then, the FD's of the subschemes are: $F_{R_1} = \{B \rightarrow DE\}$ and $F_{R_2} = \{E \rightarrow G\}$.

- δ is not join-lossless: $\vartheta_J(\delta) = 0$.
- FD's $E \rightarrow G$ and $B \rightarrow DE$ are preserved, but $A \rightarrow B$ is not: $\vartheta_P(\delta) = 2/3$.
- R_1 is in BCNF, and R_2 is not in 2NF: $\vartheta_N(R_1) = 3$, $\vartheta_N(R_2) = 0$, thus $\vartheta_N(\delta) = (3 + 0)/(3 \times 2) = 1/2$.
- δ suffers one issue: attribute A is not mentioned in any relation, thus $|\Omega| = 1$, and $\vartheta_I(\delta) = 1/2$.
- Assuming balanced weights, the total score is: $\theta = (0 + \frac{2}{3} + \frac{1}{2} + \frac{1}{2})/4 = \frac{5}{12}$.

5   Interactive Decomposition and Recommendations

In order to help a database designer make a good decomposition, s/he first needs to be able to measure the "goodness" of the decomposition, and second to better understand the consequences of different decomposition choices; thus s/he can make design decisions accordingly. We propose a semi-automatic method for decomposing a relational schema in an interactive way, where the designer can build the subschemes in a step-by-step way.

Given an original database schema R, and a set of FD's F, the designer starts with one empty subscheme (with no attributes). Then, s/he can add attributes to this subscheme or add another subscheme. Gradually, the decomposition becomes richer and other types of actions become possible, such as removing an attribute from certain subscheme or removing an entire subscheme. In general, four types of actions are possible at any step:

1. Add a new subscheme. $\delta \leftarrow \delta \cup R'$
2. Remove a subscheme. $\delta \leftarrow \delta - R'$
3. Add an attribute to a subscheme. $\delta \leftarrow (\delta - R') \cup (R' \cup \{a\})$
4. Remove an attribute from a subscheme. $\delta \leftarrow (\delta - R') \cup (R' - \{a\})$

Every time an action is taken, two things happen. First, the "goodness" measures (Sect. 4) are computed for the current decomposition δ, including the total score $\theta(\delta)$. This involves finding the FD projections and normal forms of each subscheme, and the set of structural issues, if any.

Second, a list of next-possible-actions $\Psi = \{\psi_1, \cdots, \psi_m\}$ is generated (Sect. 5.1). For each next-possible-action $\psi_j \in \Psi$, the corresponding decomposition δ_{ψ_j} (resulting if the action ψ_j is taken) is computed behind the scene, along with its goodness metrics including the total score $\theta(\delta_{\psi_j})$. The total score is used to rank the list of actions, such that the actions with higher scores are displayed first. Moreover, each action is annotated as *positive*, *equivalent* or *negative*, respectively, based on whether the action score $\theta(\delta_{\psi_j})$ is, respectively, *higher than*, *equal to*, or *less than* the score of the current decomposition $\theta(\delta)$.

This way, a ranked list of action-score pairs is presented to the designer: $\{\langle \psi_j, \theta(\delta_{\psi_j}) \rangle\}$. In this list, *positively* annotated actions form the recommendations that are given to the designer to choose one from them. Those recommendations are guaranteed to give, if taken, a better decomposition than the current one in terms of the total score of the goodness measures.

When an action (unnecessarily positive) is taken, the list of recommendations will change (re-computed). That is because the next possible actions would be different, and their corresponding decompositions will vary accordingly.

Obviously, the designer is not obligated to take the first recommendation in the list, s/he is free to take any action. However, taking top recommended actions would rapidly lead to better decomposition. It is important to note that an optimal decomposition is not always possible. Therefore, the designer may stop the decomposition process whenever s/he feels satisfied with the goodness measures s/he gets.

A prototype of this interactive method is implemented in Java as a user-friendly GUI (Fig. 1). The source code, examples, and other resources are available at: https://goo.gl/rSPOin.

5.1 Algorithm for Recommendation Generation

The list of recommendations of each step is generated using Algorithm 1. For each subscheme, we consider actions of adding an attribute that is not currently in the subscheme (lines 3:7), and removing an attribute from the subscheme (lines 8:12). If there are more than one subscheme, we also consider removing each subscheme (lines 13:17). For each key, we consider adding a subscheme that consists of the attributes of that key, if such a subscheme is not already present in the decomposition (lines 18:22). For each FD in the minimal cover of F, we consider adding a subscheme that consists of the left and right side attributes of that FD, if such a subscheme is not already in the decomposition (lines 23:28).

For each one of these actions, ψ, the corresponding decomposition δ_ψ is found and its total score $\theta(\delta_\psi)$ is computed. The pairs $\langle \psi, \theta_\psi \rangle$ are then added to the list of recommendations Ψ, which is, at the end, ordered by the values of θ_ψ and returned.

Algorithm 1. Compute a List of Recommendations

Require: R original schema, F a set of FD's, δ a decomposition of R.
Ensure: Ψ a set of recommendations.
1: $\Psi := \emptyset$
2: **for** each subscheme $R_i \in \delta$ **do**
3:      **if** $|R - R_i| > 1$ **then**
4:          **for** each attribute $A \in R - R_i$ **do**
5:              $\psi := (\text{Add } A \text{ to } R_i)$;
6:              $\delta_\psi := (\delta - R_i) \cup (R_i \cup \{A\})$;
7:              $\theta_\psi := \theta(\delta_\psi)$; $\Psi := \Psi \cup \{\langle \psi, \theta_\psi \rangle\}$
8:      **if** $|R_i| > 1$ **then**
9:          **for** each attribute $B \in R$ **do**
10:             $\psi := (\text{Remove } B \text{ from } R_i)$;
11:             $\delta_\psi := (\delta/R_i) \cup (R_i/\{A\})$;
12:             $\theta_\psi := \theta(\delta_\psi)$; $\Psi := \Psi \cup \{\langle \psi, \theta_\psi \rangle\}$
13: **if** $|\delta| > 2$ **then**
14:     **for** each subscheme $R_i \in \delta$ **do**
15:         $\psi := (\text{Remove } R_i \text{ from } \delta)$;
16:         $\delta_\psi := \delta/R_i$;
17:         $\theta_\psi := \theta(\delta_\psi)$; $\Psi := \Psi \cup \{\langle \psi, \theta_\psi \rangle\}$
18: **for** each key K of R **do**
19:     **if** $K \notin \delta$ **then**
20:         $\psi := (\text{Add } K \text{ to } \delta)$;
21:         $\delta_\psi := \delta \cup K$;
22:         $\theta_\psi := \theta(\delta_\psi)$; $\Psi := \Psi \cup \{\langle \psi, \theta_\psi \rangle\}$
23: **for** each FD $X \rightarrow Y \in E = minCover(F)$ **do**
24:     $S := X \cup Y$
25:     **if** $S \notin \delta$ **then**
26:         $\psi := (\text{Add } S \text{ to } \delta)$;
27:         $\delta_\psi := \delta \cup S$;
28:         $\theta_\psi := \theta(\delta_\psi)$; $\Psi := \Psi \cup \{\langle \psi, \theta_\psi \rangle\}$
29: order Ψ by θ_ψ desc.
30: **return** Ψ

5.2 Example

Consider the schema $R(A, B, C, D, E, G)$ with FD's $F = \{AB \rightarrow CD, A \rightarrow E, B \rightarrow G, EG \rightarrow C\}$. Let us start a new decomposition δ with a new subscheme R_1 and add attribute A to R_1. At this point, the goodness measures are: $\vartheta_J = 0$, $\vartheta_P = 0$, $\vartheta_N = 1$, $\vartheta_I = 0.125$ (due to 7 issues), and $\theta = 0.28125$. The top 3 recommendations are:

$\langle \psi_{11} :: \text{Add attribute } E \text{ to subscheme } R_1, 0.3542 \rangle$
$\langle \psi_{12} :: \text{Add new subscheme: } \{C, E, G\}, \quad 0.3542 \rangle$
$\langle \psi_{13} :: \text{Add new subscheme: } \{A, E\}, \quad\quad 0.3482 \rangle$

If we take the first recommendation, we get $\delta = \{R_1(A, E)\}$ with $F_{R_1} = \{A \rightarrow E\}$. This will make $\vartheta_P = 0.25$ (1 FD is preserved), and $\vartheta_I = 0.1667$ (issues are reduced to 5); therefore, $\theta = 0.3542$ (as the recommendation promised). The top-3 next recommendations are:

$\langle\psi_{21} ::$ Add new subscheme: $\{C, E, G\}, 0.4583\rangle$
$\langle\psi_{22} ::$ Add new subscheme: $\{B, G\}, \quad 0.425\rangle$
$\langle\psi_{23} ::$ Add new subscheme: $\{A, B, D\}, 0.3958\rangle$

If we again take the first recommendation, we get $\delta = \{R_1(A, E), R_2(C, E, G)\}$ with $F_{R_2} = \{EG \rightarrow C\}$. This will make $\vartheta_P = 0.5$ (2 FD's are preserved), and $\vartheta_I = 0.333$ (2 issues remaining: attributes B, D are not mentioned in any subscheme); therefore, $\theta = 0.4583$ (Fig. 1).

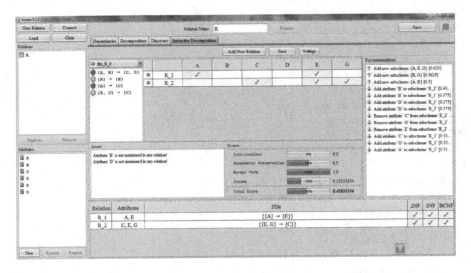

Fig. 1. Decomposition after adding subscheme $R_2\{C, E, G\}$

The top-3 next recommendations are:

$\langle\psi_{21} ::$ Add new subscheme: $\{C, E, G\}, 0.4583\rangle$
$\langle\psi_{22} ::$ Add new subscheme: $\{B, G\}, \quad 0.425\rangle$
$\langle\psi_{23} ::$ Add new subscheme: $\{A, B, D\}, 0.3958\rangle$

Continuing the decomposition process with recommended actions, we shall add $R_3(A, B, D)$ and $R_4(B, G)$, hence an optimal decomposition can be achieved: $\delta = \{R_1(A, E), R_2(C, E, G), R_3(A, B, D), R_4(B, G)\}$ which is lossless-join, all FD's are preserved, all subschemes are in BCNF, and free from structural issues; thus, $\theta(\delta) = 1$.

6   Conclusions and Future Work

We have proposed a semi-automatic method for interactive decomposition of relational databases. A database designer is guided throughout the decomposition process by a list of recommendations that tell him what are the next possible actions that can lead to a better decomposition.

Interactive decomposition seems to be a good approach to help the designer better understand the pros and cons of every possible action by quantitatively assessing the goodness of the decomposition. This method can be considered as an interesting addition to the arsenal of already established methods and tools within database design literature. As a future work, an extensive experimentation is needed to evaluate the method, and to study the impact of each goodness property on the overall quality of database design. Looking ahead, interactive decomposition can be extended to involve extra goodness measures such as: minimality, freedom from globally redundant attributes, and freedom from attribute replication. Moreover, it can be extended to consider other types of dependencies (e.g. MVDs) and higher normal forms (e.g. 4NF).

Other directions of future work concern improving the tool and making the interface more user-friendly; by introducing, for instance, graphical representation on functional dependencies in the form of FD diagrams, and graphical representation of join- diagram of the decomposition.

Finally, future work should address the usefulness of the decomposition method in real-life situations where the search space is huge. Therefore, it is necessary to conduct some experiments with real-life databases consisting of many, possibly non-normalized, relations.

References

1. Abiteboul, S., Hull, R., Vianu, V.: Foundations of Databases. Addison-Wesley, Boston (1995)
2. Aho, A.V., Beeri, C., Ullman, J.D.: The theory of joins in relational databases. ACM Trans. Database Syst. **4**(3), 297–314 (1979)
3. Armstrong, W.W.: Dependency structures of data base relationships. In: IFIP Congress, pp. 580–583 (1974)
4. Arora, A.K., Carlson, C.R.: The information preserving properties of relational database transformations. In: Proceedings of the Fourth International Conference on Very Large Data Bases, VLDB 1978, vol. 4, pp. 352–359. VLDB Endowment (1978)
5. Beeri, C., Honeyman, P.: Preserving functional dependencies. SIAM J. Comput. **10**(3), 647–656 (1981)
6. Bernstein, P.A.: Synthesizing third normal form relations from functional dependencies. ACM Trans. Database Syst. **1**(4), 277–298 (1976)
7. Biskup, J., Dayal, U., Bernstein, P.A.: Synthesizing independent database schemas. In: Proceedings of the 1979 ACM SIGMOD International Conference on Management of Data, SIGMOD 1979, pp. 143–151. ACM, New York (1979)
8. Codd, E.F.: A relational model of data for large shared data banks. Commun. ACM **13**(6), 377–387 (1970)

9. Darwen, H., Date, C.J., Fagin, R.: A normal form for preventing redundant tuples in relational databases. In: Proceedings of the 15th International Conference on Database Theory, ICDT 2012, pp. 114–126. ACM, New York (2012)

10. Date, C.J., Darwen, H., Lorentzos, N.A.: Temporal Data and the Relational Model. Elsevier, Amsterdam (2002)

11. Date, C.: An Introduction to Database Systems, 8th edn. Addison-Wesley Longman Publishing Co. Inc., Boston (2003)

12. Date, C., Darwen, H., Lorentzos, N.: Time and Relational Theory, Second Edition: Temporal Databases in the Relational Model and SQL, 2nd edn. Morgan Kaufmann Publishers Inc., San Francisco (2014)

13. Demba, M.: Algorithm for relational database normalization up to 3NF. Int. J. Database Manag. Syst. **5**, 39–51 (2013)

14. Dongare, Y., Dhabe, P., Deshmukh, S.: RDBNorma: a semi-automated tool for relational database schema normalization up to third normal form. arXiv preprint arXiv:1103.0633 (2011)

15. Du, H., Wery, L.: Micro: a normalization tool for relational database designers. J. Netw. Comput. Appl. **22**(4), 215–232 (1999)

16. Elmasri, R., Navathe, S.: Fundamentals of Database Systems, 6th edn. Addison-Wesley Publishing Company, Boston (2010)

17. Fagin, R.: Multivalued dependencies and a new normal form for relational databases. ACM Trans. Database Syst. **2**(3), 262–278 (1977)

18. Fagin, R.: Normal forms and relational database operators. In: Proceedings of the 1979 ACM SIGMOD International Conference on Management of Data, SIGMOD 1979, pp. 153–160. ACM, New York (1979)

19. Garcia-Molina, H., Ullman, J.D., Widom, J.: Database Systems: The Complete Book, 2nd edn. Prentice Hall Press, Upper Saddle River (2008)

20. Kung, H.J., Tung, H.L.: A web-based tool to enhance teaching/learning database normalization. In: Proceedings of the 2006 Southern Association for Information Systems Conference. Jacksonville (2006)

21. Maier, D.: The Theory of Relational Databases. Computer Science Press, Rockville (1983)

22. Maier, D., Mendelzon, A.O., Sadri, F., Dullman, J.: Adequacy of decompositions of relational databases. J. Comput. Syst. Sci. **21**(3), 368–379 (1980)

23. Piza-Dávila, H.I., Gutiérrez-Preciado, L.F., Ortega-Guzmán, V.H.: An educational software for teaching database normalization. Comput. Appl. Eng. Educ. **25**(5), 812–822 (2017)

24. Stefanidis, C., Koloniari, G.: An interactive tool for teaching and learning database normalization. In: Proceedings of the 20th Pan-Hellenic Conference on Informatics, PCI 2016, pp. 18:1–18:4. ACM, New York (2016)

25. Taofiki, A.A., Tale, A.O.: A visualization tool for teaching and learning database decomposition system. J. Inf. Comput. Sci. **7**(1), 003–010 (2012)

26. Tsou, D.M., Fischer, P.C.: Decomposition of a relation scheme into Boyce-Codd normal form. SIGACT News **14**(3), 23–29 (1982)

27. Ullman, J.D.: Principles of Database and Knowledge-base Systems, vol. I. Computer Science Press Inc., New York (1988)

28. Yazici, A., Karakaya, Z.: JMathNorm: a database normalization tool using mathematica. In: Shi, Y., van Albada, G.D., Dongarra, J., Sloot, P.M.A. (eds.) ICCS 2007. LNCS, vol. 4488, pp. 186–193. Springer, Heidelberg (2007). https://doi.org/10.1007/978-3-540-72586-2_27

Artificial Intelligence, Data Mining and Knowledge Discovery

Comparison Study on Convolution Neural Networks (CNNs) vs. Human Visual System (HVS)

Manuel Caldeira[1], Pedro Martins[2(✉)], José Cecílio[3], and Pedro Furtado[3]

[1] François Rabelais University, Tours, France
arscaldgral@gmail.com
[2] Department of Computer Sciences, Polytechnic Institute of Viseu, Viseu, Portugal
pedromom@estv.ipv.pt
[3] Department of Computer Sciences, University of Coimbra, Coimbra, Portugal
{jcecilio,pnf}@dei.uc.pt

Abstract. Computer vision image recognition has undergone remarkable evolution due to available large-scale datasets (e.g., ImageNet, UEC-Food) and the evolution of deep Convolutional Neural Networks (CNN). CNN's learning method is data-driven from a sufficiently large training data, containing organized hierarchical image features, such as annotations, labels, and distinct regions of interest (ROI). However, acquiring such dataset with comprehensive annotations, in many domains is still a challenge. Currently, there are three main techniques to employ CNN: train the network from zero; use an off-the-shelf already trained network, and perform unsupervised pre-training with supervised adjustments. Deep learning networks for image classification, regression and feature learning include Inception-v3, ResNet-50, ResNet-101, GoogLeNet, AlexNet, VGG-16, and VGG-19.

In this paper we exploit the use of three CNN to solve detection problems. First, the different CNN architectures are evaluated. The studied CNN models contain 5 thousand to 160 million parameters, which can vary depending on the number of layers. Secondly, the different studied CNN's are evaluated, based on dataset sizes, and spatial image context. Results showing performance vs. training time vs. accuracy are analyzed. Thirdly, based on human knowledge and human visual system (HVS) classification, the accuracy of CNN's is studied and compared. Based on obtained results it is possible to conclude that the HVS is more accurate when there is a data set with a wide variety range. However, if the dataset if focused on only niche images, the CNN shows better results than the HVS.

Keywords: Convolutional Neural Networks · CNN ·
Neural networks · GoogLeNet · Inception-v3 · ResNet · Food data-set ·
Human classification

© Springer Nature Switzerland AG 2019
S. Kozielski et al. (Eds.): BDAS 2019, CCIS 1018, pp. 111–125, 2019.
https://doi.org/10.1007/978-3-030-19093-4_9

1   Introduction

Convolutional Neural Networks (CNN's) have significantly evolved the field of image classification [18,20], and object detection accuracy [9,27]. When comparing object detection versus image classification, object detection is a more complex task, which needs more complex methods [11,33], due to the need of multistage pipelines that are slow and have low performance. Solutions compromise speed and accuracy.

There are numerous machine learning and CNN libraries, for instance: Cuda-ConvNet, Torch, Theano, and Caffe4. In this work we used MatConvNet, which is a MATLAB tool, that implements CNN's oriented to computer vision. CNN's revolutionized computer vision since [18], replacing traditional image processing pipelines using, for instance: SIFT, Jseg, DBscan. CNN's simple learning based on convolutions and rectifications are far from trivial, and the reason is related to the need to learn by adjusting coefficients (backpropagation) from massive amounts of data (i.e.: millions of images), to have a well balance trained network. Just like other existing tools, MatConvNet is optimized for this tasks by using a set of optimizations which allow automatic parallelism and support the use of GPUs (using CUDA DevKit and a suitable NVIDIA GPU) to speed-up specific mathematical computations.

In this paper, using real-world food image data sets (UEC-256 and Food-101), we analyze three Convolutional Neural Networks applied to food detection and classification, GoogLeNet [29], Inception-v3 [30], and ResNet-101 [12]. In particular, using the food dataset UECFood-256, we explored and evaluated different CNN architectures keeping the same configuration parameters, while training the networks from scratch. Based on performed training, for all tested CNN's, the performance is described and evaluated. Moreover, based on an inquired study group, comprised of approximately 100 university students, the accuracy of the CNN's computer vision approach is compared with human vision-based classification [8].

CNN results, based on a complete food data-set, all classes, 6 epoch training, show an accuracy of 70.68% vs. 80.6% HVS. However, when training the CNN for only 16 food classes (the same used for the HVS study) CNNs demonstrate an accuracy of 89.89%. Moreover, when increasing the number of epochs form 6 to 20, CNNs accuracy improves to 93.86%.

This paper is organized as follows. Section 2, makes a short resume of the related work in the field. Section 3 describes the global experimental setup used to test the CNNs. Section 4, shows obtained results and draws conclusions. Section 5, concludes the work.

2   Related Work

This study is motivated by recent studies in the field [2,3] in computer vision.

In general, object recognition involves extraction of features from images, represented according to a specific model. Then, features are fed to a classifier to recognize the image objects, in our case, food. Features can be categorized in: global (e.g.: color histograms, circular shapes), or local (e.g.: pixel color, SIFT features). The selection of such features have a key role in the final accuracy. Many combinations can be set and have been proposed. While in [26], color and texture are combined, in [32] image features are integrated, including contour, motion, texture, color. Authors in [25], work using different color spaces, such as, RGB, HSV, LAB, and texture properties.

Other works on the field, for example [24], focus on the use of Bag of Features (BoF) algorithms in order to identify food. Authors in [5], use the Pittsburgh Fast-Food Image Database (PFID), and apply using a baseline algorithm using a bag of SIFT features. Studies like [1,7,13], use and compare different BoF algorithms.

Accuracy is a very important measure to evaluate classification of objects in images, and, it can be affected by data structures and quantization methods. For instance, in [17] authors use Fisher Vectors (FV) instead of traditional BoF to optimize quantization. This work shows that finding the appropriate features that best represent images is a challenge.

The key advantage in using deep learning methods [19] is their capacity to deal with a large number of image categories and features, automatically adjusting assumptions about processed images [18]. Few studies have explored Deep Learning for Food Recognition using neural networks. In [16] features are combined using FV from 1000 food-related categories, retrieved from ImageNet. Other architectures, derived from Inception [29], are explored in [22], where modules are modified, by introducing 1×1 convolutional layers, this way reducing the input dimension for the next layer. In [10], the inception v3 network is modified to improve computational efficiency. For [21], VGG-16 network multi-task loss is exploited, and in [4] problems like food ingredients recognition are also exploited. For both previous cases a conditional random field is applied to optimize the distribution probability of ingredients identification.

During ImageNet's Large Scale Visual Recognition Challenge 2012 (ILSVRC2012), AlexNet [18] was a success, proving its efficiency. Other related works study the effectiveness of CNN for food recognition [6,31], however, it is considered that more work is necessary for those solutions to be practical. Mainly because of the necessary large number of parameters, and excessive computation cost.

In 2014 ILSVRC competition, a CNN named GoogLeNet [29], obtained very good performance recognizing general objects. This was also improved with the introduction of Inception, a multi-branch convolution module, configurable for different purposes. Compared with AlexNet (8 layers), GoogLeNet outperforms accuracy [29], and uses twelve times less parameters.

In [28], three ResNet architectures are studied and compared. First, Inception-ResNet-v1, a hybrid Inception version that has a similar computational cost to Inception-v3. Second, inception-ResNet-v2, a costlier hybrid Inception version with significantly improved recognition performance. Last, Inception-v4, a pure Inception variant without residual connections with roughly the same recognition performance as Inception-ResNet-v2. Authors study how the introduction of residual connections lead to dramatically improved training speed for the Inception architecture.

3   Experimental Setup

In this work is used the UEC FOOD 256 as use case [15]. This data-set contains more than 32.000 images, representing 256 food categories [14]. Each food image has a bounding box associated, indicating the location of the food. Most food categories in this data-set are popular in Japan, therefore, some categories might not be familiar to other people than Japanese [23].

The data-set was divided in three parts: 60% for training; 20% for testing; 20 % for validation.

The study was performed using an Intel i5 3.4 GHz processor, 16 GB RAM, Asus nvidia GeForce gtx 1070 8 GB, Windows 10 64 bits, MatLab R2018a.

For all tested CNNs (GoogleNet; Inception v3; ResNet 101), the same parameters were configured, Table 1.

Table 1. Testing configuration for all CNNs

Training cycle	
Epoch	6
Iterations	15264
Iterations per epoch	2544
Validation	
Frequency	3 iterations

Figure 1, shows the labeling for Figs. 2a, b, c and 5a, b, c.

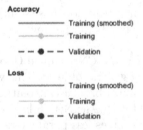

Fig. 1. Graphs labeling

To compare the CNNs with the HVS, a small test group, comprised of a total of 20 subjects was set, ten were female, and the other ten male, all in the age rank between 25 and 40 years old, all with European nationality.

Similar to CNNs, Europeans do not know Japanese food dishes. Thus, training is necessary before testing. For this reason, the inquiry was divided into two stages. A training stage was given to the subjects, consisting of showing a set of total 98 slides for the 16 food categories. A slide shows 7 images for the chosen food category and 1 images from other food categories, in random order. Then the test subject had to indicate the one that did not belong to the respective food category. This allows the test subjects to learn the names and key characteristics of the food categories.

In the second stage, image classification, a set of images is shown in random order, from the 16 trained classes, each with 16 class label options, where only one is correct. The test subjects need to select the correct label for the image. HVS data-set included 686 food photos, each class with a total of 43 photos.

4   Comparison Study

Using the UEC Food 256 full data-set, the following CNNs were trained from the scratch to classify different types of foods: GoogleNet; Inception v3; ResNet 101.

Figure 2 shows the training progress for the three used CNNs. From these charts it is possible to observe how each network evolves, epoch after epoch, to achieve its best, given the same configurations for all.

Table 2. Results comparison, validation accuracy, validation loss, and training time

	Results GoogleNet	Results Inception v3	ResNet 101
Validation accuracy	47.73%	65.43%	70.68%
Validation loss	2%	1.4%	1.2%
Training time	6105 min 12 s	5290 min 42 s	7678 min 17 s
Classification time (per image)	0.138 s	0.655 s	0.27 s

Table 2 shows a summary of the comparison of the obtained accuracy and training time results.

When comparing all charts, Fig. 2, ResNet 101 (a), and GoogleNet (c) converge faster (in fewer epochs) to the final result, while Inception V3 "climbs" to the final result/accuracy more gradually over the training time (more epochs are necessary). On the other hand, Inception v3, shows more potential to improve the accuracy performance, if extended training is performed. Never the less, with the same test conditions, ResNet 101 obtained better accuracy results.

In Fig. 3 is represented the Confusion Matrix (confusion matrix) for the CNN, resNet 101, with all 256 classes considered. Note that, most miss-classifications fell outside the 16 considered classes, therefore, the last column, sum represents

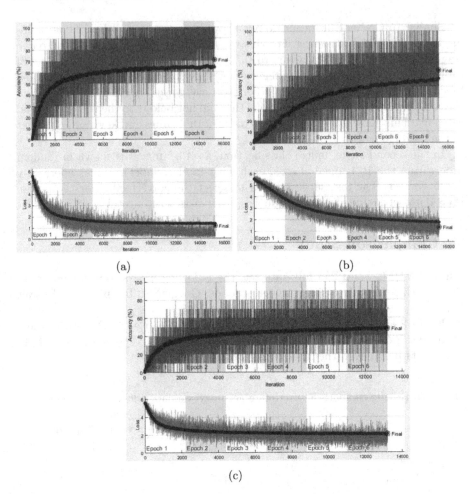

Fig. 2. CNN training and loss comparison: (a) ResNet CNN, (b) InceptionV3 CNN, (c) GoogleNet CNN

the sum of all miss-classifications. With the CNN ResNet 101, with all food classes of the data-set available to classify only the 16 classes used in the HVS survey inquiry, this CNN reached an accuracy of 67.7%.

Figure 4 shows the Confusion Matrix for the survey inquiry performed to our test group. Note that, as happens with humans, during the test/inquiry, the CNN also has a wide knowledge of other foods (256 types of food), which as in humans, can influence classification. HVS was able to reach an accuracy of 80.6%, 13.4% more than the CNN.

Table 3, resumes the Confusion Matrix in Fig. 3 (16 food classes classification, with the knowledge of all 256 food classes) and Fig. 4 (16 food classes classification of 16 food classes training, with a life time of food knowledge). From obtained results, when considering all trained food classes for both CNN and

Predict (CNN)

Truth (CNN)	#	1	2	3	4	5	6	7	8	9	10	11	12	13	14	15	16	sum
Pizza	1	4										1						8
Fried fish	2		7															14
Sashimi	3		2	5														18
Sweet and sour pork	4				5													12
Natto	5				1	10												14
Stir fried beef. peppers	6						11											18
Steamed meat dumpling	7							6										7
Chinese soup	8								3									6
Champon	9								1	9								11
Parfait	10										10							11
Roast duck	11											7						7
Glutinous rice balls	12												9			1		12
Laksa	13													10				10
Mie goreng	14														14			17
Nasi campur	15															10		12
Curry puff	16																11	18
																		131

Fig. 3. Confusion Matrix CNN ResNet 101 (256 classes trained - 16 classes classification)

human test subjects, HVS performed better than the CNN. However we must consider that both make mistakes when:

- The training data-set does not cover comprehensively the classification images.
- When the image to classify is too different from the learned ones.
- Other dishes similarities, influence the classification (i.e., over-fitting).

In the previous experiments, the CNNs were trained with 256 food categories. In the next experiment, the three CNNs were trained again from the scratch. This time only considering the training and classification images used during the HVS inquiry training. Results are shown in the following subsection.

Table 3. CNN vs. HVS results

	HVS	CNN ResNet 101
Validation accuracy	80.6%	67.2%
Training time (avg)	17 min 56 s	7678 min 17 s
Classification time (avg)	14 s (per image)	0.27 s (per image)

Truth (HVS) \ Predict (HVS)		Pizza	Fried fish	Sashimi	Sweet and sour pork	Natto	Stir fried beef, peppers	Steamed meat dumpling	Chinese soup	Champon	Parfait	Roast duck	Glutinous rice balls	Laksa	Mie goreng	Nasi campur	Curry puff	
		1	2	3	4	5	6	7	8	9	10	11	12	13	14	15	16	sum
Pizza	1	23																23
Fried fish	2		16	1								1						18
Sashimi	3		1	25					1									27
Sweet and sour pork	4		2		18		3			1				3		1		28
Natto	5				1	25												26
Stir fried beef, peppers	6				5		18			1				1		2		27
Steamed meat dumpling	7							20					1				2	23
Chinese soup	8								24	6				5				35
Champon	9						4	2		15				5	6			32
Parfait	10	1									26							27
Roast duck	11											22				1		23
Glutinous rice balls	12							2					25					27
Laksa	13				2		1			2				12	3			20
Mie goreng	14				1					1					15			17
Nasi campur	15		7			1										22		30
Curry puff	16		1											1			23	25
																		329

Fig. 4. Confusion Matrix HVS (16 classes trained - 16 classes classification)

4.1 Additional Results, CNN (16 Classes) vs. HVS

In this subsection, additional tests were performed with the three CNNs, using only 16 food classes (and the same 6 epochs). With these results it is possible to evaluate which CNN is more efficient, with a similar training as the HVS, and how the results compare with HVS in this case. For this purpose, the same images used during HVS inquiries, for training and classification were used to train the three CNNs from scratch. The CNN training data-set had the same images as the HVS training set, total 686 food photos, each class with a total of 43 photos.

Figure 5, shows the CNNs training progress (accuracy and loss). In Table 4, results are summarized. Both GoogleNet and InceptionV3 reached the same accuracy (85.92%), and ResNet101 was the most efficient (89.89%). However, ResNet101 was the CNN that took more time to train, and GoogleNet the fastest.

Tables in Fig. 6 shows the confusion matrix for the three CNNs and the HVS (Fig. 6d the same as Fig. 4) for comparison purposes. Most important conclusions are that:

- HVS errors are constant and visually/syntactically related.
- CNNs errors are much more disperse and (for us humans) without visual relation.

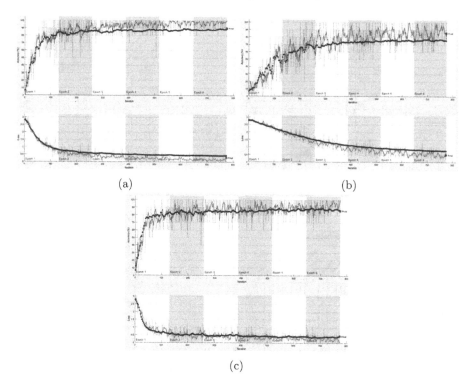

(a)   (b)

(c)

Fig. 5. CNN, 16 class training, 16 classes classification, 6 epoch: (a) ResNet CNN, (b) InceptionV3 CNN, (c) GoogleNet CNN

4.2   More (20) Epochs Impact on CNN Accuracy

Next we increased the number of training epochs from 6 to 20. The data-set was the same as in the previous and the HVS experiments (16 classes).

Figure 7 shows the training progress for the three CNNs using 20 epochs. From these images it is possible to conclude that after approximately 10 epochs the CNNs do not significantly improve their accuracy.

Table 4. Results comparison, 16 food classes, 6 epoch: validation accuracy, validation loss, and training time

	Results GoogleNet	Results Inception v3	ResNet 101
Validation accuracy	85.92%	85.92%	89.89%
Validation loss	0.45%	0.5%	0.3%
Training time	15 min 18 s	73 min 27 s	94 min 26 s

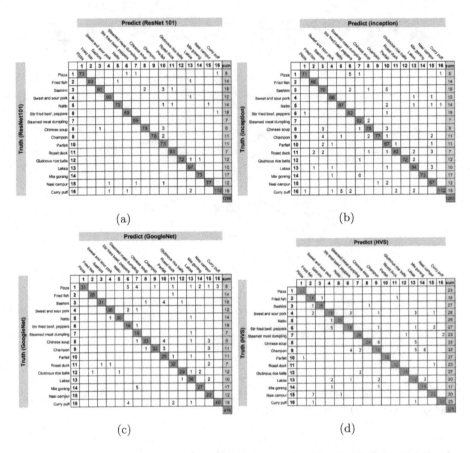

Fig. 6. Confusion matrix, 16 class training, 16 classes classification, 6 epoch: (a) ResNet CNN, (b) InceptionV3 CNN, (c) GoogleNet CNN, (d) HVS (for comparing purposes - same as Fig. 4)

Table 5 compares accuracy, loss and training times of the three CNNs. When using 20 epochs, instead of 6, CNNs can improve significantly their accuracy, almost stabilizing in a maximum around the 90% accuracy.

Tables in Fig. 8, compare the confusion matrix, of the CNNs with 20 epochs, with the HVS. Similar to what happens with 6 epochs, CNN confusion matrix's errors are dispersed and not directly related, as happens with the HVS.

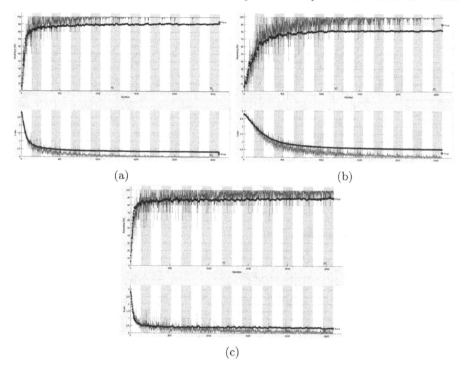

(a)   (b)

(c)

Fig. 7. CNN, 16 class training, 16 classes classification, 20 epochs: (a) ResNet CNN, (b) InceptionV3 CNN, (c) GoogleNet CNN

Table 5. Results comparison, 16 food classes, 20 epochs: validation accuracy, validation loss, and training time

	Results GoogleNet	Results Inception v3	ResNet 101
Validation accuracy	89.17%	90.25%	93.86%
Validation loss	0.30%	0.30%	0.25%
Training time	49 min 31 s	237 min 22 s	288 min 28 s

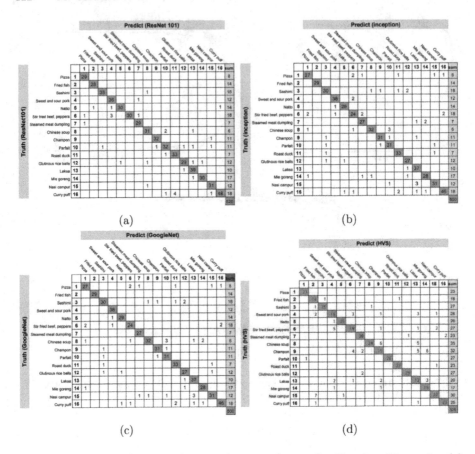

Fig. 8. Confusion matrix, 16 class training, 16 classes classification, 20 epochs: (a) ResNet CNN, (b) InceptionV3 CNN, (c) GoogleNet CNN, (d) HVS (for comparing purposes - same as Fig. 4)

5   Conclusions and Future Work

In this work we evaluated approaches to food recognition using different deep Convolutional Neural Network models introduced by Google: GoogleNet, Inception V3 and ResNet 101. These models make use of building blocks called "Inception modules", that are used to create very deep networks. We also compared these with the Human Visual System (HVS).

A problem that is worth discussing is related with the need for high computational resources to train the networks from scratch. Considering that a large number of the training session are necessary to tune some of the network parameters for reaching the best possible accuracy, it is clear that organizing and performing such experiments is not trivial. Because of this, it is also considered the need for further investigation on the settings of the training/fine-tuning parameters of the architecture for classification; we believe that deep learning and its applications, food recognition in particular, still offers many opportunities for research at all levels.

Considering all 266 food classes, when comparing the CNN with the HVS, for a limited number of classes (16 different classes), the HVS performed better than the CNN, 67.2% CNN accuracy vs. 80.6% HVS. However, when comparing the HVS with the CNNs trained with only with 16 classes, the accuracy of the CNNs improved significantly above the HVS, up to 89.89%. Additional tests show that increasing the number of epochs, 6 to 20, improves even more the accuracy, to 93.86%, when using only 16 food classes.

An important difference between CNN and HVS regarding classification is that the CNNs make errors with different food classes, of all types; HVS makes semantic errors almost always with the same food classes. Therefore, HVS, shows more consistency with the provided answers.

Acknowledgements. "This article is financed by national funds through FCT Fundação para a Ciência e Tecnologia, I.P., under the project UID/Multi/04016/2016. Furthermore, we would like to thank the Instituto Politécnico de Viseu, the University of Coimbra, the University of Lisbon, and the University of François Rabelais, for their support."

References

1. Anthimopoulos, M.M., Gianola, L., Scarnato, L., Diem, P., Mougiakakou, S.G.: A food recognition system for diabetic patients based on an optimized bag-of-features model. IEEE J. Biomed. Health Inform. **18**(4), 1261–1271 (2014)
2. Chatfield, K., Lempitsky, V.S., Vedaldi, A., Zisserman, A.: The devil is in the details: an evaluation of recent feature encoding methods. In: BMVC, vol. 2, p. 8 (2011)
3. Chatfield, K., Simonyan, K., Vedaldi, A., Zisserman, A.: Return of the devil in the details: delving deep into convolutional nets. arXiv preprint arXiv:1405.3531 (2014)
4. Chen, J., Ngo, C.W.: Deep-based ingredient recognition for cooking recipe retrieval. In: Proceedings of the 2016 ACM on Multimedia Conference, pp. 32–41. ACM (2016)
5. Chen, M., Dhingra, K., Wu, W., Yang, L., Sukthankar, R., Yang, J.: PFID: Pittsburgh fast-food image dataset. In: 2009 16th IEEE International Conference on Image Processing (ICIP), pp. 289–292. IEEE (2009)
6. Christodoulidis, S., Anthimopoulos, M., Mougiakakou, S.: Food recognition for dietary assessment using deep convolutional neural networks. In: Murino, V., Puppo, E., Sona, D., Cristani, M., Sansone, C. (eds.) ICIAP 2015. LNCS, vol. 9281, pp. 458–465. Springer, Cham (2015). https://doi.org/10.1007/978-3-319-23222-5_56
7. Farinella, G.M., Moltisanti, M., Battiato, S.: Classifying food images represented as Bag of Textons. In: 2014 IEEE International Conference on Image Processing (ICIP), pp. 5212–5216. IEEE (2014)
8. Geman, S.: Hierarchy in machine and natural vision. In: Proceedings of the Scandinavian Conference on Image Analysis, vol. 1, pp. 179–184 (1999)

9. Girshick, R., Donahue, J., Darrell, T., Malik, J.: Rich feature hierarchies for accurate object detection and semantic segmentation. In: Proceedings of the IEEE Conference on Computer Vision and Pattern Recognition, pp. 580–587 (2014)
10. Hassannejad, H., Matrella, G., Ciampolini, P., De Munari, I., Mordonini, M., Cagnoni, S.: Food image recognition using very deep convolutional networks. In: Proceedings of the 2nd International Workshop on Multimedia Assisted Dietary Management, pp. 41–49. ACM (2016)
11. He, K., Zhang, X., Ren, S., Sun, J.: Spatial pyramid pooling in deep convolutional networks for visual recognition. In: Fleet, D., Pajdla, T., Schiele, B., Tuytelaars, T. (eds.) ECCV 2014. LNCS, vol. 8691, pp. 346–361. Springer, Cham (2014). https://doi.org/10.1007/978-3-319-10578-9_23
12. He, K., Zhang, X., Ren, S., Sun, J.: Deep residual learning for image recognition. In: Proceedings of the IEEE Conference on Computer Vision and Pattern Recognition, pp. 770–778 (2016)
13. Hoashi, H., Joutou, T., Yanai, K.: Image recognition of 85 food categories by feature fusion. In: 2010 IEEE International Symposium on Multimedia (ISM), pp. 296–301. IEEE (2010)
14. Kagaya, H., Aizawa, K.: Highly accurate food/non-food image classification based on a deep convolutional neural network. In: Murino, V., Puppo, E., Sona, D., Cristani, M., Sansone, C. (eds.) ICIAP 2015. LNCS, vol. 9281, pp. 350–357. Springer, Cham (2015). https://doi.org/10.1007/978-3-319-23222-5_43
15. Kawano, Y., Yanai, K.: Automatic expansion of a food image dataset leveraging existing categories with domain adaptation. In: Agapito, L., Bronstein, M.M., Rother, C. (eds.) ECCV 2014. LNCS, vol. 8927, pp. 3–17. Springer, Cham (2015). https://doi.org/10.1007/978-3-319-16199-0_1
16. Kawano, Y., Yanai, K.: Food image recognition with deep convolutional features. In: Proceedings of the 2014 ACM International Joint Conference on Pervasive and Ubiquitous Computing: Adjunct Publication, pp. 589–593. ACM (2014)
17. Kawano, Y., Yanai, K.: FoodCam: a real-time mobile food recognition system employing fisher vector. In: Gurrin, C., Hopfgartner, F., Hurst, W., Johansen, H., Lee, H., O'Connor, N. (eds.) MMM 2014. LNCS, vol. 8326, pp. 369–373. Springer, Cham (2014). https://doi.org/10.1007/978-3-319-04117-9_38
18. Krizhevsky, A., Sutskever, I., Hinton, G.E.: ImageNet classification with deep convolutional neural networks. In: Advances in Neural Information Processing Systems, pp. 1097–1105 (2012)
19. LeCun, Y., Bengio, Y., Hinton, G.: Deep learning. Nature **521**(7553), 436 (2015)
20. LeCun, Y., et al.: Backpropagation applied to handwritten zip code recognition. Neural Comput. **1**(4), 541–551 (1989)
21. Lin, G., Shen, C., Van Den Hengel, A., Reid, I.: Exploring context with deepstructured models for semantic segmentation. IEEE Trans. Pattern Anal. Mach. Intell. **40**, 1352–1366 (2017)
22. Liu, C., Cao, Y., Luo, Y., Chen, G., Vokkarane, V., Ma, Y.: DeepFood: deep learning-based food image recognition for computer-aided dietary assessment. In: Chang, C.K., Chiari, L., Cao, Y., Jin, H., Mokhtari, M., Aloulou, H. (eds.) ICOST 2016. LNCS, vol. 9677, pp. 37–48. Springer, Cham (2016). https://doi.org/10.1007/978-3-319-39601-9_4
23. Matsuda, Y., Hoashi, H., Yanai, K.: Recognition of multiple-food images by detecting candidate regions. In: 2012 IEEE International Conference on Multimedia and Expo (ICME), pp. 25–30. IEEE (2012)
24. O'Hara, S., Draper, B.A.: Introduction to the bag of features paradigm for image classification and retrieval. arXiv preprint arXiv:1101.3354 (2011)

25. Oliveira, L., Costa, V., Neves, G., Oliveira, T., Jorge, E., Lizarraga, M.: A mobile, lightweight, poll-based food identification system. Pattern Recogn. **47**(5), 1941–1952 (2014)
26. Puri, M., Zhu, Z., Yu, Q., Divakaran, A., Sawhney, H.: Recognition and volume estimation of food intake using a mobile device. In: 2009 Workshop on Applications of Computer Vision (WACV), pp. 1–8. IEEE (2009)
27. Sermanet, P., Eigen, D., Zhang, X., Mathieu, M., Fergus, R., LeCun, Y.: Overfeat: integrated recognition, localization and detection using convolutional networks. arXiv preprint arXiv:1312.6229 (2013)
28. Szegedy, C., Ioffe, S., Vanhoucke, V., Alemi, A.A.: Inception-v4, inception-ResNet and the impact of residual connections on learning. In: AAAI, vol. 4, p. 12 (2017)
29. Szegedy, C., et al.: Going deeper with convolutions. In: CVPR (2015)
30. Szegedy, C., Vanhoucke, V., Ioffe, S., Shlens, J., Wojna, Z.: Rethinking the inception architecture for computer vision. In: Proceedings of the IEEE Conference on Computer Vision and Pattern Recognition, pp. 2818–2826 (2016)
31. Yanai, K., Kawano, Y.: Food image recognition using deep convolutional network with pre-training and fine-tuning. In: 2015 IEEE International Conference on Multimedia & Expo Workshops (ICMEW), pp. 1–6. IEEE (2015)
32. Zhu, F., et al.: The use of mobile devices in aiding dietary assessment and evaluation. IEEE J. Sel. Top. Sig. Process. **4**(4), 756–766 (2010)
33. Zhu, Y., Urtasun, R., Salakhutdinov, R., Fidler, S.: segDeepM: exploiting segmentation and context in deep neural networks for object detection. In: 2015 IEEE Conference on Computer Vision and Pattern Recognition (CVPR), pp. 4703–4711. IEEE (2015)

Multi-criteria Decision Analysis in the Railway Risk Management Process

Jacek Bagiński[✉], Barbara Flisiuk, Wojciech Górka, Dariusz Rogowski, and Tomasz Stęclik

Institute of Innovative Technologies EMAG,
ul. Leopolda 31, 40-189 Katowice, Poland
{jbaginski,bflisiuk,wgorka,drogowski,tsteclik}@ibemag.pl

Abstract. The article presents a way to use a Multi-criteria Decision Analysis (MCDA) method such as Analytic Hierarchy Process (AHP) for the assessment and selection of security and safety measures in the railway industry. The situation in the industry regarding security information exchange and the risk management process was presented briefly, as well as a proposal to support this process with the elements of the multi-criteria analysis.

Keywords: Decision support · Safety management system · Risk management · Software support · MCDA · AHP

1 Introduction

Safety Management Systems (SMS) used in the railway industry mainly focus on the risk level estimation process. However, there is a gap in that process which is the lack of decision making support regarding the selection of safety or security measures decreasing identified risk. The paper presents one of the results of a research project "Central Threats Register" (Pol. CRZ - Centralny Rejestr Zagrożeń), run by Institute EMAG, which helps to fill that gap.

The solution is based on the implementation of the AHP method into the railway industry. The method has been successfully used in many other sectors and industries for many years now [1,11,12,15]. CRZ will be a central platform for cooperation and information sharing for different railway infrastructure stakeholders according to the requirements of the Commission Regulation No. 1078/2012.

The AHP method in the CRZ platform will support the selection of security measures which are planned to reduce identified risks. It is assumed that CRZ will enable interested parties to make common assessments and to solve decision problems during the selection of proper security controls. Besides, CRZ will facilitate risk analysis and the exchange of security and threats related data by the implementation of a common knowledge base.

The paper is organized as follows. Section 2 presents the state of the art in the field of security assessment methods used in the railway industry. Section 3

S. Kozielski et al. (Eds.): BDAS 2019, CCIS 1018, pp. 126–138, 2019.
https://doi.org/10.1007/978-3-030-19093-4_10

introduces a solution to the problem and describes AHP use by a few examples. Section 4 explains the methodology and use case chosen to verify the solution. Section 5 discusses the results and Sect. 6 presents conclusions and future work.

2   Analysis of the Current Situation

In the chapter the state of the art of risk assessment methods used in the railway industry is presented.

2.1   Legal Basis

The Commission Regulation (EU) No. 402/2013 on the common safety method and risk evaluation and assessment [6] clearly defines the basic risk assessment process that is required for use in the railway industry (Fig. 1).

The industry stakeholders are obliged to perform risk management, which includes threats (hazards) identification, risk analysis, estimation for identified threats (hazards), and risk treatment process (planning and performing the activities to reduce the risk to an acceptable level).

Different entities can use different methods for risk assessment and internal reporting. The Office of Rail Transport (UTK) - the railway market regulator in Poland - has developed a set of guidelines for risk management and monitoring [17,18]. These guidelines describe several examples of such assessment methods as: Brainstorming, Checklist, Primary Hazard Analysis (PHA), Hazard and Operability studies (HAZOP), Structure What-If (SWIFT), Failure Mode Effect Analysis (FMEA), Event Tree Analysis (ETA), Fault Tree Analysis (FTA), Root Cause Analysis (RCA), and Ishikawa Diagram. The UTK guideline also includes other methods which are presented in the ISO 31010 standard [10]. The next section describes the results of a survey in which the authors try to find out which methods are the most popular among railway companies.

2.2   Survey on the Railway Risk Assessment

The survey results presented below are based on questionnaires sent to approximately 15% of railway entities in Poland. The survey was a part of the CRZ project. According to the answers the most of risk assessment activities are based on FMEA, Risk Score and Consequence/Probability matrix methods (Fig. 2).

The risk assessment and the selection of security controls are carried out by using simple documents and methods created, for instance, in Excel spreadsheets. Further, common assessment of security issues and selection of security measures are based on very popular methods, like brainstorming or "what-if" analysis.

Risk assessment methods used in the railway industry utilize predefined configurable scales of probability and consequences of threats/incidents. The methods focus on the assessment of the risk level and do not support the selection of proper security measures. Moreover, they do not facilitate the comparison of

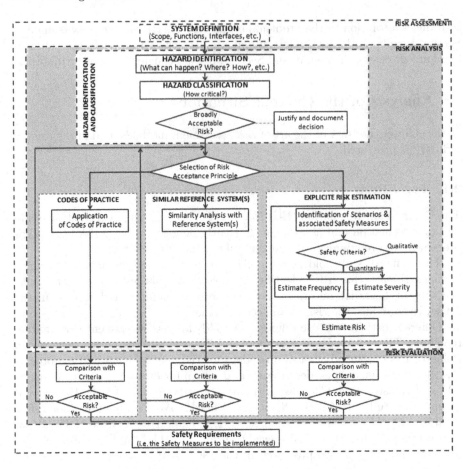

Fig. 1. Risk assessment process according to [6]

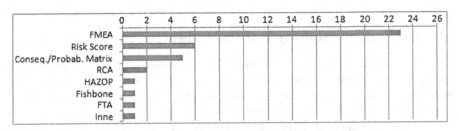

Fig. 2. The numbers of responders using selected risk assessment methods

available options and activities according to financial and non-financial criteria while choosing the best safeguards at the moment.

Little attention has been paid to using decision support methods (including MCDA-type methods [19]) in the Polish railway industry so far.

2.3   Related Works

Some of MCDA methods (AHP, Analytic Network Process - ANP) are used in the transport industry to plan transport routes for passengers or cargo trains [16]. Sometimes they concern development problems of transport networks (including railway) in specific areas [5,7]. For instance, they can be used for finding the best transport alternatives in conurbations or simulating the development of tram ways in a city. A good example of MCDA application to railway safety was described in paper [3]. Here MCDA was used for the assessment of a safety status of such railway infrastructure components as rolling stocks, rails and automatic switches, level crossings, railway stations, signaling systems and automatic devices, and others.

Literature research showed no studies about safety management systems with built-in decision support methods, neither for risk/threat assessment nor for security/safety control measures assessment and selection. In the next section a solution is proposed to overcome that gap.

3   Proposed Solution

Due to the complexity and extent of a typical railway infrastructure there can be various types of its possible consequences to human health, train passengers, users of infrastructure elements (e.g. road-rail level crossings, railway stations), natural environment (e.g. transport of hazardous materials like chemicals, oil, radioactive materials), local communities (e.g. railway lines or infrastructure elements under repair or construction).

The possible impact of planned security measures as well as the importance of consequences and impacted areas can be evaluated differently. The priorities of evaluation criteria change and depend on the evaluators' group, i.e. some aspects will be important for security managers while others for local communities, governments, management boards of railway companies and their financial departments.

This is why, in the CRZ project, we propose a software tool with a built-in MCDA method capable of supporting the assessment and selection of safety/security measures.

The MCDA method is used to assess and prioritize various options from the most to the least preferred ones. The method is often used to assess complex financial and non-financial issues. Overall preference to an assessed item, asset or object is determined thanks to the results obtained from the method. The results consider assessments and comparisons made with respect to the chosen individual criteria.

The manual [19] covers a range of multi-criteria analysis techniques and provides guidance on how to use them for the assessment of options or decisions.

The multi-criteria analysis techniques group includes Analytical Hierarchy Process (AHP) – a method presented in 1980 by Saaty [13]. It is a technique in which a set of parameters are evaluated using the absolute scale (for both measurable and non-measurable criteria). The method relies on the decomposition

of the problem into constituent elements and repeating the evaluation sequence.
In the sequence the elements are compared in pairs.

The decision problem is modeled as a tree (Fig. 3) where the first level
presents a general goal (objective) and the successive levels contain criteria and
sub-criteria of the assessment. The lowest level includes available options, alternative actions, or decisions.

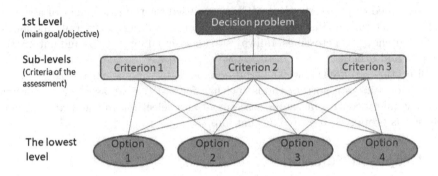

Fig. 3. Example of decision problem tree

In order to determine the weight of individual criteria, they are compared in
pairs (pair-wise comparisons) using the scale proposed by Saaty (Table 1).

Decision makers express their preferences using the above scale and the result
of comparisons is recorded in the form of a square preference matrix. The resulting matrix is a square matrix with a dimension equal to the number of compared
elements.

$$A_k = \begin{bmatrix} a_{11} & a_{1j} & a_{1n} \\ a_{i1} & a_{ij} & a_{in} \\ a_{n1} & a_{nj} & a_{nn} \end{bmatrix}$$

where a_{ij} is a value which informs how much more important the element i is in
relation to the element j with respect to the k criterion. The matrix A is recip-
rocal, which means that elements of the evaluation matrix have the property:
$a_{ji} = \frac{1}{a_{ij}}$, $a_{ij} > 0$, for each i, $j = 1, 2, \ldots, n$. Table 2 presents an example of
such a matrix.

The decision makers should perform the comparison (using the scale presented in Table 1) for each pair of criteria (on the particular level of criteria).
Having the comparison matrices ready, the priority vector for each matrix must
be found. The priority vector expresses the preferences between elements (such as
criteria or decision options, depending on the level of the performed assessment).

Various methods of approximate calculation of the priority vector are used.
The following are the two main methods [4, 14] for calculation of w_i - elements
of the priority vector.

Table 1. Evaluation ranking scale for the pairwise comparison [4]

Importance	Definition	Explanation
1	Equal importance	Two activities contribute equally to the objective
2	Weak or slight	
3	Moderate importance	Experience and judgement slightly favour one activity over another
4	Moderate plus	
5	Strong importance	Experience and judgement strongly favour one activity over another
6	Strong plus	
7	Very strong or demonstrated importance	An activity is favoured very strongly over another; its dominance demonstrated in practice
8	Very, very strong	
9	Extreme importance	The evidence favouring one activity over another is of the highest possible order of affirmation

Table 2. Example of preference matrix

Elements (a_{ij}) of A matrix	a_{i1}	a_{i2}	a_{i3}
a_{1j}	1	4	$\frac{1}{2}$
a_{2j}	$\frac{1}{4}$	1	$\frac{1}{4}$
a_{3j}	2	4	1

Method I: Utilize the geometric mean to calculate the elements of priority vector w_i.

$$a_i^* = \sqrt[n]{\prod_{j=1}^{n} a_{ij}} \tag{1}$$

$$a^* = \sum_{i=1}^{n} a_{ij}^* \tag{2}$$

$$w_i = \frac{a_i^*}{a^*} \tag{3}$$

Method II: The matrix A is transformed to A^* using formulas (4), (5). Then the priority vector is performed as the arithmetic averages from the row of the normalized comparison matrix (A^*)

$$\alpha_j^* = \sum_{i=1}^n a_{ij} \tag{4}$$

$$a_{ij}^* = \frac{a_{ij}}{\alpha_j^*} \tag{5}$$

$$w_i = \frac{1}{n} \sum_{j=1}^n a_{ij}^* = \frac{1}{n} \sum_{j=1}^n \frac{a_{ij}}{\alpha_j^*} \tag{6}$$

The next step in the classic approach to AHP is to create comparison matrices for each option, against each criteria. The global priorities for each option can be calculated as the sum of the products of priorities in each branch, from a single option up to the main goal (decision problem).

In order to ensure better quality of results, the AHP method assumes the calculation of Consistency Ratio (CR), which is determined as the ratio of the Consistency Index (CI) of the pairwise pairing matrix to the Random Index (RI) determined for random matrix values of the same size.

$$CI = \frac{\lambda_{max} - n}{n - 1} \tag{7}$$

$$CR = \frac{CI}{RI} \tag{8}$$

where n is the size of the matrix, λ_{max} is the largest eigenvalue of the matrix, RI is a consistency index of matrices, the values of which are completely random. [2, 9,13] present tables of typical RI values for matrices of different sizes, determined on the basis of many simulations (100, 500, 1000, 100000).

4   Methodology

There are various hazards, e.g. railway crossing accident, failure in railway transmission lines, collision of trains, equipment failures which were registered and are a matter of big concern in railway companies. The AHP method was verified on one chosen problem presented below.

The problem refers to proper selection of security controls to counter the threat of breaking overhead electrical transmission lines due to icing. The use case analysis was performed in two steps:

1. First, the traditional brainstorming method was used by the experts to assess and select safety measures in the way it was done up till now.
2. Second, the assessment and selection of the same safety measures were performed using the AHP method.

Such an approach allowed to assess whether the results are consistent with the experts' predictions and whether the method can be used in this area.

5   Results and Discussion

In this section we present and compare the results achieved in the brainstorming and AHP methods.

5.1   Brainstorming Results

For the selected use case the following controls were indicated by the experts as possible alternatives for the assessment:

1. Mechanical de-icing using devices mounted on trolleys.
2. Mechanical de-icing using devices mounted to pantographs of traction vehicles.
3. Mechanical de-icing by hitting the wires with rods.
4. Chemical de-icing (application of a $CaCl_2$ to the traction network).
5. Electrical de-icing (melting ice by heating the electric traction).
6. Reduction of adhesion of ice or rime by using fat (e.g. transformer oil or chemicals) in the period preceding the occurrence of icing.
7. Covering a short section of cables with anti-frost measures in zones where frequent starting of traction vehicles occur.
8. More frequent train drives including manual clearing of traction from accumulated snow and ice.
9. Driving trains with raised two panthographs, the first of them facing the moving direction should not draw power.
10. More frequent inspections of electrified railway lines to observe overhead lines, icing, snow covering of insulators, condition of tensioning devices.

For these measures, a set of evaluation criteria has been defined: Impact on safety/security, Impact on the environment, Effectiveness, Financial costs, Possibility of implementation, Time of implementation.

According to the brainstorming method which is currently used by experts, the most important assessment criteria are the cost of implementation and the impact on safety/security measures. The experts, according to their best knowledge and experience, selected options: 1, 2, 4, 7, 10 from the list of 10 alternatives (safety measures). These options are considered by railway experts as the most appropriate for implementation.

Next, the experts used the AHP method to assess the same options according to the same criteria.

5.2   AHP Process

In this step previously selected criteria and safety measures were assessed. The results from the AHP method were compared to the results gained from the brainstorming method. Special Excel spreadsheets were prepared to assess the chosen use case according to the AHP method.

A few chosen railway experts were engaged in the assessment. Each of them had to make a full AHP assessment including the comparison of criteria and the comparison of security measures. Then the results from all experts were aggregated into the final ranking of criteria and security measures.

The evaluation of criteria and measures was performed using a typical AHP 9-value scale (Table 1). Table 3 presents an example of the criteria assessment results achieved by one of the experts.

Table 3. Example of criteria evaluation made by one expert

	Impact on safety/security	Impact on environment	Effectiveness	Financial costs	Possibility of implement	Time of implementation
Impact on safety/security	1.000	5.000	0.500	0.333	0.333	0.500
Impact on environment	0.200	1.000	0.250	0.500	0.333	0.500
Effectiveness	2.000	4.000	1.000	0.250	0.333	0.500
Financial costs	3.000	2.000	4.000	1.000	0.333	0.333
Possibility of implement	3.000	3.000	3.000	3.000	1.000	0.333
Time of implementation	2.000	2.000	2.000	3.000	3.000	1.000

Safety measures were assessed by each expert in the same way like criteria. Next, all security measures were compared to each other considering defined criteria.

The full version of AHP method assumes checking the consistency of result matrices. For inconsistent assessments, the problem is the inaccuracy of results [4]. This is why additionally a consistency ratio (CR) was calculated for each resulting matrix. It allowed to check how many answers were thought-out, well-judged, and consistent.

Summing up, the whole AHP process consists of the following steps:

1. Criteria assessment
2. Calculating CR for the results of criteria assessment
3. Safety measures assessment
4. Calculating CR for the results of safety measures assessment
5. Aggregating results from all experts and ranking safety measures.

5.3   AHP Results and Discussion

AHP assessments were made by a few railway experts coming from maintenance of electric traction networks and individual carriers. In the beginning, the results had their consistency ratios (CR) values between 0.05 and 0.57. Such a wide range was caused by the lack of the experts' experience in using the method. It was the very first time they used AHP.

Such inconsistency in CRs can be minimized by re-evaluations but this increases the complexity and time-consuming nature of the overall assessment.

Another possibility of decreasing inconsistencies in CRs is the rejection of ratings (matrices) whose consistency index exceeds the acceptable value, which in turn may lead to discarding many results of other experts.

Due to the small number of experts, the first solution of a re-evaluation was chosen. Then the results collected from several experts were aggregated. The aggregation of assessment results was made in the form of a geometric mean.

The analysis of the results revealed that AHP results were not entirely consistent with brainstorming method results. During brainstorming, railway experts judged the financial costs and impact on safety as the most important criteria. While the AHP method revealed that criteria related to the possibility of implementation and time of implementation of security measures were assessed higher by the same experts (Fig. 4).

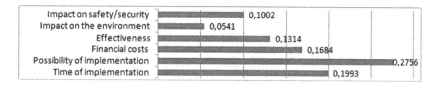

Fig. 4. Ranking of criteria (aggregated results)

It does not mean the AHP method gives bad results. It rather means the method delivers a broader view of security measures. In that case the chosen security measures should not be only cheap and effective but also feasible for implementation in the current environment and in acceptable time.

As a result of brainstorming, the following five alternatives were selected as recommended for implementation: 1, 2, 4, 7, 10. While, according to the AHP results, the following safety measures got the top five scores: 9, 2, 1, 10 and 4 (Fig. 5), and alternative No. 7 goes right after alternative No. 4.

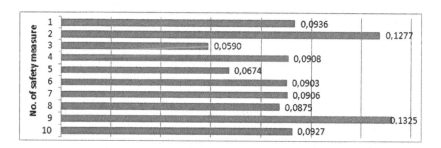

Fig. 5. Ranking of safety measures (aggregated results)

The inconsistency regarding safety measure 9 may result from the fact that the evaluators represented different types of railway entities and the selected test case did not affect all the entities to the same extent. The maintenance of the electric traction network is the responsibility of the infrastructure owner and not individual carriers, while the experts' group included also representatives of the carriers. Therefore the discrepancies in the safety measures assessment were greater than in the criteria assessment. Railway carriers actually use solution No. 9 as the most effective even in strong frosts, although reluctantly, because it often causes damage to the pantograph.

6   Conclusions and Future Work

The paper presents the AHP multi-criteria method which can be implemented into a software tool. It can replace a simple brainstorming method giving more options of assessment according to the defined criteria. The paper presents shortly the basis of the AHP method and compares this method to the brainstorming method.

Literature review and our own survey demonstrated that little attention had been paid to using MCDA methods in the railway industry. This is why the CRZ project can be the solution to that problem by offering software support for AHP method. The main goals of the project are: (1) supporting information exchange about railway threats, and (2) supporting railway experts in risk management and solving security problems within the entire railway industry (not only on the level of one particular railway enterprise).

The AHP multi-criteria method allows common assessments by many experts at the same time. This capability was successfully examined with railway experts during the project. The assessment criteria can be prioritized individually by each expert as well as separate assessments and comparisons of all alternatives can be made individually.

The proposed method has also got some weaknesses the users must be aware of. However there are some possibilities to diminish them. The results depend on the proper selection of assessment options and their coherence with the decision maker's point of view. We made an assumption in the project that a person who asks for support is responsible for the decision problem definition (including the assessment criteria and possible variants to be assessed). Thus the problem definition phase is a crucial point to reduce the lack of adequacy of criteria and variants.

There is a risk that some evaluators may act as false/fake experts or act in collusion. To eliminate such cases we can consider a ranking system to analyze how many solutions are proposed by a given expert which reflect actual implementation and to list the results (positive, negative, neutral). In the AHP method the experts assess each solution on the basis of their own experience. But in the end it is the decision maker's responsibility to select the right solution while experts are only an advisory board.

There is a possibility of imprecise or random assessment too. But AHP can indicate these inappropriate or random responses by determining a Consistency Ratio for each comparison matrix. The higher value of the indicator the greater uncertainty about the result.

It happens that the ratings made by different evaluators' groups (managers, decision-makers) have higher weights than others. Such cases were also reported by the railway experts. They found out that the assessments could be different depending on the evaluators' groups within the same railway company. It also depends on the type of a railway company: passengers carrier, freight carrier, an entity in charge of maintenance (EMC). Therefore the future implementations of AHP in a software tool should include some improvements and modifications considering the level of importance of individual evaluators [8]. To avoid discrepancies resulting from the type of a railway company (passengers/freight carrier, EMC), and to increase the accuracy of experts' responses it is planned to apply in the CRZ system the mechanism of assigning experts to decision-making problems based on specialization (infrastructure owners, carriers).

The presented results on AHP application in the railway industry constitute only preliminary work in the CRZ project. One of its elements will be the decision support and evaluation of alternatives using AHP. The software implementation of a well-defined risk assessment and decision support method is an added value for the CRZ project and its prospective users - railway security professionals.

The next stage of the project will be the implementation of a software tool and making it available on the network platform. The assessment of the method usability and the impact of its application on the efficiency of the operator's actions will be possible after gathering more data from the users. The effectiveness of selected measures and correctness of decisions made with AHP support will be analyzed. It will enable to evaluate how the use of the AHP method affects the performance of railway operators.

Acknowledgements. The work was carried out within the CRZ project, co-financed by the European Regional Development Fund, Grant No. POIR.04.01.02-00-0024/17-00.

References

1. Algarin, C.A.R., Llanos, A.P., Castro, A.O.: An analytic hierarchy process based approach for evaluating renewable energy sources. Int. J. Energy Econ. Policy **7**(4), 38–47 (2017)
2. Alonso, J., Lamata, M.: Estimation of the random index in the analytic hierarchy process. In: Proceedings of Information Processing and Management of Uncertainty in Knowledge-Based Systems, Perugia, vol. 1, pp. 317–322 (2004)
3. Bureika, G., Bekintis, G., Liudvinavičius, L., Vaičiūnas, G.: Applying analytic hierarchy process to assess traffic safety risk of railway infrastructure. Eksploatacja i Niezawodność - Maint. Reliab. **15**(4), 376–383 (2013)
4. Cabała, P.: Using the analytic hierarchy process in evaluating decision alternatives. Oper. Res. Decis. **20**(1), 5–23 (2010)
5. Ergun, M., Iyinam, S., Iyinam, A.: An assessment of transportation alternatives for Istanbul Metropolitan City for year 2000. In: Proceedings of the 8th Meeting of the Euro Working Group Transportation, Rome, pp. 183–190 (2000)
6. European Commission: Commission implementing regulation (EU) no 402/2013 of 30 April 2013 on the common safety method for risk evaluation and assessment and repealing regulation (EC) no 352/2009 (2013). https://eur-lex.europa.eu/legal-content/EN/TXT/PDF/?uri=CELEX:32013R0402

7. Fierek, S., Szarata, A.: Wykorzystanie symulacji modeli podróży i wielokryterialnej metody rankingowej do projektowania rozbudowy sieci tramwajowej. In: Materiały konferencyjne: Modelowanie podróży i prognozowanie ruchu, vol. 2, no. 98, pp. 183–190. ZNT SITK Rzeczpospolitej Polskiej, Kraków (2012)
8. Forman, E., Peniwati, K.: Aggregation individual judgements and priorities with the analytic hierarchy process. Eur. J. Oper. Res. 11, 165–169 (1998)
9. Golden, B., Wang, Q.: An alternate measure of consistency. In: Golden, B., Wasil, E., Harker, P. (eds.) The Analytic Hierarchy Process, pp. 68–81. Springer, Berlin (1989). https://doi.org/10.1007/978-3-642-50244-6_5
10. IEC/ISO 31010: 2009 risk management - risk assessment techniques (2010)
11. Książek, M., Nowak, P., Rosłon, J., Wieczorek, T.: Multicriteria assessment of selected solutions for the building structural walls. Procedia Eng. 91, 406–411 (2014)
12. Nataraj, S.: Analytic hierarchy process as a decision-support system in the petroleum pipeline industry. Issues Inf. Syst. V I(2), 16–21 (2005)
13. Saaty, T.: The Analytic Hierarchy Process: Planning, Priority Setting, Resource Allocation. McGraw-Hill, New York (1980)
14. Saaty, T., Hu, G.: Ranking by eigenvector versus other methods in the analytical hierarchy process. Appl. Math. Lett. 11(4), 121–125 (1998)
15. Sojda, A., Wolny, M.: Zastosowanie metody AHP w ocenie projektówinwestycyjnych Kopalni Węgla Kamiennego. Studia Ekonomiczne 207, 212–222 (2014)
16. Stoilova, S., Nikolova, R.: An application of AHP method for examining the transport plan of passenger trains in Bulgarian railway network. Transp. Probl. 13(1), 37–48 (2018)
17. The Office of Rail Transport (UTK): Guideline: Ekspertyza dotyczącapraktycznego stosowania przez podmioty sektora kolejowego wymagań i metodybezpieczeństwa w zakresie oceny ryzyka (2015)
18. The Office of Rail Transport (UTK): The office of rail transport (UTK)guideline: Ekspertyza dotycząca praktycznego stosowania przez podmiotysektora kolejowego wymagań i metody oceny bezpieczeństwa w zakresiemonitorowania (2015)
19. Velasquez, M., Hester, P.T.: An analysis of multi-criteria decision making methods. Int. J. Oper. Res. 10(2), 56–66 (2013)

NFL – Free Library for Fuzzy and Neuro-Fuzzy Systems

Krzysztof Siminski[(✉)]

Institute of Informatics, Silesian University of Technology,
ul. Akademicka 16, 44-100 Gliwice, Poland
krzysztof.siminski@polsl.pl

Abstract. The paper presents «Neuro-Fuzzy Library» (NFL) – a free library for fuzzy and neuro-fuzzy systems. The library written in C++ is available from the GitHub repository. The library implements data modifiers (for complete and incomplete data), clustering algorithms, fuzzy systems (descriptors, t-norms, premises, consequences, rules, and implications), neuro-fuzzy systems (precomposed MA, TSK, ANNBFIS, and subspace ANNBFIS for both classification and regression tasks). The paper is accompanied by numerical examples.

Keywords: Fuzzy system · Neuro-fuzzy system · Clustering · Incomplete data · C++

1 Introduction

Fuzzy and neuro-fuzzy system have proved their usefulness in many practical applications. Fuzzy systems incorporate fuzzy sets and fuzzy logic into artificial and computational intelligence. They share with humans a common representation of knowledge – fuzzy IF-THEN rules. Fuzzy rules are easy to interpret both by humans and machines. Human experience expressed as rules can be easily incorporated in a fuzzy system. Neuro-fuzzy systems add the ability of automatic identification of rules from presented data. They can extract knowledge hidden in the data and present it in form of easily readable rules. Both fuzzy and neuro-fuzzy systems have many applications in real life problems.

Our objective is to provide a reliable up-to-date open library for fuzzy and neuro-fuzzy systems to wide community of scientists and all interested in applications of neuro-fuzzy systems.

«Neuro-Fuzzy Library» (NFL) is available from the GitHub repository https://github.com/ksiminski/neuro-fuzzy-library.

2 Description

NFL library is written in the C++11 language. This language has been chosen because it is popular and widely known, builds to fast machine code, and its

© Springer Nature Switzerland AG 2019
S. Kozielski et al. (Eds.): BDAS 2019, CCIS 1018, pp. 139–150, 2019.
https://doi.org/10.1007/978-3-030-19093-4_11

compilers are available for many platforms. The library is commented in the Doxygen style, so the technical documentation is easily available from the library source code.

Figure 1 presents a UML diagram of essential classes in the library.

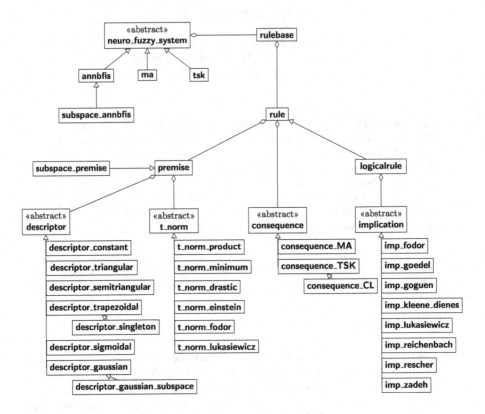

Fig. 1. UML diagram of essential classes

2.1   Data Modifiers

NFL library implements some data modifiers (class **data_modifier**):

- for complete data
    1. **data_modifier_normaliser**: normalisation to minimum 0 and maximum 1,
    2. **data_modifier_standardiser**: standardisation to average 0 and standard deviation 1,
- for incomplete data
    1. **data_modifier_marginaliser**: marginalisation – removal of incomplete data items,

2. **data_modifier_imputer**: imputation with a constant,
3. **data_modifier_imputer_average**: imputation with average of an incomplete attribute,
4. **data_modifier_imputer_median**: imputation with median of an incomplete attribute,
5. **data_modifier_imputer_knn_average**: imputation with average of an incomplete attribute from k nearest neighbours,
6. **data_modifier_imputer_knn_median**: imputation with medians of an incomplete attribute from k nearest neighbours,
7. **data_modifier_imputer_values_from_knn**: imputation with values of an incomplete attribute from k nearest neighbours,

2.2 Fuzzy Systems

With NFL library you can compose your own fuzzy system. The library implements a variety of descriptors, t-norms, implications, and consequences.

Descriptors. The available descriptors in the NFL library are:

- constant value (**descriptor_constant**)
- singleton (**descriptor_singleton**)
- triangular (**descriptor_triangular**) with core c and support interval (s_{min}, s_{max}) (Fig. 2)

$$\mu(x) = \begin{cases} \frac{x - s_{min}}{c - s_{min}}, & s_{min} < x < c \\ 1, & x = c \\ \frac{s_{max} - x}{s_{max} - c}, & c < x < s_{max} \\ 0, & \text{otherwise} \end{cases} \tag{1}$$

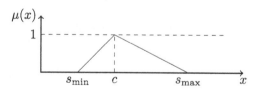

Fig. 2. Example of a triangular descriptor

- semitriangular
  - leftmost semitriangular (**descriptor_semitriangular**) descriptor (Fig. 3)

$$\mu(x) = \begin{cases} 1, & x \leqslant c \\ \frac{s - x}{s - c}, & c < x < s \\ 0, & x \leqslant x \end{cases} \tag{2}$$

- rightmost semitriangular descriptor (Fig. 3)

$$\mu(x) = \begin{cases} 0, & x \leqslant s \\ \frac{x-s}{c-s}, & s < x < c \\ 1, & c \leqslant c \end{cases} \tag{3}$$

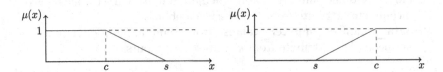

Fig. 3. Example of a leftmost (left) and rightmost (right) semitriangular descriptors

- trapezoidal (**descriptor_trapezoidal**) with core interval $[c_{\min}, c_{\max}]$ and support interval $(s_{\min}, s_{\max})$ (Fig. 4)

$$\mu(x) = \begin{cases} \frac{x - s_{\min}}{c_{\min} - s_{\min}}, & s_{\min} < x < c_{\min} \\ 1, & c_{\min} \leqslant x \leqslant c_{\max} \\ \frac{s_{\max} - x}{s_{\max} - c_{\max}}, & c_{\max} < x < s_{\max} \\ 0, & \text{otherwise} \end{cases} \tag{4}$$

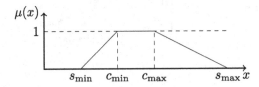

Fig. 4. Example of a trapezoidal descriptor

- sigmoidal (**descriptor_sigmoidal**) (Fig. 5)

$$\mu(x) = \frac{1}{1 + \exp(-\alpha(x - \beta))} \tag{5}$$

- gaussian (**descriptor_gaussian**) with mean m and standard deviation s

$$\mu(x) = \exp\left[-\frac{(x - m)^2}{2s^2}\right]; \tag{6}$$

- subspace gaussian (**descriptor_gaussian_subspace**) similar to gaussian descriptor, but a descriptor has an weight.

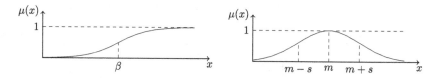

Fig. 5. Example of a sigmoidal (left) and gaussian (right) descriptors

Table 1. T-norms

T-norms	Class	$T(a, b)$
product	**t_norm_product**	ab
minimum	**t_norm_minimum**	$\min(a, b)$
drastic	**t_norm_drastic**	$\begin{cases} \min(a,b), & a = 1 \vee b = 1 \\ 0, & \text{otherwise} \end{cases}$
Einstein	**t_norm_einstein**	$\frac{ab}{2-(a+b-ab)}$
Fodor	**t_norm_fodor**	$\begin{cases} \min(a,b), & a + b > 1 \\ 0, & \text{otherwise} \end{cases}$
Łukasiewicz	**t_norm_lukasiewicz**	$\max[a + b - 1, 0]$

T-Norms. The AND operator joining the descriptors is modelled with T-norms. The T-norms provided by the NFL library are listed in Table 1.

Implications. Some fuzzy systems use logical interpretation of fuzzy rules. The rules are treated as true mathematical fuzzy implication. They are listed in Table 2.

Conclusions. The NFL library provides three types of conclusions:

– Mamdani-Assilan [9] type based on triangular fuzzy sets (**consequence_MA**),
– Takagi-Sugeno-Kang [15, 16] type based on a linear combination of input values (**consequence_TSK**),
– Czogała-Łęski type [1] – parametrised consequences (moving consequences) (**consequence_CL**).

2.3 Neuro-Fuzzy Systems

A neuro-fuzzy system adds a new feature to fuzzy systems. A neuro-fuzzy system elaborates its fuzzy rule base from presented data. Commonly identification of fuzzy rule base is done in two steps: first the input domain of the task is partitioned into several regions that are transformed into premises of rules, then in the second step the rule base is tuned to better fit the presented data.

One of methods of initial partition of the input domain is clustering. The NFL library contains new implementations of several clustering algorithms:

Table 2. Implications

Implication	Class	$x \to y$
Fodor	**imp_fodor**	$\begin{cases} 1, & x \leqslant y \\ \max\left[1 - x, y\right], & \text{otherwise} \end{cases}$
Gödel	**imp_goedel**	$\begin{cases} 1, & x \leqslant y \\ y, & \text{otherwise} \end{cases}$
Goguen	**imp_goguen**	$\begin{cases} 1, & x \leqslant y \\ \frac{y}{x}, & \text{otherwise} \end{cases}$
Kleene-Dienes	**imp_kleene_dienes**	$\max[1 - x, y]$
Łukasiewicz	**imp_lukasiewicz**	$\min[1, 1 - x + y]$
Reichenbach	**imp_reichenbach**	$1 - x + xy$
Rescher	**imp_rescher**	$\begin{cases} 1, & x \leqslant y \\ 0, & \text{otherwise} \end{cases}$
Zadeh	**imp_zadeh**	$\max[1 - x, \min(x, y)]$

– fuzzy c-means (FCM) [3] (class **fcm**)
– possibilistic fuzzy c-means (possibilistic FCM) [5] (**fcm_possibilistic**),
– conditional fuzzy c-means (conditional FCM) [10] (**fcm_conditional**),
– subspace fuzzy c-means (SFCM) [4] (**sfcm**),
– fuzzy c-ordered-means (FCOM) [7] (**fcom**),
– rough fuzzy c-means (RFCM) [11] (**rfcm**),
– subspace fuzzy c-ordered-means (SFCOM) [14] (**sfcom**),
– rough subspace fuzzy c-means (RSFCM) [13] (**rsfcm**).

The NFL library implements three types of neuro-fuzzy systems with gradient tuning paradigm based on Mamdani-Assilan (class **ma**), Takagi-Sugeno-Kang (**tsk**), and Czogała-Łęski fuzzy systems (**annbfis** and **subspace_annbfis**).

3   Examples

In the repository we prepared 5 experiments to show the features and usage of the NFL library.

3.1   exp-1: Data Modifiers

This experiment shows all data modifiers enumerated in Sect. 2.1 and an example of a chain of modifiers. Modifications are shown both for complete and incomplete data. For the last example (a chain of modifiers) the original incomplete data set is first imputed with constant value, normalised, imputed with average of attributes, and finally standardised.

3.2   exp-2: Clustering of Complete Data

The experiments runs 5 clustering algorithms:

1. fuzzy C-means (FCM)
2. possibilistic FCM
3. conditional FCM
4. subspace FCM (SFCM)

   This experiment presents a subspace FCM algorithm. This algorithm assigns weights to attributes in each cluster separately. Thus the algorithm extracts fuzzy subspaces from the task domain. The data set has two clusters but the importance of attributes is not the same. In the first cluster important attributes are: 1, 3, and 5; in the second cluster – 2, 4, and 5. The 6th attribute has low importance in both clusters. The weights assigned to attributes are presented in Table 3. The informative attributes have high weights and non-informative ones – low. The constraints of the algorithm on attribute weights makes them sum up to 1 in each cluster.

Table 3. Weights assigned to attributes in clusters by the SFCM algorithm

Attribute	Cluster 1	Cluster 2
1	0.28201	0.00265
2	0.00258	0.34119
3	0.32973	0.00243
4	0.00279	0.26542
5	0.38000	0.38606
6	0.00289	0.00225
	1.00000	1.00000

5. fuzzy C-ordered means (FCOM)

   This experiment runs on data set with three informative clusters at locations $(10, 10)$, $(10, -10)$, $(-10, 10)$. Each cluster has 200 points with dispersion 1. Additional outliers cluster at $(-10, -10)$ built with 20 points with dispersion 0.1. The FCOM algorithm assigns typicality to each data item. The algorithm elaborated three clusters with centres at $(10.3735, 9.96222)$, $(9.94037, -9.99038)$, and $(-9.99487, 10.0413)$. A histogram of typicalities is presented in Fig. 6.

3.3   exp-3: Clustering of Incomplete Data

The experiment shows an example of clustering of incomplete data with specialized algorithms:

1. rough fuzzy C-means (RFCM),

2. rough subspace fuzzy C-means (RSFCM) – the algorithm elaborates attribute weights,
3. rough fuzzy C-ordered means (RFCOM) – the algorithm elaborates typicalities of data items.

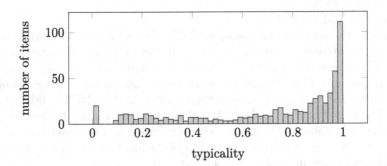

Fig. 6. Typicalities assigned to data items. The left bin with almost zero typicality represents outliers.

3.4   exp-4: Fuzzy System

The experiments shows an example of a fuzzy system with precomposed fuzzy rule base. The input domain of the system has two attributes x_1 and x_2. The former attributes is partitioned with two triangular fuzzy sets, the latter – with two semitriangular fuzzy sets presented in Fig. 7. The descriptors in rules are joined with a product T-norm. The consequences follow the Mamdami-Assilan paradigm and are composed of four triangular fuzzy sets. The rule base of the system defines a surface presented in Fig. 8.

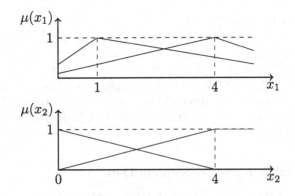

Fig. 7. Partition of domain of the x_1 and x_2 attribute.

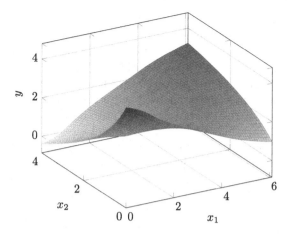

Fig. 8. Surface defined by the fuzzy rule base in exp-4

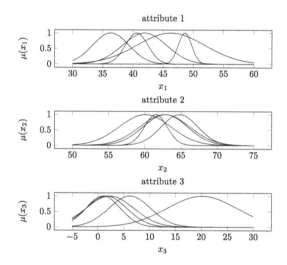

Fig. 9. Partition of the input domain for the 'haberman' data set elaborated by the ANNBFIS system

3.5   exp-5: Neuro-Fuzzy System

Four types of neuro-fuzzy systems have been precomposed:

1. Mamdani-Assilan (class **ma**)
2. Takagi-Sugeno-Kang (**tsk**)
3. Czogała-Łęski in two versions: ANNBFIS (**annbfis**) and subspace ANNBFIS (**subspace_annbfis**) [12].

All systems have Gaussian fuzzy sets in premises joined with a product T-norm. For Czogała-Łęski neuro-fuzzy systems with logical interpretation of fuzzy rules the user can any of implications listed in Table 2. All neuro-fuzzy systems use a gradient tuning method. For all systems methods for both regression and classification tasks have been implemented. For classification task a threshold for separation of positive class from negative one is calculated in three ways: with Youden index [17], minimal distance threshold or mean of both classes.

The examples present the precomposed neuro-fuzzy systems in regression and classification tasks. For regression the 'leukocytes' data set is used. It is a data set describing concentration of leukocytes in blood modelled with Mackey-Glass equation [8]. The equation was solved with condition $x(0) = 0.1$ with Runge-Kutt method with step $k = 0.1$ [6]. The data series was the base for creation of tuples with template $[x(t), x(y - 6), x(t - 12), x(t - 18), x(k + 6)]$. The train data set holds items 501 to 700, the test data set – items 701 to 1000. The results are gathered in Table 6.

For classification the 'habermas' data set is used. This data set contains survival data of patients who had undergone surgery for breast cancer [2]. This is a two class problem. The first 153 items are used as a train set, the following 153 – as a test set. Table 4 presents precision and recall elaborated by the precomposed neuro-fuzzy systems with the three thresholds for separation of the positive class from the negative one. Table 5 presents a confusion matrix for the 'haberman' test data set elaborated by the ANNBFIS system with Youden index. Figure 9 presents an example of a partition of the input domain for the 'haberman' data set.

Table 4. Results elaborated for the 'haberman' data set

System	Threshold Type	Value	Train Precision	Recall	Test Precision	Recall
TSK	Mean	1.5000	0.6315	0.6153	0.2763	1.0000
	Minimal distance	1.2708	0.4603	0.7435	0.5476	0.5476
	Youden	1.2728	0.4915	0.7435	0.2763	1.0000
MA	Mean	1.5000	0.4285	0.1538	0.3666	0.5238
	Minimal distance	1.2535	0.4307	0.7179	0.3135	0.8809
	Youden	1.2702	0.4814	0.6666	0.3333	0.7380
ANNBFIS	Mean	1.5000	0.7500	0.3846	0.5806	0.4285
	Minimal distance	1.2938	0.4677	0.7435	0.2770	0.9761
	Youden	1.2455	0.5373	0.9230	0.5200	0.6190
Subspace ANNBFIS	Mean	1.5000	0.6666	0.2564	0.6111	0.2619
	Minimal distance	1.2640	0.4705	0.6153	0.3333	0.9285
	Youden	1.2414	0.4354	0.6923	0.4905	0.6190

Table 5. Confusion matrix for test data for the 'haberman' data set elaborated with the ANNBFIS neuro-fuzzy system with Youden index

Total population (POP) = 153	Original positive (OP) = 42	Original negative (ON)= 111	Prevalence $= \frac{OP}{POP}$ $= 0.27450$	Accuracy (ACC) $= \frac{TP+TN}{POP}$ $= 0.73856$
Predicted positive = 50	True positive, TP, power = 26	False positive, FP, type I error = 24	Positive predictive value, PPV, precision $= \frac{TP}{TP+FP}$ $= 0.52000$	False discovery rate, FDR $= \frac{FP}{TP+FP}$ $= 0.48000$
Predicted negative = 103	False negative, FN, type II error = 16	True negative, TN = 87	False omission rate, FOR $= \frac{FN}{FN+TN}$ $= 0.15533$	Negative predictive value, NPV $= \frac{TN}{FN+TN}$ $= 0.84466$
	True positive rate, TPR, recall, sensitivity $= \frac{TP}{TP+FN}$ $= 0.61904$	False positive rate, FPR, fall-out $= \frac{FP}{TN+FP}$ $= 0.21621$	Positive likelihood ratio (LR+) $= \frac{TPR}{FPR}$ $= 2.86309$	Diagnostic odds ratio, DOR $= \frac{LR+}{LR-}$ $= 3.92708$
	False negative rate, FNR, miss rate $= \frac{FN}{TP+FN}$ $= 0.57142$	True negative rate, TNR, specificity, SPC $= \frac{TN}{TN+FP}$ $= 0.78378$	Negative likelihood ratio (LR−) $= \frac{FNR}{TNR}$ $= 0.72906$	F1 score = $\frac{2TPR\times PPV}{TPR+PPV}$ $= 0.56521$

Table 6. Results elaborated for the 'leukocytes' data set with the precomposed neuro-fuzzy systems. RSME stands for root means square error, MAE – for mean absolute error.

	RMSE		MAE	
System	Train	Test	Train	Test
TSK	0.000650	0.001079	0.000522	0.000715
MA	0.398608	0.401140	0.346149	0.349690
ANNBFIS	0.000639	0.001016	0.000513	0.000685
Subspace ANNBFIS	0.000683	0.003824	0.000537	0.001105

4 Conclusions

The paper presents the NFL – Neuro-Fuzzy Library for fuzzy and neuro-fuzzy systems. The library is free and aims at providing the community of scientists with reliable and fast fuzzy and neuro-fuzzy systems. It implements data modi-fiers for complete and incomplete data, clustering algorithms, fuzzy and neuro-fuzzy systems. The library is written in the C++ language and is available from the GitHub repository.

Acknowledgements. The research has been supported by the Rector's Grant for Research and Development (Silesian University of Technology, grant number: 02/020/RGJ19/0165).

References

1. Czogała, E., Łęski, J.: Fuzzy and Neuro-Fuzzy Intelligent Systems. Series in Fuzziness and Soft Computing. Springer, Heidelberg (2000). https://doi.org/10.1007/978-3-7908-1853-6
2. Dheeru, D., Karra Taniskidou, E.: UCI machine learning repository (2017). http://archive.ics.uci.edu/ml
3. Dunn, J.C.: A fuzzy relative of the ISODATA process and its use in detecting compact, well separated clusters. J. Cybern. **3**(3), 32–57 (1973)
4. Gan, G., Wu, J.: A convergence theorem for the fuzzy subspace clustering (FSC) algorithm. Pattern Recogn. **41**(6), 1939–1947 (2008). https://doi.org/10.1016/j.patcog.2007.11.011
5. Krishnapuram, R., Keller, J.: A possibilistic approach to clustering. IEEE Trans. Fuzzy Syst. **1**, 98–110 (1993)
6. Leski, J.: Systemy neuronowo-rozmyte (in Polish: Neuro-fuzzy systems). Wydawnictwa Naukowo-Techniczne, Warszawa (2008)
7. Leski, J.M.: Fuzzy c-ordered-means clustering. Fuzzy Sets Syst. **286**, 114–133 (2014). https://doi.org/10.1016/j.fss.2014.12.007
8. Mackey, M.C., Glass, L.: Oscillation and chaos in physiological control systems. Science **197**(4300), 287–289 (1977)
9. Mamdani, E.H., Assilian, S.: An experiment in linguistic synthesis with a fuzzy logic controller. Int. J. Man-Mach. Stud. **7**(1), 1–13 (1975)
10. Pedrycz, W.: Conditional fuzzy clustering in the design of radial basis function neural networks. IEEE Trans. Neural Netw. **9**(4), 601–612 (1998)
11. Siminski, K.: Clustering with missing values. Fundamenta Informaticae **123**(3), 331–350 (2013)
12. Siminski, K.: Neuro-fuzzy system with weighted attributes. Soft Comput. **18**(2), 285–297 (2014). https://doi.org/10.1007/s00500-013-1057-z
13. Siminski, K.: Rough fuzzy subspace clustering for data with missing values. Comput. Inform. **33**(1), 131–153 (2014)
14. Siminski, K.: Fuzzy weighted C-ordered means clustering algorithm. Fuzzy Sets Syst. **318**, 1–33 (2017). https://doi.org/10.1016/j.fss.2017.01.001. http://www.sciencedirect.com/science/article/pii/S0165011417300180
15. Sugeno, M., Kang, G.T.: Structure identification of fuzzy model. Fuzzy Sets Syst. **28**(1), 15–33 (1988)
16. Takagi, T., Sugeno, M.: Fuzzy identification of systems and its application to modeling and control. IEEE Trans. Syst. Man Cybern. **15**(1), 116–132 (1985)
17. Youden, W.J.: Index for rating diagnostic tests. Cancer **3**(1), 32–35 (1950)

Detection of Common English Grammar Usage Errors

Luke Immes[✉] and Haim Levkowitz

Computer Science Department,
University of Massachusetts at Lowell,
Lowell, MA, USA
lpimmes@townisp.com,
{limmes,haim}@cs.uml.edu

Abstract. Our research aims to provide writers with automated tools to detect grammatical usage errors and thus improve their writing. Correct English usage is often lacking in scientific and industry papers. [16] has compiled 130 common English usage errors. We address the automated detection of these errors, and their variations, that writers often make. Grammar checkers, e.g., [9] and [11], also implement error detection. Other researchers have employed machine learning and neural networks to detect errors. We parse only the part of speech (POS) tags using different levels of generality of POS syntax and word-sense semantics. Our results provide accurate error detection and are feasible for a wide range of errors. Our algorithm specifies precisely the ability to increase or decrease the generality in order to prevent a large number of false positives. We derive this observation as a result of using The Brown corpus, which consists of 55, 889 untagged sentences, covering most genres of English usage, both fiction and non-fiction. This corpus was much larger than any corpus employed by related researchers. We implemented 80 of Swan's most common 130 rules; and detected 35 true positives distributed among 15 of Swan's rules. Such a low true positive rate, 35/55889, had been expected. No false positives were detected. We employed a separate, smaller, test suite of both true positive and true negative examples. Our system, as expected, correctly detected errors in all the true positive examples, and ignored all the true negative ones. The Language-Tool system had a detection rate of $28/130 = 22\%$; Grammarly had a detection rate of $60/130 = 46\%$. Our results show significant improvement in the detection of common English usage errors.

Keywords: Big data · English grammar · Part of speech tag · Heuristic · Rule · Converge · Machine learning · Neural network · English usage · POS · NER · Common error · Uncommon error · Context · Grammatical usage rule · False positive · True positive · Brown corpus

© Springer Nature Switzerland AG 2019
S. Kozielski et al. (Eds.): BDAS 2019, CCIS 1018, pp. 151–167, 2019.
https://doi.org/10.1007/978-3-030-19093-4_12

1   Introduction, Problem Addressed

Correct English writing is a difficult task, even for a college educated, native speaker of English—in other words poor grammatical usage does not reflect well upon the writer. The intended writing styles we had in mind were cover letters, marketing reports, short stories, and legal briefs. (Chatbots—generating intelligent English sentences given an arbitrary English query—involve sentence generation, not error detection.) We made this claim after carefully studying Swan's Practical English usage, [16], with 130 common rules. We are not just referring to, e.g., subject-verb agreement or the proper use of *its* versus *it's*.

Our research was also motivated by the fact that studies have shown that foreign language students have difficulty with English grammar, [1]. Specifically, Iraqi high school students have serious problems with the usage of verb tenses, articles, and prepositions. Frequent errors involved missing words and additional words. Furthermore, mastering prepositions is onerous for the English language learner, [5]—they are just too numerous.

We discuss some Swan rules as a linguist would view them. In error rule 595.2 (too, too much) we have the following errors, i.e., the true positives:

It's too much hot in this house.
I arrived too much early.

The corrections, i.e., the true negatives:

It's too hot in this house.
I arrived much too early.

Obviously, one can substitute hot, cold, dark, light, etc. This involves simple syntax checking on the adjective.

For error rule 325.1 (like: verb), one has to be careful using simple replacement to correct an error:

You look as John.

The correction is:

You look like John.

However, substituting *like* for *as* will not work in the following case, because the following sentence is correct.

He whistled as he locked the office.

We discuss how our algorithm did not consider this a false positive shortly. An explanation of what exactly is wrong—syntax as well as semantics—is provided by Swan, as well as positive and negative sentence examples. Since our algorithm implements Swan's rules, there necessarily must be a close correspondence between our implementation of the rule and Swan's original examples of the rule. Grammar checkers on the market may have, in fact, used Swan's rules.

To reiterate, Swan's rules are both syntactic, and semantic—implicitly capturing the user's intention. At first glance—when examining the more difficult and uncommon kinds of errors that occur—one knows that there is an error of some kind, but not the reason for it. We decided to automate the process, while being mindful to adjust the generality of rule detection in order to avoid a large number of false positives.

Given 130 of Swan's common examples, we thoroughly tested 80 of these (due to time constraints). For each of these rules, there are examples of both true positives and true negatives. Our algorithm detected all the true positives and ignored all the true negatives, as expected. These examples are independent of any corpus (or document).

We tested a large corpus covering major genres of English, including fiction, non-fiction, new stories, business news, and academic papers, using the Brown Corpus [2]. This corpus has 55, 889 sentences—of correct grammatical usage. Therefore, it was no surprise that the number of true positives was very low, 35 out of 55, 889, distributed among 15 of Swan's rules. This corpus was much larger than any corpus mentioned in related research (Sect. 3); total run time was one minute 34 s. Finding false negatives is a separate research endeavor, which will be discussed (Sect. 4.2).

Each word in the Brown corpus has an adjacent part of speech (POS) tag, which we removed because our algorithm requires a tagged input sentence in the form of XML. We employed Stanford's POS tagging [15] during pre-processing, which provides for the correct input format; it also provides for named entity recognition (NER). Our solution parses only the part of speech tags, e.g., root verbs or nouns (within that tag), during Swan rule invocation.

The details of our contribution are discussed next.

2 Contribution, Intellectual Merit

Our algorithm detects common and uncommon English usage errors via POS (part of speech) tags, and root words denoting context. Specifically, we have:

- The Brown Corpus was used as a test for false positives. Thirty-five true positives distributed among fifteen Swan rules resulted. This was a desirable low rate, 35/55,889, as expected.
- Separately, independent testing (apart from any corpus) of all 80 of Swan's rules resulted in the detection of all true positives as well as ignoring all true negatives.

Next, we review detecting English grammatical errors.

3 Related Work

Some authors used a large corpus of correct and incorrect sentences; measures of success are retrieval percentage rates. *Precision* measures accuracy, and *recall* measures thoroughness of search—Are any true positives missed?

[9] is an on-line grammar checker[1] with a free and premium edition. Testing all of Swan's 130 errors, it detected 35 errors and 25 additional errors (if using the premium edition). Performance was $60/130 = 46\%$; we expected at least 85%. Some error explanations were sufficient (to make the correction), but in many cases not helpful.

[11] is another on-line grammar checker[2] with free and premium editions. Testing all of 130 of Swan's errors, the free edition detected 24 errors, and another 4 errors (if using the premium edition). Performance was $28/130 = 22\%$, i.e., very low in our opinion. Language-Tool reported an 'Out of memory' exception when tested under the Brown corpus.

A neural network for machine translation, [18], finds errors in a target[3] foreign language during statistical machine translation. The architecture employs Gated Recurrent Units. Inputs are the n-gram surface syntax (including POS tagging), parse trees of concomitant siblings, children, and parents. Outputs are two word classes: good/bad word, with corresponding fluency[4] error types: grammar, lexicon, orthography, multiple errors, and other errors. Training employed the SCATE corpus [18] of 2067 sentence pairs (e.g., English-Dutch). Fluency OK score was .81, and fluency BAD score was .47; these are good results in our opinion. They employed a *recurrent* neural network (which converged). Furthermore, they used many more examples than we did. (Either or both of these factors may the reason for our neural network not converging, see Sect. 3.1.) However, they considered only broad types of errors, which are not as specific as 80 of Swan's error rules.

[7] implemented a recurrent neural network to detect article (determiner) errors. Precision was 76% and recall was 41% which is passable in our opinion. The NUS Corpus of Learner English was employed which included a subset of article insertion, deletion and substitution errors (total: 5782). Our own algorithm had some difficulty with article detection using Swan's rules. It would be prudent for us to consider the NUS examples in future development and testing of our algorithm.

[4] developed CyWrite which detected quantifier, subject-verb agreement, article, and run-on-sentence error types—only four kinds of errors, whereas our algorithm covered 80 of Swan's rules. CyWrite employed the statistical parser from [15] using a corpus of 5500 sentences. Given the error types, precision varied between .75 and .62, and recall varied between .30 and .63, which is a good result (in terms of accuracy). False positives and false negatives occurred within all error types. For example, false positives arose because their uncountable list of nouns was not complete. Our algorithm has the same constraint—lists of uncountable nouns (or other types) must be augmented to prevent both false

---

[1] Rated first out of 20 grammar checking applications.

[2] Rated fifth out of 20 grammar checking applications.

[3] The target foreign language is the language one is translating into, e.g., from Dutch into English.

[4] Fluency, in machine translation, means how natural the resultant translation is to a native speaker of the language.

positives and false negatives. Another reason for `CyWrite`'s false positives is that the POS tagger tagged some word improperly. Our algorithm encountered the same problem—but it overrode the tagger's defaults in some cases (for particular Swan rules). Similar to our algorithm, `CyWrite` is a rule based approach. However, we did not employ the statistical parser [15], as we did in a prior independent work on machine translation, because resultant parse trees were inconsistent.

The system presented in the doctoral dissertation of [19] helps tutors of English (as a second language) by providing automated corrections and the correction's rationale. Specifically, the system finds redundant expressions, and misused or confusing words (e.g., prepositions, articles). Testing for redundant words involves supervised machine learning on a corpus of sentences. The scorer is trained on which words are likely to be redundant—words are dropped according to *fluency*: using an n-gram model, and *meaning*: alignment between words and their translation into another foreign language. Accuracy is 37% for the most redundant word; but 70% to determine if a sentence has redundant words at all. Testing for misused words involves simulating how similar words are used in different contexts, during the time the student had been learning English. If words are used in similar contexts they are likely to be confused. Swan does not have any rules, that we know of, which consider redundant words. But, confusing word combinations are covered indirectly throughout many of Swan's rules. Our algorithm handled the syntactic cases (of confusing words) easily, but discerning the user's intention via surrounding context may be tricky (see Sect. 4.2).

The [12] text book thoroughly analyses the kinds of co-locations that proofreaders of English commonly miss. The book covers prepositional constructions that foreign language learners find difficult. In this respect it is comparable to the reference text that we employ—[16]. But, the text also discusses various implementation solutions for error detection, e.g., machine learning, neural networks, and rule based systems.

[8] has system that corrects N+N (combined noun phrase) for French users of English. In the ICLE corpus of 226,922 words 5.52% had N+N errors. Perhaps the noun modifier is used incorrectly (too general), or ambiguity arises with two or more nouns together. [8] developed a small hand-coded grammar to run against the corpus. Parse trees are compared with the N+N error. That approach is similar to ours—it is a rule based. But, we covered significantly more errors, 80 of Swan's rules. We analyzed the POS tags themselves versus parse trees [15]).

[6] take an interesting approach to aid machine learning algorithms for error detection. Their algorithm, GenERRate, generates errors automatically—errors that would normally have been made. We are not generating errors, we detect them in the given input sentence.

In the doctoral dissertation of [3], that system automatically acquires models of prepositional and determiner usage, for the purpose of correcting non-native English speakers. The system performs well on correct preposition selection, but less than optimal on missing or extraneous determiners. No precision or recall rates were reported. Our algorithm covers many more grammatical errors. We take symbolic approach which facilitates developing (and debugging) one rule at a time. De-Felice takes a machine learning approach which looks at preposition and determiner examples as a whole—which is harder to debug.

[17] use a maximum entropy machine learning algorithm to detect errors in non-native speakers of English—prepositions in particular. Precision was 84%, recall was 19%. English prepositional usage (or its absence) is quite varied with users' intention, e.g., *on, near, at, by* the beach. We cover many more error types, and our accuracy is significantly higher, and we can point out the exact error.

Considering the contributions of the authors above, our approach is completely different: we attempt to detect the error explicitly by means of grammatical rules, parsing only POS tags. To reiterate, we take a symbolic approach which facilitates development versus a non-symbolic one. However, there is a probabilistic component in our approach, since the POS tagger relies upon n-grams which is not rule-based. Refer to Sect. 4.2 our handling of false negatives.

3.1   Our Own Related Neural Network Approach

Our motivation for using a non-rule based approach, concurrently with our rules based approach, is twofold: to reduce the knowledge engineering time for building rules, and to improve the accuracy of error detection—minimize false positives (occurring within the rules). See our neural network details in Appendix B; we will only summarize the results here.

During our neural network training phase, *no* convergence occurred—because each of Swan's rules has on average five error examples—not nearly enough examples. If the network does not converge during training, one cannot run that neural network.

Many examples are required to train a neural network as evidenced by the following text processing applications. `TensorFlow` (machine learning using a neural network), [10], learns the context of a given word via its surrounding words. Massive amounts of text are used when training the network. Documents are then classified based upon context. Another `TensorFlow` application, [14], uses chatbots to generate intelligent English text. That application also requires an enormous amount of data, i.e., the Movie Dialog Corpus (Cornell University) of $20K$ conversations.

In the future use many more examples during the training phase. For *each* Swan rule (described in Sect. 1), a linguist would write p true positives and n true negatives so that the network converges using all of Swans rules. p and n would need to be much larger than five. In order to make the process faster, one could generate sentences automatically, filling in nouns, verbs, adjectives, while still conforming to the rule's semantics. The minimum number of Swan's rules to train with until network convergence could then be determined.

Next, we discuss our detection of usage errors.

4   Examples of English Grammar Usage, Error Detection

We employed some of the following POS tags [15]:

1. CC coordinating conjunction; and, but, or
2. CD cardinal number; one, two
3. DT Determiner; a, the
4. IN preposition, subordinating conjunction; of, in, by
5. JJ, yellow
6. MD Modal, can, should
7. NN Noun, singular or mass; llama
8. NNS Noun, plural; llamas
9. NNP Proper noun, singular; IBM
10. PRP Personal pronoun; I, you, he, they
11. PRP$ Possessive pronoun; your, one's
12. TO to, a preposition
13. VB Verb, base form; eat
14. VBD Verb, past tense; ate
15. VBG Verb, gerund or present participle; eating
16. VBN Verb, past participle; eaten
17. VBP Verb, non-3rd person singular present; they eat
18. VBZ Verb, 3rd person singular present; he eats
19. WDT Wh-determiner; which, that
20. WP Wh-pronoun; what, who
21. WRB Wh-adverb; how, where

We list some examples shown in Sects. 4.1 and 4.2. LEMMA refers to the root word, POS refers to the part of speech (tag)—all of which are shown above. NER refers to named entity recognition; 0 under NER means not applicable.

Needless to say, our algorithm has no trouble with explicit syntax, but it does not understand semantics, unless there is explicit marking. Thus, checking for a usage error using syntax—POS tags—is straightforward. However, discerning semantics requires examining surrounding words to determine context without being too general.

Swan actually created 634 rules of English grammar usage, of which 130 are most common. Most of these rules just mention true positives, without any true negatives. Swan's number scheme is rather arbitrary. For each rule there is a sub-rule which we denoted with an underscore and a number. Several rules (xxx_xx), and examples—both correct and incorrect—are discussed next.

4.1   Swan Error 595_2.

Consider rule 595_2. '... too much hot ...' should be 'too hot ...' The actual errors are in the comments. The source code easily detects the problem via the too function. Detection of other errors is not this straightforward, as we will see next.

```
/*
 * It's too much hot in this house.
 * I arrived too much early.
 */
def swan_595_2_find(sen: Array[Array[String]]): Int = {

  def too(p: Int, sent: List[Array[String]]): Int = {
    val v: Int = verb(p, sent)
    val a: Int = if (v != -1) find_LEM(v + 1, "too", sent) else -1
    val b: Int = if (a != -1) find_LEM(a + 1, "much", sent) else -1
    val c: Int = if (b != -1) find_POS(b + 1, "JJ", sent) else -1

    val va: Boolean = a - v == 1
    val ab: Boolean = b - a == 1
    val bc: Boolean = c - b == 1

    val hits: List[Int] = List(v, a, b, c).filter(_ > -1)
    if (va && ab && bc && hits.size == 4) c else -1
  }

  val sent: List[Array[String]] = sen.toList
  too(0, sent)
}
```

Here our algorithm implements the following heuristic (as well as the second one in Sect. 4.2):

Find combinations of words that constitute a grammatical usage error.

4.2   Swan Error 326_1.

Now consider error 326_1. The error is using *as* when one should have used *like*. Our algorithm detects the following errors:

You look as John.
My sister looks as me.
My brother looks as John.
He ran as the wind.
She is good at subjects as mathematics.
In warm countries as Turkey, the climate is warm.

Our algorithm, correctly, did not detect errors in the following sentences.

I had another place which I used as a studio.
He whistled as he locked the office.

Occasionally, we had no choice but to use the root word of the verbs, e.g., *run*, *look* and the nouns *country*, *subject* as contextual markers, otherwise many false positives occur. One can consider these additional words as a form of generalizing the error detection. Thus, the last two sentences above were, correctly, not considered errors. What makes our algorithm *innovative* is that it implements the following heuristic (as well as the first one in Sect. 4.1):

Find combinations of surrounding words (that are not errors in themselves) which circumscribe a grammatical usage error, while being careful not to induce false positives.

Note that one would have to add synonyms for these words in a future version. See Appendix A for the actual code; start from the `comb` function (bottom of page).

Minimizing False Negatives. Automated testing for false negatives in the Brown Corpus was deemed impractical—one must manually examine a sentence and determine whether or not a Swan rule should have been invoked.

Some of Swan's rules require discerning semantics from verbs and their surrounding argument syntax, e.g., *'I am boring with the lecture.'* should be *'I am bored with the lecture.'* Implicitly we have *x verbing with y*. A linguist would have to determine, beforehand (apart from our algorithm), all such root verbs with attendant arguments that could be used incorrectly. Our algorithm handled specific true positive cases from Swan's rules, but more generality is possible. Machine learning techniques (e.g., classification), and neural networks (if convergent) may assist in this endeavor (Sect. 3.1).

4.3   Algorithm, and Rule Base Maintenance

Our algorithm is seemingly straightforward—detect the usage error from combinations of POS syntax. LEMMA, i.e., root words, and NER, i.e., named entities, all provide additional information. However, there are some subtleties when implementing each detection rule. For example, one has to account for separation between words: Usually it is one word apart, but it could be exactly two words or perhaps less than or equal to two words. Failure to account for word separation results in the error not being detected, or false positives.

Furthermore, sometimes it is necessary to search for a portion of the rule, then search backwards (to the left). This occurs when looking for errors involving verbs. If the verb form is encountered too early during the parsing, that rule will not be detected. For example, the algorithm may encounter a non-relevant verb before the relevant verb is found. By looking backwards at verb arguments our algorithm can determine the relevant verb.

Another heuristic we employed was searching for *arguments* before and after the error itself. Arguments were typically subjects and direct objects of some verb. If this is not done, false positives occur. The POS for that argument must be examined; it might be a noun, pronoun or person, but not a possessive pronoun. Also, the *ing* form of a verb, POS encoding: VBG, may be encoded as JJ, an adjective. If this is not accounted for in the error detection, that rule will not be triggered.

As mentioned before, in rule 326_1, sometimes it is necessary to use actual root words to provide a context to prevent an overwhelming number of false positives. However, one then needs to add synonyms, or analogs of these words to the code.

The Brown Corpus was essential when refining these error detection rules. When developing a new rule, one may find that both true positive and true negative examples yield expected results, only to find false positives when testing under the Brown Corpus.

All the other rules, 80 total, were implemented in a similar fashion using the above heuristics.

5   Conclusions

We have created an algorithm that allows 80 of the 130 common, and uncommon English usage errors to be detected. When testing, roots of nouns or verbs may have to be added to determine a context—thereby preventing false positives. The Brown corpus, 55, 889 sentences, was tested and 35 true positives distributed among 15 Swan rules were detected. Such a low rate of false positives, 35/55,889, was not surprising.

Our algorithm's declarative heuristics, i.e., a comprehensive explanation of why our algorithm works is:

Find combinations of words that constitute a grammatical usage error. (Sect. 4.1)
Find combinations of surrounding words (that are not errors in themselves) which circumscribe a grammatical usage error, while being careful not to induce false positives. (Sect. 4.2)

Circumscribing grammatical usage in context, i.e., the second heuristic, makes our algorithm *innovative*. Section 4.3 discusses the heuristic details for proper error detection and avoiding false positives.

The sample sentences shown in the comment section of the source code provide a separate test suite for each individual rule. Our algorithm detected all true positives and ignored all true negatives in this suite. Recall from Sect. 3 that [11] had a detection low rate of $28/130 = 22\%$; [9] had a low detection rate of $60/130 = 46\%$.

Common applications might be a cover letter or sales report. Non-native English speakers would be prime users, but native speakers often need this kind of help.

6   Broader Implications

Our system should help writers significantly reduce the number of usage errors. As more errors are detected by our system the user's writing skill should improve. Our application would be useful to English speakers, as well as non-native English foreign language speakers.

There are Big Data implications with respect to human factors combinations—using gradual, scaled development. Different document types (e.g., fiction, nonfiction) and different authors, as well as different works by the same author may give rise to different classes Swan errors.

7   Future Directions

Error detection would be more thorough if *all* of Swan's 130 rules were implemented making our existing algorithm more robust.

On going testing would involve other documents, e.g., newspapers, novels, letters, marketing collateral, news reports.

An alternative to adjusting a rule's generality (to prevent false positives) is a domain specific language (DSL) based on Swan's rules. This DSL could be constructed to allow a user to write English clauses and sentences correctly in the first place. This solution would take much more time than just implementing error detection rules.

If a foreign language analog of Swan's rules was to be implemented we would do the following. A linguist would map Swan's rules to some foreign language, in the meta-language of English, if books or other text sources are not available. A POS tagger would have to be found, or written, which parses that foreign language. Then, we would update/rewrite each of our Swan rule implementations using this new scheme. This would be a separate undertaking.

Knowledge engineering efficiency, in terms of adding new Swan rules, may be improved provided that convergence of out neural network (Sect. 3.1), during the training phrase, is attained.

Minimizing false negatives was discussed in Sect. 4.2.

Appendix

A   Swan Error 326_1.

```
/*
Sentence: 0
ID WORD        LEMMA        POS  NER
0  You         you          PRP  O
1  look        look         VBP  O    < look
2  as          as           IN   O    <
3  John        John         NNP  PERSON   <
4  .           .            .    O


Sentence: 1
ID WORD        LEMMA        POS  NER
0  My          my           PRP$ O
1  sister      sister       NN   O
2  looks       look         VBZ  O
3  as          as           IN   O
4  me          I            PRP  O
5  .           .            .    O


Sentence: 2
ID WORD        LEMMA        POS  NER
0  My          my           PRP$ O
```

```
1  brother      brother        NN   O
2  looks        look           VBZ  O
3  as           as             IN   O
4  John         John           NNP  PERSON
5  .            .              .    O
```

Sentence: 3
```
ID WORD         LEMMA          POS  NER
0  He           he             PRP  O
1  ran          run            VBD  O    < ran
2  as           as             IN   O      as
3  the          the            DT   O
4  wind         wind           NN   O
5  .            .              .    O
```

Sentence: 4
```
ID WORD         LEMMA          POS  NER
0  She          she            PRP  O
1  is           be             VBZ  O
2  good         good           JJ   O
3  at           at             IN   O
4  subjects     subject        NNS  O        nns
5  as           as             IN   O
6  mathematics mathematics NNS  O        nns
7  .            .              .    O
```

Sentence: 5
```
ID WORD         LEMMA          POS  NER
0  In           in             IN   O                prep
1  warm         warm           JJ   O
2  countries    country        NNS  O        nns
3  ,            ,              ,    O
4  as           as             IN   O                as
5  Turkey       Turkey         NNP  LOCATION  nnp
6  ,            ,              ,    O
7  the          the            DT   O
8  climate      climate        NN   O
9  is           be             VBZ  O
10 warm         warm           JJ   O
11 .            .              .    O
```

Sentence: 6
```
ID WORD         LEMMA          POS  NER
0  I            I              PRP  O
1  had          have           VBD  O
2  another      another        DT   O
3  place        place          NN   O
4  which        which          WDT  O
5  I            I              PRP  O
6  used         use            VBD  O
```

```
7   as          as          IN   O
8   a           a           DT   O
9   studio      studio      NN   O
10  .           .           .    O

Sentence: 7
ID  WORD        LEMMA       POS  NER
0   He          he          PRP  O
1   whistled    whistle     VBD  O
2   as          as          IN   O
3   he          he          PRP  O
4   locked      lock        VBD  O
5   the         the         DT   O
6   office      office      NN   O
7   .           .           .    O

 *
 */
def swan_326_1_find(sen: Array[Array[String]]): Int = {

  def arg_det_arg(p: Int, sent: List[Array[String]]): Int = {

    def det_arg(p: Int, sent: List[Array[String]]): Int = {
      val a: Int = find_POS(p, "DT", sent)
      if (a != -1) arg(a + 1, sent) else -1
    }
    val a: Int = det_arg(p, sent)
    if (a != -1) a else arg(p, sent)
  }

  def v_as_n_pr(p: Int, sent: List[Array[String]]): Int = {
    val lms
      : List[String] = List("look", "run")
        //need to reduce false positives!
    val w: List[String] = List("if", "it")
    // vrb    a
    val b: Int = find_W(p, "as", sent)
    val c: Int = if (b != -1) arg_det_arg(b + 1, sent) else -1
    val a: Int = if (c != -1 && b - 1 >= 0) verb(b - 1, sent) else -1

    val lem: Int = if (a != -1) find_LEMS(b - 1, lms, sent) else -1
    val lm: Boolean = b - lem == 1 && lem != -1

    val wr: String = if (a != -1) find_nextW(sent.size, b + 1, sent)
        else ""
    val badWrd: Boolean = w.contains(wr)

    val ab: Boolean = b - a == 1
    val bc: Boolean = c - b <= 2
    val hits: List[Int] = List(a, b, c).filter(_ > -1)
```

```
        if (lm && !badWrd && ab && bc && hits.size == 3) c else -1
}


/*
Sentence: 2
ID WORD        LEMMA        POS  NER
0  She         she          PRP  O
1  is          be           VBZ  O
2  good        good         JJ   O
3  at          at           IN   O
4  subjects    subject      NNS  O   arg
5  as          as           IN   O   <
6  mathematics mathematics  NNS  O   <
7  .           .            .    O


ID WORD        LEMMA        POS  NER
0  In          in           IN   O
1  warm        warm         JJ   O
2  countries   country      NNS  O   arg
3  ,           ,            ,    O
4  as          as           IN   O   <
5  Turkey      Turkey       NNP  LOCATION  <
6  ,           ,            ,    O
7  the         the          DT   O
8  climate     climate      NN   O
9  is          be           VBZ  O
10 warm        warm         JJ   O
11 .           .            .    O
*
*/
def nns_as_nn(p: Int, sent: List[Array[String]]): Int = {
  val nn: List[String] = List("NNS", "NNP")
  val lms: List[String] = List("subject", "country")
      //need to reduce false positives

  val a: Int = find_POS(p, "IN", sent)
  val b: Int = if (a != -1) find_POS(a + 1, "NNS", sent) else -1
  val lm: Int = if (a != -1) find_LEMS(a + 1, lms, sent) else -1
  val c: Int = if (b != -1) find_W(b + 1, "as", sent) else -1
  val d: Int = if (c != -1) find_POSs(c + 1, nn, sent) else -1

  val lmChk: Boolean = b-lm <= 2 && lm != -1
  val ab: Boolean = b - a <= 2
  val bc: Boolean = c - b <= 2
  val cd: Boolean = d - c == 1
  val hits: List[Int] = List(a, b, c, d).filter(_ > -1)
  if (lmChk && ab && bc && cd && hits.size == 4) d else -1
}
```

```
def comb(p: Int, sent: List[Array[String]]): Int = {

  val a: Int = v_as_n_pr(p, sent)
  if(a != -1) a else nns_as_nn(p, sent)
}

val sent: List[Array[String]] = sen.toList
comb(0, sent)
}
```

B   Our Own Related Neural Network

Lets first review the mechanics of our off-the-shelf neural network, [13], that we employed.

First, the network's input and output nodes are chosen with values—these are concrete examples with and without errors. Second, the network is trained using these examples. Finally, the network is tested with similar (but not identical) examples. If the network does not converge during training, use a different architecture—hidden nodes, but retain with the original training examples.

Swan's errors (634 of them) can be listed in any numerical order, and our subset (at the time) of 52 errors was mapped into binary digits. 000000—0—constitutes no error in the input string. Six binary digits were enough to cover 52 errors—11110 would be Swan's error #30. Text strings were converted, internally, into decimal numbers; padding to the right occurred to make the sentences of equal length. These converted text strings constitute our training input. Our final training input and output is in Table 1. The last error example, 463.2 (Swan), involves the weather—the simple present tense should not be used for talking about a temporary condition.

The training algorithm [13] was back-propagation; we chose a slow training rate to minimize errors-0.01. While training, the network failed to converge. Because of this failure, we tried separate digits for each error (using significantly more nodes)—to no avail. Hidden layers were varied from three to eight layers; the number of nodes per layer varied from nine to 100. In other words, many combinations were tried—none of the architectures converged for our given (training) inputs and outputs. Convergence would decrease and then stop, and stay constant. Thus, no testing could be performed on the neural network.

However, testing the error detection for exactly two rules, and a non-error state did work as expected. We are not surprised—neural networks require many examples and counter examples—but each Swan rule usually has on average five examples, and at most ten examples which is not nearly enough for training purposes.

Table 1. Neural Network training inputs; 0=OK, outputs

Training output	Training input
Binary index into Swan's error #	String converted into decimal #s—internally
0 0 0 0 0 1	When I was ten I was drinking
0 0 0 0 0 0	When you were ten you drank
0 0 0 0 1 0	John went to beach
0 0 0 0 0 0	He went to the beach
0 0 0 0 0 0	My mother went to the beach
0 0 0 0 0 0	My mother's sister went to the beach
0 0 0 0 1 1	You have done an error
0 0 0 0 0 0	You have made an error
0 0 0 0 1 1	He has done an error
0 0 0 0 0 0	He has made an error
0 0 0 0 1 1	You have done a mistake
0 0 0 0 0 0	You have made a mistake
0 0 0 1 0 0	Look it rains
0 0 0 0 0 0	Look it's raining
0 0 0 1 0 0	Look it hails
0 0 0 0 0 0	Look it's hailing
0 0 0 1 0 0	Look it sleets
0 0 0 0 0 0	Look it's sleeting

References

1. Al-Shujairi, Y., Tan, H.: Grammar errors in the writing of Iraqi English language learners (2017). http://journals.aiac.org.au/index.php/IJELS/article/view/3925. Accessed 9–10 2018
2. BrownCorpus: Brown Corpus (2018). https://www1.essex.ac.uk/linguistics/external/clmt/w3c/corpus_ling/content/corpora/list/private/brown/brown.html. Accessed 9–10 2018
3. De-Felice, R.: Automatic error detection in non-native English. Oxford University, Ph.D. thesis (2008)
4. Feng, H., Saricaoglu, A., Chukharev-Hudilainen, E.: Automated error detection for developing grammar proficiency of ESl learners. Calico J.: Comput. Assist. Lang. Instr. Consortium **33**, 49 (2016)
5. Fernando, S., Dias, A.: Mastering language use and usage: repositions perceived according to cognitive linguistics (2017). http://www.serie-estudos.ucdb.br/index.php/serie-estudos/article/view/1087. Accessed 9–10 2018
6. Foster, J., Andersen, Ø.E.: GenERRate: generating errors for use in grammatical error detection. In: 2009 Proceedings of the Fourth Workshop on Innovative use of NLP for Building Educational Applications, EdAppsNLP (2009)

7. Garimella, M.: Detecting article errors in English learner essays with recurrent neural networks. Department of Computational Linguistics (M.A.), Brandeis University (2016)

8. Garnier, M.: Correcting erroneous N+N structures in the productions of French users of English. Nottingham University (2011)

9. Grammarly: Grammarly (2019). https://app.grammarly.com. Accessed 1 2019

10. Hope, T., Resheff, Y., Lieder, I.: Learning Tensor Flow. O'Reilly, Beijing (2017)

11. LanguageTool: Language-Tool (2018). https://languagetool.org/. Accessed 9–10 2018

12. Leacock, C., Chodorow, M., Gamon, M., Tetreault, J.: Automated Grammatical Error Detection for Language Learners, 2nd edn. Morgan and Claypool Publishers, San Rafael (2014)

13. NeurographStudio: Neurograph Studio (2018). http://neuroph.sourceforge.net/download.html/. Accessed 5–6 2018

14. Skukla, N.: Machine Learning with Tensor Flow. Manning Publications, New York (2018)

15. StandfordNLPGroup: Standford NLP Group (2018). http://nlp.stanford.edu/projects/stat-parsing.shtml. Accessed 6–7 2017

16. Swan, M.: Practical English Usage, 3rd edn. Oxford University Press, Oxford (2015)

17. Tetreault, J., Chodorow, M.: The ups and downs of preposition error detection in ESL writing. In: Proceedings of the 22nd International Conference on Computational Linguistics 1, COLING 2008 (2008)

18. Tezcan, A., Hoste, V., Macken, L.: A neural network architecture for detecting grammatical errors in statistical machine translation. Prague Bull. Math. Linguis. **108**, 133–145 (2017)

19. Xue, H.: Computational models of problems with writing of English as a second language learners. Ph.D. thesis, Pittsburgh University (2015)

Link Prediction Based on Time Series of Similarity Coefficients and Structural Function

Piotr Stąpor[✉], Ryszard Antkiewicz, and Mariusz Chmielewski

Institute of Computer and Information Systems, Military University of Technology,
ul. gen. Sylwestra Kaliskiego 2, 00-908 Warsaw, Poland
piotr.stapor@wat.edu.pl

Abstract. A social network is a structure whose nodes represent people or other entities embedded in a social context while its edges symbolize interaction, collaboration or exertion of influence between these fore-mentioned entities [3]. From a wide class of problems related to social networks, the ones related to link dynamics seems particularly interesting. A noteworthy link prediction technique, based on analyzing the history of the network (i.e. its previous states), was presented by Prudêncio and da Silva Soares in [5]. In this paper, we attempt to improve the quality of edges' formation prognosis in social networks by proposing a modified version of aforementioned method. For that purpose we shall compute values of certain similarity coefficients and use them as an input to a supervised classification mechanism (called *structural function*). We stipulate that this function changes over time, thus making it possible to derive time series for all of its parameters and obtain their next values using a forecasting model. We might then predict new links' occurrences using the forecasted values of similarity metrics and supervised classification method with the predicted parameters. This paper contains also the comparison of ROC charts for both legacy solution and the novel method.

Keywords: Social network · Dynamic graph · Link prediction · Structural function

1 Introduction

Currently, networks are a commonly used tool used to describe a wide range of real-world phenomena [6]. A great amount of attention has been devoted to social networks analysis. The problem of link prediction has been already described extensively in literature. Liben-Nowell and Kleinberg [3] provide useful information and insights for regarding that issue, with references to some classical prediction measures based on topological features of analyzed network [7]. Lu and Zhou summarized, in [4], popular algorithms used for linkage is inside complex networks. Another interesting paper worth mentioning is [7]. In

© Springer Nature Switzerland AG 2019
S. Kozielski et al. (Eds.): BDAS 2019, CCIS 1018, pp. 168–179, 2019.
https://doi.org/10.1007/978-3-030-19093-4_13

this publication, authors provide a comprehensive and systematic survey[1] of the link prediction problem in social networks. The topics discussed there cover both classical and latest link prediction techniques, their applications, and active research groups [7]. Many solutions described there either make use of network's various topological metrics or perform data mining in order to reveal new or apply already existing structural patterns. The paper, however, neglected the analysis of how does these topological metrics evolve over time.

Especially interesting link prediction technique was proposed by Prudêncio and da Silva Soares in [5]. Authors proposed there calculating similarity scores for each pair of disconnected nodes at different time-frames, thus building a separate time series for each such pair. Subsequently, a forecasting model is applied to the series in order to predict their next values, which are then going to be used as input to unsupervised and supervised link prediction methods [5].

In this paper, we present modifications to the original method proposed by Soares and Prudêncio. Predicted values of similarity metrics are treated as an input to a supervised classification method. In further parts of this article, we will refer to this classification mechanism by a term *structural function*, as its value decides for whether a link exists for any given pair of nodes. We stipulate that this so called structural function changes over time, thus making is possible to derive time series for all parameters of a given structural function, and obtain their next values using the forecasting model. We might then predict new links' occurrences using the forecasted values of similarity metrics and supervised classification methods with predicted parameters.

The paper is organized as follows. In the Sect. 2 we will present preliminaries—the link prediction problem along with basic definitions. In Sect. 3 we will present the new method of link prediction. Section 4 contains a short description of a conducted experiment and its results. Conclusions are contained in Sect. 5.

2   Prerequisites

Definition 1 (Dynamic graph). *Let $G = (V, E)$ represent a graph containing vertices from set V and edges from E. Additionally, let $\mathbb{T}$ denote a set of moments in time, such that $\mathbb{T} = \{1, 2, ..., T, T + 1, T + 2, ..., \mathfrak{T}\}$, where $T > 1$ stands for the actual time. Through the term "dynamic graph" we shall understand an indexed family of graphs with t as a running index:*

$$\mathcal{G} = (G_t)_{t \in \mathbb{T}}, \tag{1}$$

where $G_t = (V_t, E_t)$ such that $V_1 = V_2 = ... = V_{\mathfrak{T}}$.

The link prediction problem can be formulated (based on [7]) as follows: Consider a social network of structure $\mathcal{G} = (G_t)_{t \in \mathbb{T}}$. The link prediction aims at: (a) forecasting a creation or disappearance of links between nodes in the future

---

[1] The authors cite 131 papers in their publication.

time-frame $t^* : t^* > T$ or (b) finding missing or unobserved links in current state of the network.

Definition 2 (Cumulative Dynamic graph). *A dynamic graph* $\mathcal{G} = (G_t)_{t \in \mathbb{T}}$, *where* $G_t = (V_t, E_t)$, *is a "cumulative dynamic graph" if additionally the following requirement is fulfilled:*

$$\forall t_1, t_2 \in \mathbb{T} : t_1 \leqslant t_2 \implies E_{t_1} \subseteq E_{t_2}. \tag{2}$$

In this paper, we will limit our scope to sole prediction of new edges in graph G_{T+1}, while assuming that already-existing links are not deleted. Hence, whenever $\mathcal{G}$ appears, it symbolizes a cumulative dynamic graph.

Definition 3 (Graph's coefficient). *A coefficient* C *in the context of* $\mathcal{G}$ *can be thought as a function* $C_{\mathcal{G}} : V^2 \times \mathbb{T} \to \mathbb{R}$, *that returns a certain value for a pair of vertices and time frame* t, *according to the structure of graph* G_t.

2.1 Similarity Coefficients

In order to be able to compare our method against the one proposed in [5], we have decided to focus our preliminary research around measures used therein. Let $\Gamma_{\mathcal{G}}(v, t)$ denote a set of neighbors of a given vertex v in graph G_t. The common neighbors (CN) measure, for a pair of two vertices (v and w) can be defined as follows:

$$CN_{\mathcal{G}}(v, w, t) = |\Gamma_{\mathcal{G}}(v, t) \cap \Gamma_{\mathcal{G}}(w, t)| \tag{3}$$

According to CN measure suggests that the higher the mutual neighbors count, for a given pair of nodes, the higher the possibility that a connection between that pair should exist, yet it remains hidden or will exist. By its definition, CN is closely tied with Jaccard's coefficient (JC), known also as Link Relevance measure, which in fact is a CN value divided by analyzed pair's all neighbors count.

$$JC_{\mathcal{G}}(v, w, t) = \frac{|\Gamma_{\mathcal{G}}(v, t) \cap \Gamma_{\mathcal{G}}(w, t)|}{|\Gamma_{\mathcal{G}}(v, t) \cup \Gamma_{\mathcal{G}}(w, t)|} \tag{4}$$

JC is used to measure connection strength and thus plays an important role in the process of hidden links discovery inside graph knowledge-bases [1,2].

The Preferential Attachment (PA) is another measure that we will make use of. It assigns higher link materialization possibility to pairs of vertices with greater adjacent nodes count product. Though simple, the results obtained from experiments assert its ability to predict link formation.

$$PA_{\mathcal{G}}(v, w, t) = |\Gamma_{\mathcal{G}}(v, t)| \times |\Gamma_{\mathcal{G}}(w, t)| \tag{5}$$

Lastly, [5] proposes Adamic-Adar (AA) measure.

$$AA_{\mathcal{G}}(v, w, t) = \sum_{z \in |\Gamma_{\mathcal{G}}(v, t) \cap \Gamma_{\mathcal{G}}(w, t)|} \frac{1}{\log |\Gamma_{\mathcal{G}}(z, t)|} \tag{6}$$

We will leave JC and AA and utilize CN and PA for now. The reason behind this decision is to reduce the time complexity during the proof of concept phase. Furthermore, we would also like to avoid probable interdependence between applied measures. Thus, we refrain from using the JC as it seems to be correlated to a certain degree with CN. Once the proposed method yields positive results, we shall include more coefficients in future tests.

2.2 Description of the Algorithm Proposed by Prudêncio and da Silva Soares

The algorithm of link prediction proposed in [5] has the following steps:

I. For each pair of non-connected (v, w) nodes, create a time series $\overline{C}_T(v, w)$ of similarity coefficients' vector

$$\overline{C}_T(v, w) = (s_t^{v,w})_{t=1,..,T}, \tag{7}$$

$$s_t^{v,w} = \left[C_{\mathcal{G}}^1(v, w, t) \, C_{\mathcal{G}}^2(v, w, t) \, ... \, C_{\mathcal{G}}^N(v, w, t) \right]^\top, \tag{8}$$

where $C_{\mathcal{G}}^i(v, w, t)$ is the value of i-th similarity coefficient for pair of nodes (v, w) at moment t; the following metrics can be used as a similarity coefficient: Common Neighbors (CN), Preferential Attachment (PA), Adamic-Adar (AA) and Jaccard's Coefficient (JC);

II. Using time series $\overline{C}_T(v, w)$ and one of the forecasting methods (Moving Average, Average, Random Walk, Linear Regression, Simple Exponential Smoothing or Linear Exponential Smoothing) compute the future $T + 1$ values:

$$s_{T+1}^{*v,w} = \left[C_{\mathcal{G}}^{*1}(v, w, T+1) \, C_{\mathcal{G}}^{*2}(v, w, T+1) \, ... \, C_{\mathcal{G}}^{*N}(v, w, T+1) \right]^\top. \tag{9}$$

III. Basing on the value of $s_{T+1}^{*v,w}$, use either unsupervised or supervised method to predict new links.

In the unsupervised methods, the pairs of disconnected nodes are ranked according to their scores defined by a chosen similarity coefficients. It is assumed, that the top ranked pairs have highest probability of being connected in the future. The link prediction is treated as a classification task in the supervised approach. As a classifier, *Support Vector Machine* (SVM) is used in [5]. Data from the family $\mathcal{G}' = (G_1, G_2, ..., G_T)$ play the role of a training set, while data from graph $G_T + 1$ are used as a test network.

3   The Novel Method Proposition

The method of link prediction, proposed in this paper, is a modification of the one presented in [5]. We assume that not only values of similarity coefficients are changing in time but also the relation (approximated by a structural function) between values of a given similarity coefficient and probability of new link appearance.

For every $t \in \mathbb{T}$, the task of finding whether a given edge exists can be expressed as a classification problem. A pair of vertices $(v, w) \in V_t$ shall be assigned either to class 1 if there exists an edge $(v, w) \in E_t$ or to class 0 otherwise—i.e. $(v, w) \notin E_t$. Due to the randomness of the edge occurrence and the randomness[2] of similarity coefficients' values, we can use a classifier in the form of a regression function:

$$E\left\{I_{\{(v,w)\in E_t\}} | \boldsymbol{s}_t^{v,w} = \boldsymbol{val}\right\}, \tag{10}$$

where $\boldsymbol{val} \in \mathbb{R}^N$ (N is the number of used similarity coefficients) and also:

$$I_{\{(v,w)\in E_t\}} = \begin{cases} 1, & (v, w) \in E_t \\ 0, & (v, w) \notin E_t \end{cases}. \tag{11}$$

Due to the character of (11) the classifier takes the form of

$$P\left\{I_{\{(v,w)\in E_t\}} = 1 | \boldsymbol{s}_t^{v,w} = \boldsymbol{val}\right\}. \tag{12}$$

Definition 4 (Structural function). *A structural function for a given $\mathcal{G} = (G_t)_{t\in\mathbb{T}}$, N coefficients and time step t is a function of a signature*

$$f_{\mathrm{str}}^{C,\mathcal{G}} : \mathbb{R}^N \times \mathbb{T} \to [0, 1], \tag{13}$$

that maps coefficients' values $\boldsymbol{val} \in \mathbb{R}^N$ obtained from graph G_t to the conditional probability of a link existence:

$$f_{\mathrm{str}}^{C,\mathcal{G}}(\boldsymbol{val}, t) = P\left\{I_{\{(v,w)\in E_t\}} = 1 | \boldsymbol{s}_t^{v,w} = \boldsymbol{val}\right\} \tag{14}$$

It might be helpful to note that a realization of $f_{\mathrm{str}}^{C,\mathcal{G}}$ for one coefficient $C_\mathcal{G}$ is a mapping:

$$val, t \mapsto \frac{\left|\{(v, w) \in V_t^2 | C_\mathcal{G}(v, w, t) = val \wedge (v, w) \in E_t\}\right|}{\left|\{(v, w) \in V_t^2 | C_\mathcal{G}(v, w, t) = val\}\right|}. \tag{15}$$

3.1 Forecasting Links with a Structural Function of the Last Known Moment

To predict edges for time $t^* = T + 1$, evaluate coefficients for every vertices' pair and each time-step $t \in \{1, 2, ...T\}$ in series. The employment of polynomial regression mechanism allows to obtain forecasted values $\boldsymbol{s}_{T+1}^{*v,w}$ for each pair of nodes. Now, we may assess the probability that a link exists between a pair of vertices (v, w) in $\mathcal{G}$, with respect to selected coefficients, by inserting forecasted values into the structural function obtained from the last known moment - T. The final decision whether a link exists involves choosing a certain threshold value $\alpha \in [0, 1]$. If the obtained probability value is higher or equal to the threshold value, we assume that the link materializes.

---

[2] Similarity coefficients' values are random variables as they are functions of a random graph structure.

3.2   Overview of Our Derived Method

Our version of algorithm has the following form:

I. For each pair of non-connected nodes create a time series of similarity coefficients' vector as in (7);

II. For each $t = \overline{1,T}$ approximate a structural function for a mapping between similarity coefficients' vector for any pair of vertices to the existence or absence of a link $s_t^{v,w} \mapsto \{1,0\}$. In other words, estimate parameters of the structural function for each time period $t = 1,2,...,T$ and therefore obtain time series of these parameters;

III. Calculate $s_{T+1}^{*v,w}$ using time series $\overline{C}_T(u,v)$ and polynomial regression (9);

IV. Calculate the values of the structural function's parameters for $T+1$ by applying a polynomial regression to time series of structural function parameters;

V. Finally, predict new links existence by passing $s_{T+1}^{*v,w}$ as an argument to the structural function and comparing its result with a threshold value.

3.3   Forecasting Links with the Predicted Structural Function

The structural function depends on time interval via its second argument. By restricting it to some $\tau \in \mathbb{T}$, we may observe how coefficient C relates to the probability of edge existence in that snapshot. One way of observing, how this behavior changes through time is to look at $f_{\text{str}}^{C,\mathcal{G}}$ as a sequence of its own restrictions.

$$f_{\text{str}}^{C,\mathcal{G}}\Big|_{t=1}, f_{\text{str}}^{C,\mathcal{G}}\Big|_{t=2}, ..., f_{\text{str}}^{C,\mathcal{G}}\Big|_{t=\tau}, ..., f_{\text{str}}^{C,\mathcal{G}}\Big|_{t=T} \qquad (16)$$

To simplify future formulas we shall apply here some syntactic sugar and treat $f_{\text{str}}^{C,\mathcal{G}}|_{t=1}$ as f_1^C, $f_{\text{str}}^{C,\mathcal{G}}|_{t=2}$ as f_2^C and so on, keeping the obvious $\mathcal{G}$ context in mind.

$$f_\tau^C(val) \triangleq f_{\text{str}}^{C,\mathcal{G}}(val,t)\Big|_{t=\tau} \qquad (17)$$

The changes occurring in dynamic graph's structure throughout its subsequent phases may cause f_τ^C to return quite different value than $f_{\tau+1}^C$ for the very same pair of nodes and coefficient C. If the analyzed net alters in a particular manner, the obtained values may reveal a certain trend. For example, during our research we have found out that a network showing collaborations between authors of scientific publications exhibits a characteristic of a logistic function. This corollary led us to the idea of prognosing the structural function values at time $t^* = T+1$, that is what f_{T+1}^{*C} would have looked like at time t^*. Performing logistic regression (or any that fits the trend) for each function from $f_1^C, f_2^C, ..., f_T^C$ sequence will leave us with T corresponding vectors of logistic models coefficients: $B_1, B_2, ..., B_T$. To discovery of their behavior can be achieved by running polynomial regression for each position in obtained vectors.

If $B_i = \begin{bmatrix} b_{i1} & b_{i2} & ... & b_{in} \end{bmatrix}^\top$ then applying the polynomial regression for each of its position will allow us to predict a coefficients' vector for the time $t^* = T+1$, which we will denote it as $\boldsymbol{B}_{T+1}^* \left[b_1^* \ b_2^* \ ... \ b_n^* \right]^\top$.

$$b_{11}, b_{21}, b_{31}, ..., b_{T1} \xrightarrow{\text{pred. by reg.}} b_1^*$$

$$b_{12}, b_{22}, b_{32}, ..., b_{T2} \xrightarrow{\text{pred. by reg.}} b_2^*$$

$$\vdots$$

$$b_{1n}, b_{2n}, b_{3n}, ..., b_{Tn} \xrightarrow{\text{pred. by reg.}} b_n^*$$

The predicted coefficients $\boldsymbol{B}_{T+1}^*$ can then be inserted into logistic function formula, hence unfolding the expected shape of structural function f_{t*}^C. For vector $\boldsymbol{s}_{T+1}^{*v,w} = \left[s_1^* \ s_2^* \ ... \ s_N^* \right]^\top$:

$$f_{T+1}^{*C} \left(\boldsymbol{s}_{T+1}^{*v,w} \right) = \frac{\exp \left(\boldsymbol{B}_{T+1}^* \cdot \bar{\boldsymbol{s}} \right)}{1 + \exp \left(\boldsymbol{B}_{T+1}^* \cdot \bar{\boldsymbol{s}} \right)}, \tag{18}$$

where $\bar{\boldsymbol{s}} = \left[1 \ s_1^* \ s_2^* \ ... \ s_N^* \right]^\top$.

For example, the application of logistic regression for one coefficient C will result in a series of vectors $\boldsymbol{B}_i = \left[b_1 \ b_2 \right]^\top$ and a prediction: $\boldsymbol{B}_{T+1}^* = \left[b_1^* \ b_2^* \right]^\top$. In this case, the probability value can be evaluated with the formula:

$$f_{T+1}^{*C} \left(C_{\mathcal{G}}^*(v, w, T+1) \right) = \frac{\exp \left(b_1^* + b_2^* C_{\mathcal{G}}^*(v, w, T+1) \right)}{1 + \exp \left(b_1^* + b_2^* C_{\mathcal{G}}^*(v, w, T+1) \right)}. \tag{19}$$

Again, as in Sect. 3.1, a link (v, w) will materialize when $f_{T+1}^{*C} \left(\boldsymbol{s}_{T+1}^{*v,w} \right) \geqslant \alpha$, where α is the chosen threshold.

3.4   Extending the Method for N Coefficients

The method can be accommodated to take any positive number of measures into account. This can be accomplished by inserting each measure's values into a set of even-length intervals. The number of divisions and their size may vary for each coefficient. Let $\mathbb{C} = \left\{ C_{\mathcal{G}}^1, C_{\mathcal{G}}^2, ..., C_{\mathcal{G}}^N \right\}$ be a set of N coefficients' evaluation functions. Now, let $\mathfrak{D} \colon \mathbb{C} \times \mathbb{N}^+ \to 2^{\mathbb{R}}$ denote a function that, for a given coefficient $C_{\mathcal{G}}$, divides space $[0, \max C_{\mathcal{G}}(v, w, t)]$ into a set of consecutive, equal-length, d intervals: $(C_{\mathcal{G}}, d) = \left\{ \Delta_1^C, \Delta_2^C, ..., \Delta_d^C \right\}$. Through $\max C_{\mathcal{G}}(v, w, t)$ we marked the highest value achieved for a given $C_{\mathcal{G}}$ up to the predicted time-frame.

To every interval we will now assign its representative value (via $\mathfrak{R} \colon 2^{\mathbb{R}} \to \mathbb{R}$ function)—in our study we decided to use interval's average value.

Having got through the definitions we may now construct, for a given $\mathcal{G}$, an indexed family of tables: $\mathcal{T} = (tabl_t)_{t \in \{1,2,...,T\}}$ (one table per each time frame) that will constitute a data to be consumed by logistical regression mechanism while searching for structural function coefficients. This calls for evaluating all of N coefficients for every pair of vertices in each time step.

The table contains information on how many pairs of vertices can be found in a given N-dimension space fragment and how many of them are actually linked. A given pair of nodes (v, w) belongs to the space fragment represented by $\left(\Re(\Delta^{C_{\mathcal{G}}^1}), \Re(\Delta^{C_{\mathcal{G}}^2}), ..., \Re(\Delta^{C_{\mathcal{G}}^N})\right)$ if $\forall i \in 1, 2, ..., N : C_{\mathcal{G}}^i(v, w, t) \in \Delta^{C_{\mathcal{G}}^i}$.

Table 1. Data of $tabl_t$ from time frame t, used for finding the coefficients of logistic structural function that utilizes n measures. Column designations: LP stands for *linked pairs*, AP – *all pairs*, C_i^{rep} – a representative value for a given C_i's interval.

LP	AP	C_1^{rep}	C_2^{rep}	...	C_N^{rep}
lp_1	ap_1	$\Re(\Delta_1^{C_{\mathcal{G}}^1})$	$\Re(\Delta_1^{C_{\mathcal{G}}^2})$	...	$\Re(\Delta_1^{C_{\mathcal{G}}^N})$
lp_2	ap_2	$\Re(\Delta_1^{C_{\mathcal{G}}^1})$	$\Re(\Delta_1^{C_{\mathcal{G}}^2})$	...	$\Re(\Delta_2^{C_{\mathcal{G}}^N})$
$\vdots$	$\vdots$	$\vdots$	$\vdots$	...	$\vdots$
lp_j	ap_j	$\Re(\Delta_1^{C_{\mathcal{G}}^1})$	$\Re(\Delta_2^{C_{\mathcal{G}}^2})$	...	$\Re(\Delta_1^{C_{\mathcal{G}}^N})$
lp_{j+1}	ap_{j+1}	$\Re(\Delta_1^{C_{\mathcal{G}}^1})$	$\Re(\Delta_2^{C_{\mathcal{G}}^2})$	...	$\Re(\Delta_2^{C_{\mathcal{G}}^N})$
$\vdots$	$\vdots$	$\vdots$	$\vdots$	...	$\vdots$
lp_k	ap_k	$\Re(\Delta_2^{C_{\mathcal{G}}^1})$	$\Re(\Delta_1^{C_{\mathcal{G}}^2})$	...	$\Re(\Delta_1^{C_{\mathcal{G}}^N})$
lp_{k+1}	ap_{k+1}	$\Re(\Delta_2^{C_{\mathcal{G}}^1})$	$\Re(\Delta_1^{C_{\mathcal{G}}^2})$	...	$\Re(\Delta_2^{C_{\mathcal{G}}^N})$
$\vdots$	$\vdots$	$\vdots$	$\vdots$	...	$\vdots$

Let us now introduce a logit function [8]: $L(p_i) = \ln(p_i/(1 - p_i))$, where, $p_i = lp_i/ap_i$. This lets us apply the generalized least squares (GLS) technique from [8] to find structural functions' coefficients. The algorithm continues then as shown in Sect. 3.3.

3.5 A Detailed Pseudo-Code for the Proposed Method

Let $\mathbb{X}_t$ and $\mathbb{Y}_t$ be mutable integer maps (for time frame t)—i.e. mappings of type $\mathbb{N}_0 \to \mathbb{N}_0$, such that:

- initially every $n \in \mathbb{N}_0$ is associated with zero—$n \mapsto 0$,
- every INCREMENTMAPPINGVALUEFORKEY($\mathbb{M}, n$) call, where $\mathbb{M}$ is a mapping, increments the value returned for n by 1 (E.g. After two such calls with 3 as the second argument $\mathbb{M}(3) = 2$.)

Initially, a value of coefficient C is computed ($C_{\mathcal{G}}(v, w, t)$) for every pair of nodes in the graph at every historical time step $1, 2, ..., T$. The results form a matrix X, such that its every row contains a series of sequential values obtained at different moments of time. At line 6 we increase a number of pairs with similar result by one, while at line 8 only existing links with that value are accounted for. Line 9 is responsible for fitting a regression curve of a structural function at time t. At line 9 a structural function is predicted for time $T+1$. At line 13 we than obtain a forecast of coefficient C at $T+1$. Finally the link occurrence prediction can be assessed.

Algorithm 1. The algorithm for one coefficient C returning $\mathbb{N}_0$.

1: **procedure** PREDICT($\mathcal{G}$, $C_\mathcal{G}$, α)
2:     **for** every $t = \overline{1,T}$ **do**
3:         **for** every $(v,w) \in V_t^2$ of graph $G_t = (V_t, E_t)$ **do**
4:             $c \leftarrow C_\mathcal{G}(v,w,t)$
5:             $\boldsymbol{X}[\text{GETROWFORNODEPAIR}(v,w),t] \leftarrow c$
6:             INCREMENTMAPPINGVALUEFORKEY($\mathbb{X}_t, c$)
7:             **if** $(v,w) \in G_t$ **then**
8:                 INCREMENTMAPPINGVALUEFORKEY($\mathbb{Y}_t, c$)
9:         $B_t \leftarrow$ LOGISTICREGRESSIONFIT($\mathbb{Y}_t, \mathbb{X}_t$)
10:     $B^* \leftarrow$ PREDICTNEXTUSINGPOLYNOMIALREGRESSION($\boldsymbol{B}_1, \boldsymbol{B}_2, ..., \boldsymbol{B}_T$)
11:     **for** every row r in $\boldsymbol{X}$ **do**
12:         $\boldsymbol{x} \leftarrow$ GETROWFROMMATRIX($\boldsymbol{X}, r$)
13:         $x^* \leftarrow$ PREDICTNEXTUSINGPOLYNOMIALREGRESSION($\boldsymbol{x}$)
14:         $p \leftarrow$ VALUEOFLOGISTICFUNWITHPARAMSAT($\boldsymbol{B}^*, x^*$)
15:         **if** $p \geqslant \alpha$ **then**
16:             $\boldsymbol{P}[r] \leftarrow 1$
17:         **else**
18:             $\boldsymbol{P}[r] \leftarrow 0$
19:     **return** P

The next algorithm also requires some commentary. The $\boldsymbol{D}$ vector contains numbers of divisions (intervals) for each coefficient found in a list $\mathbb{C}$. The function GETMAXIMUMINCOLUMN at line 12 returns a maximum value ever returned by a given coefficient. PREPAREDIVTABLES (line 13) creates a table for each coefficient containing intervals and their representative values. The last function call that may appear obscure to the reader is PREPAREDATATABLE() from line 15. Its purpose is to create a table of a form presented by Table 1 in Sect. 3.4.

4   The Experiment

In order to evaluate the novel method an experiment was conducted in which a prediction about future collaboration of authors in Arxiv[3] publications' database was to be attained. Like in the case of [5], the scope included all articles in *High Energy Physics – Lattice archive* (hep-lat[4]) published between 1993 and 2010 with an accuracy to a month. Each time-frame corresponded to one year. (In order to gain some reduction in algorithms' execution time, yearly data, that

---

[3] https://arxiv.org.
[4] https://arxiv.org/archive/hep-lat.

Algorithm 2. The algorithm for n coefficients in a list $\mathbb{C}$.

1: **procedure** PREDICT($\mathcal{G}, \mathbb{C}, \boldsymbol{D}, \alpha$)
2:     $N \leftarrow$ LENGTH($\mathbb{C}$)
3:     **for** every $t = \overline{1, T}$ **do**
4:         **for** every $(v, w) \in V_t^2$ of graph $G_t = (V_t, E_t)$ **do**
5:             **for** $i \leftarrow 1$ **to** N **do**
6:                 $\boldsymbol{X}_t[$GETROWSFORNODEPAIR$(v, w), i] \leftarrow \mathbb{C}[i](v, w, t)$
7:             **if** $(v_1, v_2) \in G_t$ **then**
8:                 $\boldsymbol{Y}_t[$GETROWFORNODEPAIR$(v, w)] \leftarrow 1$
9:             **else**
10:                $\boldsymbol{Y}_t[$GETROWFORNODEPAIR$(v, w)] \leftarrow 0$
11:    **for** $i \leftarrow 1$ **to** N **do**
12:        $\boldsymbol{Max}[i] \leftarrow$ GETMAXIMUMINCOLUMN$(i, \langle \boldsymbol{X}_1, \boldsymbol{X}_2, ..., \boldsymbol{X}_T \rangle)$
13:    $divTables \leftarrow$ PREPAREDIVTABLES$(\boldsymbol{Max}, \boldsymbol{D})$
14:    **for** every $t = \overline{1, T}$ **do**
15:        $dataTable_t \leftarrow$ PREPAREDATATABLE$(divTables, \boldsymbol{X}_t, \boldsymbol{Y}_t)$
16:        $\boldsymbol{B}_t \leftarrow$ LOGISTICREGRESSIONFIT$(dataTable_t)$
17:    $\boldsymbol{B}^* \leftarrow$ PREDICTNEXTUSINGPOLYNOMIALREGRESSION$(\boldsymbol{B}_1, \boldsymbol{B}_2, ..., \boldsymbol{B}_T)$
18:    **for** every row r in $\boldsymbol{X}_1$ **do**
19:        **for** $i \leftarrow 1$ **to** N **do**
20:            $\boldsymbol{x}_i \leftarrow$ CREATEVECTOR$(\boldsymbol{X}_1[r, i], \boldsymbol{X}_2[r, i], ..., \boldsymbol{X}_T[r, i])$
21:            $x_i^* \leftarrow$ PREDICTNEXTUSINGPOLYNOMIALREGRESSION$(\boldsymbol{x}_i)$
22:        $\boldsymbol{x}^* \leftarrow$ CREATEVECTOR$(x_1^*, x_2^*, ..., x_N^*)$
23:        $p \leftarrow$ VALUEOFLOGISTICFUNWITHPARAMSAT$(\boldsymbol{B}^*, \boldsymbol{x}^*)$
24:        **if** $p \geqslant \alpha$ **then**
25:            $\boldsymbol{P}[r] \leftarrow 1$
26:        **else**
27:            $\boldsymbol{P}[r] \leftarrow 0$
28:    **return** $\boldsymbol{P}$

---

was actually taken into account and processed, had been limited only to records from four months: 3rd, 6th, 9th and 12th.) It should be noted that edges are added in a cumulative manner - i.e. once a pair of vertices is bond by an edge, it remains connected. Two approaches were confronted:

- using the last structural function to obtain prognosis (AP1) and
- utilizing the predicted structural function for same purpose (AP2).

The computation has been done using a dedicated software. In order to reduce prognosis complexity, the developed software divides the analyzed multi-graph into weak connected components with respect to its last structure in time series so that for any pair of vertices coming from different components neither PA nor CN may yield output other than zero. The comparison of the quality of both solutions was assessed with the help of ROC (*Receiver Operating Characteristic*) charts.

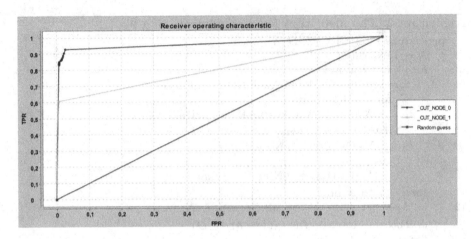

Fig. 1. ROC chart for single coefficient model—CN. The upper series curve (_CUT_NODE_0) illustrates the effectiveness of AP2 approach, while _CUT_NODE_1 of AP1. The diagonal line across represents expected the prognostic ability of a random guess.

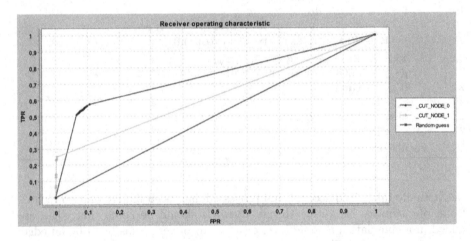

Fig. 2. ROC chart for single coefficient model—PA

A note on interpreting the ROC charts. The mechanism, that forecasts of weather the link should or should not exist, can be viewed as kind of a binary classifier, hence the usage of ROC charts as the assessment tool. The presented ROC charts shows how well the classifier performs for different levels of threshold α (please refer to Sect. 3.3 for α). The TPR (*True Positive Rate*) value assesses how well the mechanism performs for positives—i.e. how many observations categorized belong there truthfully. The higher the value, the better. On the other hand, the FPR (*False positive Rate*) measures how poorly the classification works for negatives—i.e. how many observations categorized as negatives, have been wrongly assigned. The lower the value, the better. In conclusion, the closer the ROC curve passes by the upper-left corner, the better the classifier works.

As it can be seen in Fig. 1, the usage of the forecasted structural function with CN improves the classification results. When it comes to PA (Fig. 2), the proposed modification causes FPR to increase slightly, but on the other hand it performs a better job when classifying positives

5   Conclusion and Future Work

As shown in the results section, performing prediction with the help of a forecasted structural function improves the classification rate of true positives (TPR). Although obtained false positive ratio (FPR) seems inferior to last-step structural function forecast, the gain from the improved TPR classification surpasses by far that loss, thus making the usage of forecasted structural function sensible and advisable. In near future we plan to experiment with other similarity measures (such as AA or JC) and running a series of experiments for multi-coefficient version of the algorithm. Further research would concentrate on: (a) experimentation with other kinds of social networks, (b) proposal of recommended set of uncorrelated coefficients, (c) taking into account that some pairs of nodes may become disconnected and (d) the introduction of cooperation intensity concept.

References

1. Barthélemy, M., Chow, E., Eliassi-rad, T.: Knowledge representation issues in semantic graphs for relationship detection. In: AAAI Spring Symposium on AI Technologies for Homeland Security, pp. 91–98. AAAI Press (2005)
2. Chmielewski, M., Stąpor, P.: Protégé based environment for DL knowledge base structural analysis. In: Jędrzejowicz, P., Nguyen, N.T., Hoang, K. (eds.) ICCCI 2011. LNCS (LNAI), vol. 6922, pp. 314–325. Springer, Heidelberg (2011). https://doi.org/10.1007/978-3-642-23935-9_31
3. Liben-Nowell, D., Kleinberg, J.: The link-prediction problem for social networks. J. Am. Soc. Inf. Sci. Technol. 58(7), 1019–1031 (2007)
4. Lu, L., Zhou, T.: Link prediction in complex networks: a survey. Phys. A: Stat. Mech. Appl. 390(6), 1150–1170 (2010)
5. da Silva Soares, P.R., Prudêncio, R.B.: Time series based link prediction. In: Proceedings of the International Joint Conference on Neural Networks, June 2012
6. Rossetti, G., Guidotti, R., Miliou, I., Pedreschi, D., Giannotti, F.: A supervised approach for intra-/inter-community interaction prediction in dynamic social networks. Soc. Netw. Anal. Min. 6(1), 86 (2016)
7. Wang, P., Xu, B., Wu, Y., Zhou, X.: Link prediction in social networks: the state-of-the-art. CoRR abs/1411.5118 (2014). http://arxiv.org/abs/1411.5118
8. Zeliaś, A., Pawełek, B., Wanat, S.: Prognozowanie ekonomiczne: teoria, przykłady, zadania. Wydawnictwo Naukowe PWN (2003)

The Analysis of Relations Between Users on Social Networks Based on the Polish Political Tweets

Adam Pelikant[✉][iD]

Institute of Mechatronics and Computer Systems, Lodz University of Technology,
Łódź, Poland
adam.pelikant@p.lodz.pl

Abstract. The article presents a method of acquiring selected topical data from Twitter conversations and storing it in the relational database schema with the use of a scheduled cyclical process. This kind of storage creates the opportunity to analyze the dependencies between users, hashtags, mentions etc. based on SQL or its procedural extension. Additionally, it is possible to construct data views, which facilitates the creation of front end applications leading to efficient generation of cross sections of data based on different features: time of creation, strength of relations etc. Taking all of this into consideration, an application was developed that allows for graphical representation of relations with the use of the two algorithms: a Fruchterman-Reingold and a radial one. This program accepts the visual analysis and additionally it creates the opportunity to manipulate parts of a graph or separate nodes, and obtain descriptions of their features. Some conclusions about the relations between people conversing about Polish political scene were presented.

Keywords: Social media · Social networks · Collecting tweets ·
Graphs visualization · Fruchterman-Reingold algorithm ·
Radial algorithm · User to user relations

1 Investigation Background

Social media, especially Twitter, have always been a popular subject of scientific research. As it was mentioned [25] the years 2007–2012 were very fruitful. Nearly 400 indexed scientific publications were printed or distributed online. That happened due to the fact that there was a huge, continuous growth of the research material. The recently published statistics by Omnicore

Research project - RPLD.01.02.02-IP.02-10-030/17 - "ASM Smart Data System (ASM SDS) - Badania nad systemem gromadzenia, przetwarzania i dystrybucji danych z wykorzystaniem algorytmów uczenia maszynowego do modernizacji procesów realizacyjnych i zarządczych oraz wdrażania nowych i zasadniczo udoskonalonych produktów i usług badawczych w ASM-Centrum Badan i Analiz Rynku Sp z o.o" Regionalny Program Operacyjny Województwa Łódzkiego na lata 2014–2020.

S. Kozielski et al. (Eds.): BDAS 2019, CCIS 1018, pp. 180–191, 2019.
https://doi.org/10.1007/978-3-030-19093-4_14

(https://www.omnicoreagency.com/twitter-statistics/), shows that the total number of monthly active Twitter users has already reached 330 million and the total number of tweets sent per day is 500 million. This statistics is continuously increasing. What is more, the content of tweets applies to all types of social activities, thus it creates the opportunity for research in many fields. The majority of the subjects tackled refer to computer science [21,25], but among them there are also numerous ones dedicated to business, communications, economics, medicine [3] political science [1,5,9,23] and many others. In many cases it is difficult to assign them to only one of the disciplines. One of the most attractive and frequently appearing activities is politics. Due to its character, many of the publications describe local dependencies limited to a single country [1,5,9] or region [23]. Owing to the regional information exchange in the study of social networks, it is advisable to refer to the features of the national languages. It effectively eliminates the possibility of building a universal tool for the full analysis of social networks, including Twitter.

2   Methods of Data Acquisition and Storage

There are various methods of data acquisition. Many researchers use network repositories of tweets, some collect data manually or prepare online aggregators for example TwapperKeeper [25] or agents [12]. But to obtain a really interesting subset of data it is much better to create a unique tool based on given Twitter APIs. This method is the most popular among specialists in the field of computer science [25]. This API offers many methods and features which can be used to build applications, unfortunately the information is stored in Jason tag text format, which is sometimes difficult to parse. To omit this inconvenience it is possible to use one of the templates available on a freeware license. In this work the TweetSharp template was used https://www.nuget.org/packages/TweetSharp/. It determines the programming language used to collect tweets systematically.

To understand the organization of storage itself, it is necessary to describe what kinds of reactions can be associated with a single tweet (see Fig. 1). After the post is published, the users who read the tweet may react to it in one of the two ways. The first one is to retweet the post, which means they agree with its content and, what is more, they like it. This activity does not require any special additional involvement of the user and can be done mechanically. The second type of reaction to the post published is to reply. In this case users write their own comments which relate to the tweet they have just read. Consequently, these can be both positive or negative replies with relation to the original content. Such activity needs more effort as it is necessary, not only to read the message, but additionally understand it and only then write your own content. It is obvious, that each such a response can generate further reactions. The set of these actions builds a chain of replies. It is possible that in such a series the answer originally entered by the author will appear. In both cases the number of users actively involved in the conversation is unlimited.

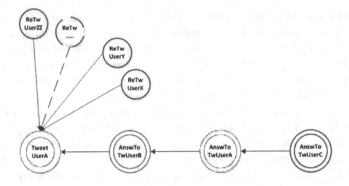

Fig. 1. Possible interactions for a single tweet

The build system (see Fig. 2) consists of: an application which searches for tweets and is triggered according to the schedule; a database server storing obtained results in relational schema; a set of SQL scripts and procedures for analytic processing; and an application for graphical representation of relations between users or other features described in tweets such as hashtags or mentions. The parts for the analysis can be supervised by the operator.

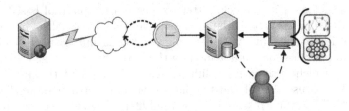

Fig. 2. Schema of tweets collecting and analyzing system

Such a system can used for general purpose, however, it was decided to include it in the analysis of the political conversations in Poland. The choice of the topic had its origins in the conviction that the amount of tweets related to it will be large and varied in terms of opinions expressed in them. In addition, like any other citizen, the author of the paper is interested in what is happening in his country. It is possible to search for tweets using some of their features but to use hashtags connected to them seems to be much more interesting. The search results are saved in the database with the diagram shown in the Fig. 3.

The main table in the schema is Tweets, which stores the content of the tweet itself, the time of its publication online, as well as three foreign keys. The users table is the dictionary which contains data describing the authors of tweets. A similar solution of storing data as Facebook users' metadata was used in [13] The table contains three columns which define the account owner. Name and ScreenName are compulsory short names. The third, optional one, stores longer

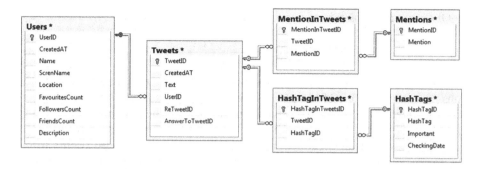

Fig. 3. Relational database diagram for collecting and analyzing tweets system

description and can be used to examine user's preferences and opinions. Due to the fact that one tweet can contain more than one hashtag (tweet keywords) and mentions (reference to people, users) these tables are connected to Tweets table using additional connecting tables (relation many to many). Hashtags table plays a dual role: a dictionary and a list of the hashtags that are searched periodically – the Important field is set to True. In an extreme case, all newly discovered tags can be automatically set to this value. In the proposed solution only the selected ones have this value set manually. In this case the key hashtags represented the most popular political parties in Poland.

3   The Analysis of User to User Relations

In most works, all relations between users of social networks are treated equally – whether they exist or not [4,9,18]. However, according to the author of this paper they have to be varied, defining the strength of influence for each node. Because retweeting needs much less activity it should have less impact than replying to a tweet which needs not "only click" but create your own comment correlating with the original statement. So that vertex strength, representing user in the graph, can be calculated as a weighted sum of connections with other users given by the formula (1)

$$\vartheta(x_i) = \sum_{j=1, x_i \neq x_j}^{n} \alpha + \sum_{j=1, x_i \neq x_j}^{n} \beta f\big(d(x_i, x_j)\big) \tag{1}$$

The α is a weight for retweets and β is a weight for replies, and $\alpha < \beta$. Additionally the impact of the reply should depend on the distance from the original entry. For example this function can be the inverse of the distance measured by the number of levels in the hierarchy between entries (2)

$$f\big(d(x_i, x_j)\big) = \frac{1}{d(x_i, x_j)} \tag{2}$$

Of course it is possible to use a less restrictive function (3).

$$f\big(d(x_i, x_j)\big) = \frac{1}{1 + \log\big(d(x_i, x_j)\big)} \tag{3}$$

It is necessary to remove the tweets written by original authors from both sums. Because the time of the publication is given, it is possible to omit tweets between which there is too little space, on the basis of the assumption that they could have been created by the bot or as a result of user interaction (collusion).

The necessity of unequal treatment of replies and retweets also results from the proportion of their number. Of course, at different moments of data collection, the proportions changed slightly. However, we can assume that they were the following: the ratio of responses to all tweets 2%, the ratio of retweets to all tweets 75%, the remaining 23% are e tweets that potentially start the discussion. It should be noted that only slightly over 50% of start tweets cause any reactions at all, others remain orphans.

To obtain a hierarchical structure of Tweet Answers it is possible to build a query which uses a defined ad hoc view. Such a solution gives the opportunity to a defined recursion based on UNION ALL operator and self-join with the view - Listing 1.

Listing 1. The query which builds hierarchical structure of tweets

```
WITH AT AS
(SELECT 1 AS LEVEL, TweetID, AnswerToTweetID, TweetID AS Root
FROM Tweets
WHERE AnswerToTweetID IS NULL
UNION ALL
SELECT LEVEL + 1, tw.TweetID, tw.AnswerToTweetID, Root
FROM AT JOIN
Tweets AS tw ON AT.TweetID = tw.AnswerToTweetID)
SELECT LEVEL, TweetID, AnswerToTweetID, Root
FROM AT
WHERE LEVEL > 1
ORDER BY LEVEL DESC
```

For the data collected so far, the largest set of answers consisted of 21 elements. But for more tweets there is only one answer. Exemplary set of results is presented in Table 1.

By joining the results of this query and the Users table we obtained connections between the people actively engaged in the online conversations. Counting the number of responses and adding the appropriate weight to them for each user, the value of strength given by the formula (1) can be set.

4   Graphical Representation of Relations Between Users

By default, the nodes representing users (hashtags or mentions) do not contain any information determining their position in space. That is why they can be

Table 1. Hierarchical relations between answers and tweets (set of 11 first rows)

Level	TweetID	AnswerToTweetID	Root
21	968807622924226560	968806055038914560	968774606223036416
20	968806055038914560	968793441927823360	968774606223036416
19	968793441927823360	968791847098298368	968774606223036416
18	968791847098298368	968790369285615617	968774606223036416
17	968790369285615617	968789334525702144	968774606223036416
16	968789334525702144	968788236687630336	968774606223036416
15	968788236687630336	968787827185025024	968774606223036416
14	968787827185025024	968786955650641920	968774606223036416
13	968786955650641920	968785992420380674	968774606223036416
12	968785992420380674	968784976757035008	968774606223036416
11	958187700099731456	958069758196412418	957968443629867009

arranged randomly. The graph drawn in this way will certainly be illegible. Its presentation can be done using many ways described in [14]. In order to improve its quality, the force directed Fruchterman–Reingold algorithm has been used [8]. In this method there are two opposing forces between each pair of connected nodes: repulsive analogous to electrostatic force (4)

$$f_r(d) = -\frac{k}{d} \tag{4}$$

and attracting the corresponding spring force (5).

$$f_a(d) = \frac{d^2}{k} \tag{5}$$

The parameter k appearing in both expressions is defined as the optimal distance between vertexes (6)

$$k = C\sqrt{\frac{XY}{n}} \tag{6}$$

where: X, Y are dimensions of the drawing area, n represents the number of nodes in a graph and C is an arbitrary chosen constant. To improve the convergence of the method for subsequent iterations of determining the position of vertices, the simulated annealing method can be used. The application that implements the F-R algorithm has been developed [20]. It enables graphical presentation of any, also inconsistent graphs. Because the result is strongly dependent on the initial location of the nodes, the possibility of randomly setting them was introduced. In addition, it was possible to introduce weights for both forces affecting the nodes and change the optimal distance. All illustrations presented in this article are the result of this software. For an arbitrary chosen moment of data collection, the graph presenting the connections between the participants (users) not the discussion (answers and retweets) looked like in the Fig. 4.

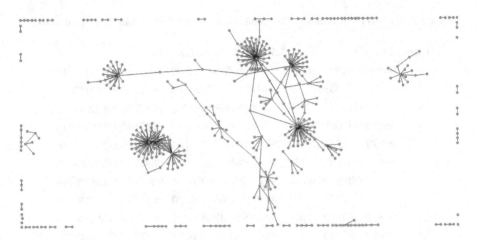

Fig. 4. Graphical representation of user to user relations for Polish political tweets using Fruchterman–Reingold algorithm

Due to the fact that the graph is inconsistent, there are several subgraphs of varied degrees of complexity. Many of them are gathered on the boundaries of a drawing region. This applies to subgraphs with a small number of nodes, for which the repulsive force (4) is dominant. When we reduce the number of subgraphs, the readability of the drawing is improved and the complexity of the relationship between the conversation users is visible. An example for the two separated subgraphs is shown in Fig. 5. The inconsistency of the graph resulted from the fact that despite the fact that only tweets concerning politics were collected, sympathizers of one option rarely took part in the conversation initiated by another of them. As a rule, the answers were positive. This is also underlined by the dominant number of retweets that constitute a quick answer to yes. It should be emphasized that the graph concerns the user to user relationship. We would get different illustrations presenting relations: hashtag to hashtag or mention to mention. All of them can be presented in a developed application by defining other view representing such a relation.

Of course for a consistent graph, the Fruchterman–Reingold method also gives correct results. However, in this case it is possible to use a different visualization – the radial method. This idea was also developed in the original computer software [20]. The radius of the most outer circle is calculated based on the dimensions of graphics area (7).

$$R = \frac{min(X, Y)}{2} \tag{7}$$

The difference of the radius between the following circles radii results from the maximal number of levels appearing in the graph (8)

$$\Delta R = \frac{R}{max(Level)} \tag{8}$$

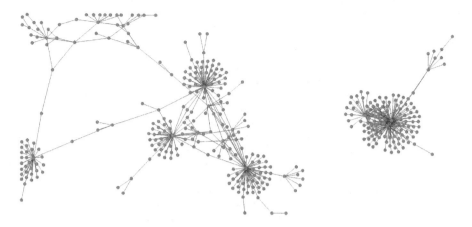

Fig. 5. Graphical representation of the two chosen user to user graphs for Polish political tweets Fruchterman–Reingold algorithm

It is possible to start drawing from the arbitrary chosen graph node. In the developed application, the start node selected was the one, which had the most number of children (directly connected nodes). It is located in the centre of the graph. Each next level of the nodes is placed at the subsequent circle. The angular position of node cl which belongs to the same parent p is described by the following Eq. (9)

$$\gamma_{cl} = \gamma_p + \frac{(l-1)\delta_p}{m} \tag{9}$$

where: l is the number of the subsequent node belonging to the parent p, m is the number of all children of that parent, and δ_p is the location of the first node of this parent. The Starting angle of l-th parent is given by (10)

$$\delta_{pl} = \gamma_{p(l+1)} - \gamma_{p(l)} \tag{10}$$

As the initial values of the recursion for the central node (root), the angle values (11) were assumed

$$\gamma_{root} = 0 \quad and \quad \delta_{root} = 2\pi \tag{11}$$

The exemplary illustration of a consistent subgraph being part of the graph representing the dependencies between the authors of tweets is shown in the Fig. 6.

For many kinds of human relationships, we can formulate a rule of small worlds, introduced by Milgram [19,22]. It has been developed into a theorem of "Six degrees of separation." It has less serious interpretations: Six Degrees of Kevin Bacon (for actors) – a Bacon number; an Erdős number (for mathematicians), a Morphy number (for chess players) or mix of them. But even these examples show the strength of the idea of a short distance between people.

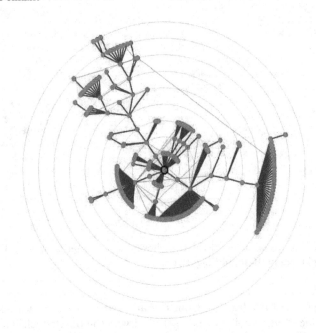

Fig. 6. Radial representation of a chosen connected graph for the Polish political tweets with a central node having the highest number of children

Also in scientific research on cooperation and communication this concept has a strong confirmation [11,17]. For graphs presented on Figs. 5 or 6 the average distance between nodes is certainly smaller than 6, but on the other hand, it is easy to point out many pairs of nodes where distance is much greater than this value. It can be said that all the participants of the discussion are connected by an overarching label, which is the hashtag they talk about [24]. However, in this case, people from several inconsistent graphs will be connected. This means that people who have never met on the network will be connected with one another. There seems to be a much better solution, namely to divide a consistent graph into clusters based on the selected features. Of course it is possible to use only the information about the number of connections between cluster members or to apply additional features described in the nodes such as: mentions, emotions of comment etc. [6,7,9,10,15,16,23]. This operation is aimed at dividing a coherent graph (small world) into subgraphs (handshake worlds). This leads to the creation of a group of close friends who exchange information and generally agree with the issues raised. The software application allows to indicate any node as the central element of the graph. Example of such a situation is presented in Fig. 7. As you can see, the person represented by the central node has only three direct connections to the child nodes. However, each of its children (child nodes) has a significant amount of direct connections to the next level. In this case, we can assign a label to the central node - the head of all bosses, which communicates only with a limited number of direct subordinates, and only they issue instructions to their executors. In this case, we can talk about connections of a mafia character or others with a very strong hierarchical structure. In the case of social networks, the person represented by the central node initiates the dialogue, and his or her subordinates either share their opinion further or initiate support for it.

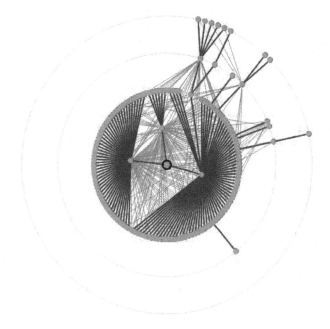

Fig. 7. Radial representation of a chosen connected graph for Polish political tweets with a central node whose children have a big number of children

5   Conclusions and Future Works

The paper presents a system for collecting and analyzing conversation data on a social network. It also contains a set of scripts that performs a database server dependency analysis and an application that allows you to visualize connections, as well as operations on this graphical representation. The next stage of work will be to build an application that analyzes the relationships between users of the portal at various levels. These levels will be defined using hashtags belonging to different categories, for example: sport, fashion, travel, etc. This will allow to verify the thesis that the relations between various levels are transferred and only to a small extent do they build relationships between representatives of small worlds from other layers. The use of experience from the analysis of emotions [2,10], and text [6] processing (taking into account the specificity of the inflection of the Polish language) will also allow to broaden the analysis of interdependencies between nodes, taking into account the emotional expression of the message. This will allow you to make adjustments in determining the attraction and repulsion forces between users.

References

1. Al-Khalifa, H.S.: A first step towards understanding saudi political activities on Twitter. Int. J. Web Inf. Syst. **8**(4), 390–400 (2012)
2. Avots, E., Sapiński, T., Bachmann, M., Kamińska, D.: Audiovisual emotion recognition in wild. In: Machine Vision and Applications, pp. 1–11 (2018)
3. Bouslimi, R., Ayadi, M.G., Akaichi, J.: Content modelling in radiological social network collaboration. In: Kozielski, S., Mrozek, D., Kasprowski, P., Małysiak-Mrozek, B., Kostrzewa, D. (eds.) BDAS 2015-2016. CCIS, vol. 613, pp. 499–506. Springer, Cham (2016). https://doi.org/10.1007/978-3-319-34099-9_38
4. Byun, C., Lee, H., Kim, Y., Ko Kim, K.: Twitter data collecting tool with rule-based filtering and analysis module. Int. J. Web Inf. Syst. **9**(3), 184–203 (2013)
5. Cheng, Y.C., Chen, P.L.: Global social media, local context: a case study of Chinese-language tweets about the 2012 presidential election in Taiwan. Aslib J. Inf. Manag. **66**(3), 342–356 (2014)
6. Cieślewicz, J., Pelikant, A.: Zastosowanie sieci językowych w reprezentacji dokumentów tekstowych (Application of language networks in the representation of text documents). Stud. Informatica **32**(2A), 517–526 (2011)
7. Fagnan, J., Zaïane, O., Goebel, R.: Visualizing community centric network layouts. In: 2012 16th International Conference on Information Visualisation, pp. 321–330. IEEE (2012)
8. Fruchterman, T.M., Reingold, E.M.: Graph drawing by force-directed placement. Softw.: Pract. Experience **21**(11), 1129–1164 (1991)
9. Gunnarsson Lorentzen, D.: Polarisation in political Twitter conversations. Aslib J. Inf. Manag. **66**(3), 329–341 (2014)
10. Kamińska, D., Sapiński, T., Pelikant, A.: Recognition of emotion intensity basing on neutral speech model. In: Gruca, D.A., Czachórski, T., Kozielski, S. (eds.) Man-Machine Interactions 3. AISC, vol. 242, pp. 451–458. Springer, Cham (2014). https://doi.org/10.1007/978-3-319-02309-0_49
11. Kardes, H., Sevincer, A., Gunes, M.H., Yuksel, M.: Six degrees of separation among US researchers. In: Proceedings of the 2012 International Conference on Advances in Social Networks Analysis and Mining (ASONAM 2012), pp. 654–659. IEEE Computer Society (2012)
12. Kijas, W.: Facebook crawler as software agent for business intelligence system. Stud. Informatica **35**(4), 89–110 (2014)
13. Kijas, W., Kozielski, M.: Integration of Facebook online social network user profiles into a knowledgebase. In: Kozielski, S., Mrozek, D., Kasprowski, P., Małysiak-Mrozek, B., Kostrzewa, D. (eds.) BDAS 2015. CCIS, vol. 521, pp. 245–255. Springer, Cham (2015). https://doi.org/10.1007/978-3-319-18422-7_22
14. Kobourov, S.G.: Force-Directed Drawing Algorithms. Handbook of Graph Drawing and Visualization, pp. 383–408 (2013)
15. Konopka, E., Pelikant, A.: Zastosowanie metod grupowania w analizie sieci społecznościowych (Application of grouping methods in the analysis of social networks). Zeszyty Naukowe Wyższej Szkoły Informatyki **13**(1), 13–37 (2014)
16. Leydesdorff, L.: Clusters and maps of science journals based on bi-connected graphs in journal citation reports. J. Documentation **60**(4), 371–427 (2004)
17. Lunze, J.: Six degrees of separation in multi-agent systems. In: 2016 IEEE 55th Conference on Decision and Control (CDC), pp. 6838–6844. IEEE (2016)

18. Mars, A., Gouider, M.S., Saïd, L.B.: A new big data framework for customer opinions polarity extraction. In: Kozielski, S., Mrozek, D., Kasprowski, P., Małysiak-Mrozek, B., Kostrzewa, D. (eds.) BDAS 2015-2016. CCIS, vol. 613, pp. 518–531. Springer, Cham (2016). https://doi.org/10.1007/978-3-319-34099-9_40
19. Milgram, S.: The small world problem. Psychol. Today **2**(1), 60–67 (1967)
20. Pilny, P., Pelikant, A.: Wizualizacja i reorganizacja grafów z zastosowaniem relacyjnej bazy danych (Graphs visualisation and reorganisation using relational database). Zeszyty Naukowe Wyższej Szkoły Informatyki **13**(1), 38–65 (2014)
21. Pomorova, O.: The commercial information leak detection technology based on the analysis of professional discussions on social networks. Stud. Informatica **39**(1), 43–64 (2018)
22. Travers, J., Milgram, S.: An experimental study of the small world problem. In: Social Networks, pp. 179–197. Elsevier (1977)
23. Udanor, C., Aneke, S., Ogbuokiri, B.O.: Determining social media impact on the politics of developing countries using social network analytics. Program **50**(4), 481–507 (2016)
24. Wang, R., Liu, W., Gao, S.: Hashtags and information virality in networked social movement: examining hashtag co-occurrence patterns. Online Inf. Rev. **40**(7), 850–866 (2016)
25. Zimmer, M., Proferes, N.J.: A topology of Twitter research: disciplines, methods, and ethics. Aslib J. Inf. Manag. **66**(3), 250–261 (2014)

Image Analysis and Multimedia Mining

Poincaré Metric in Algorithms for Data Mining Tools

Alenka Trpin[1], Biljana Mileva Boshkoska[1,2(✉)], and Pavle Boškoski[1,2]

[1] Faculty of Information Studies in Novo Mesto,
Ljubljanska cesta 31 a, Novo Mesto, Slovenia
biljana.mileva@fis.unm.si
[2] Jožef Stefan Institute, Jamova 39, Ljubljana, Slovenia

Abstract. Today we cannot imagine life without computers. The massive use of the information communication technologies has produced large amounts of data that are difficult to interpret and use. With data mining tools and machine learning methods, large data sets can be processed and used for prediction and classification. This paper employees the well known classification algorithm the k nearest neighbour and it modified use the Poincaré measurment distance instead of traditional Euclidean distance. The reason is that in different industries (economy, health, military ...) it increasingly uses and stores databases of various images or photographs. When recognizing the similarity between two photographs, it is important that the algorithm recognizes certain patterns. Recognition is based on metrics. For this purposes an algorithm based on Poincaré metric is tested on a data set of photos. A comparison was made on algorithm based on Euclidean metric.

Keywords: Poincaré metric · Weka · K nearest neighbors

1 Introduction

Face recognition is an important role in our lives. Even a baby can recognize her mother's face. Adults recognize friends by their faces, and our perception about how we feel about someone can be seen by our face characteristics (sad, angry, happy ...). Face recognition can be exploited in numerous areas some if which are:

- security (control buildings, airports, ...),
- network security (email authentication on multimedia),
- identity verification (electoral registration, banking, ...),
- criminal justice systems [6].

Face recognition is an automatic way to identify persons based on their faces intrinsic characteristics. It usually involves: detecting the facial area, normalizing the detected faces, extracting facial features from appearance or facial geometry,

© Springer Nature Switzerland AG 2019
S. Kozielski et al. (Eds.): BDAS 2019, CCIS 1018, pp. 195–203, 2019.
https://doi.org/10.1007/978-3-030-19093-4_15

and finally classifying facial images based on the extracted features [9]. Biometric technologies include identification based on physiological characteristics (such as face, fingerprints, finger geometry, hand geometry, palm, iris, ear, voice, ...) and behavioral traits (gait, signature, keystroke dynamic) [7].

In this paper we focused on the well known classification algorithm k nearest neighbor, which we modified such that we changed the Euclidean distance metric with the Poincaré distance metric and tested it on a database of grey images of faces. The remainder of this paper is organized as follows. In Sect. 2 we define the Poincaré metric, and in Sect. 3 we discuss the used kNN classification method and its modification. The preparation of data and experimental results are reported in Sect. 4. Conclusions are given in Sect. 5.

2   Poincaré Metric

Poincaré's metric is useful in hyperbolic geometry. Unlike the Euclidean geometry, where it holds that through a given point we can draw exactly one parallel to a given line, in hyperbolic geometry holds that through a given point there are infinite parallels to a given line [13]. In this research we used Poincaré's disk model, which is in n-dimensional space defined as $B^n = \{x \in \mathbb{R}, \|x\| < 1\}$.

Let P be an arbitrary point in the unit disk and the v vector defined at this point.

Let the γ be a continuous path that maps the interval $[0, 1]$ to the disk D. The definition of Poincaré length of path is:

$$\ell_\rho(\gamma) = \int_0^1 |\dot{\gamma}(t)|_{\gamma(t)} \, dt = \int_0^1 \frac{1}{1 - |\gamma(t)|^2} \cdot \|\dot{\gamma}(t)\| \, dt. \tag{1}$$

Then Poincaré metrics between two different points $P, Q \in D$ is defined as

$$d_\rho(P, Q) = \inf\{\ell_\rho(\gamma) : \gamma \in \mathcal{C}_D(P, Q)\}. \tag{2}$$

The path, which length is defined by the infimum, is called geodesic. This is the shortest distance between two points. If the points P, Q are collinear with the starting point, the shortest path through these two points is a line. In other cases, they are circular arcs in standard Euclidean geometry. The circle on which such a circular arc is located, or a geodesic, can be precisely determined. Using the analytical method we can calculate the radius and center of this circle because we know three points P, Q and $\frac{1}{P}$. The third point $\frac{1}{P}$ is the point P, which is reflection over the edge of the disk.

A hyperbolic disk with the center a and a radius r in a complex space is denoted by $D_\rho(a, r) = \{z \in D; d_\rho(a, z) < r\}$. That is a set of all points whose distance from the center is less than r, which is exactly an Euclidean disk whose closure is contained in D [8].

If P is the point that lies in the unit disk, then Poincaré distance between starting point and point P is equal to

$$d_\rho(0, P) = \frac{1}{2} \cdot \ln\left(\frac{1 + |P|}{1 - |P|}\right). \tag{3}$$

Poincaré distance between two different points $P, Q \in D$ on a single disc is defined as

$$d_\rho(P,Q) = \frac{1}{2} \ln \left(\frac{1 + \left\| \frac{P-Q}{1-\overline{P}Q} \right\|}{1 - \left\| \frac{P-Q}{1-\overline{P}Q} \right\|} \right). \tag{4}$$

3 Classification Method

In this paper we used a modified k nearest neighbor (kNN) classifier that represents each example as a data point in a d-dimensional space, where d is the number of attributes. KNN involves a two-step process:

- an inductive step for constructing a classification model from data, and
- a deductive step for applying the model to test examples [12].

In the inductive step, the kNN classifier uses distance-based comparisons that intrinsically assign equal weight to each attribute. They therefore can suffer from poor accuracy when given noisy or irrelevant attributes. The method, however, has been modified to incorporate attribute weighting and the pruning of noisy data tuples. The choice of a distance metric can be critical. Usually the Eucledian distance, the Manhattan (city block) distance, or Minkowski metrics are used [3]. In this paper we use instead the Poincare distance as given with Eqs. 3–4.

In kNN, k is a natural number that tells how many adjacent attributes affect the classification of a given instance. Choosing the appropriate k affects the classification of a given instance into a class. If k is too small, the results are too general. If k is small, there are more boundaries and fragments, so the algorithm is more sensitive to noise. If it is too large, local features do not get to the expression, it follows that the classifier classifies the experimental attribute incorrectly. The larger k, leads to more inaccurate and time-consuming models.

4 Data Processing and Results

Face recognition is a specific and hard case of object recognition. The difficulty of this problem stems from the fact that in their most common form (the frontal view) faces appear to be roughly alike and the differences between them are quite subtle. Furthermore, the human face is not a unique, rigid object [6]. A number of conditions are affected by the perception of success:

- expression on the face (cheerful, sad, angry),
- face cover (wearing a scarf, sunglasses ...),
- face view (different profiles),
- camera features (lens, lighting),
- facial features (mustache, beard, glasses),
- the size of the face (the face closer to the camera is bigger),
- face illumination.

We obtained 640 images from the UCI Machine Learning Repository, which include faces in various positions and conditions [4]. The images are given in gray scale and all share the same size and format. Images differ in the facial positions, facial expressions, and the appearances (long or short hair, beard, sunglasses). In the pre-processing step we converted each image into a format that is suitable for processing in WEKA where data are presented as vectors with numerical values. Hence we converted the images into jpg format and then we created an arff file with two attributes: image name and class (Fig. 1).

Fig. 1. Photos in different positions

Next we processed the data with WEKA using different photo filters that were suitable for black and white photographs: Binary patterns pyramid filter, Edge histogram filter, FCTH filter, Gabor filter, JPEG coefficient filter and PHOG filter.

Local binary patterns pyramid (BPP) filter creates 3×3 pixel blocks of an image. Each pixel in this block is thresholded by its center pixel value which is a binary number. Next it compares the values of the point and its neighbors: if the value of the point is greater than the value of its neighbor, the algorithm records 0, otherwise it records 1. This creates an eight-digit binary number. Next the histogram of these numbers is created and normalized. The result is a 256-bit vector used for classification [10].

Edge histogram filter (MPEG) is used to compare images at the edges. It describes the spatial distribution of edge parts, non edge parts and four directional edge parts. The image is partitioned in 16 sub-regions. A histogram of five bins (colors) is created for each sub-region. The image is divided into smaller regions, where the edges are categorized. Each bin is normalized and quantized (reduces the range of values). To measure the similarity of two edge histograms, the sum of the absolute differences of individual bins is used [11].

The FCTH filter (fuzzy colour and texture histogram) is limited to 72 bytes per image, so it is suitable for large databases. It uses three channels and forms ten bins. Each part represents a certain color. This filter is not sensitive to noise and deformation [1].

Gabor filter uses Gabor waves to remove attributes from the image. Each wave captures energy at a given frequency and in a certain direction. It provides a local description of the frequency, thereby capturing the local characteristics of the signal. This filter is useful for surface analysis [15].

A JPEG (Joint Photographic Experts Group) coefficient acquires attributes from a sequence of quantized coefficients that a person can not detect when converting to a JPEG format. The filter acquires coefficients with discrete cosine transformation. This is a well-established standard for compression of color and gray-scale images for the purposes of storage and transmission [2].

PHOG filter encodes information on the orientation of the gradient intensity to the image. It is based on a local image set and is effective in spatial distribution of the edge. It divides the image into a net, then compensates for the histogram of the gradient for each network [5].

The result of the processing is an .arff file where the first attribute represents the name of the photo, followed by the numeric attributes, as represented in Fig. 2.

In the processing phase we set up a test and learning sets and used 10 - fold transversal validation [14].

During preprocessing, we removed the attribute image name, selected the kNN classifier, implemented the Poincaré distance as a metric in Java, set up the metric for usage in WEKA. The number of neighbors were selected inductively. To find the best k with k - nearest neighbor, we selected the Linear NN search algorithm.

Table 1 shows the performance of the classifier in percentages in terms of their accuracies, which are rounded up to two decimal places. The variable k is the number of closest neighbors that were searched with the Linear NN search algorithm. In Table 1, the results differ both depending on the filters that were used and on the set k number in the kNN. In Table 1 PM stands for Poincaré metric and EM for Eeuclidean based kNN while the variable k is the number of closest neighbors. The required execution time of the modified kNN with PM was higher than the EM algorithm, since the calculation of the Poincaré distance requires calculation of three different Euclidean distances as given in Eq. 4. However, the complexity of both algorithms is the same.

Table 1. Results

Metrics	PM			EM		
Filters	$k = 1$	$k = 5$	$k = 10$	$k = 1$	$k = 5$	$k = 10$
MPEG	**98, 44%**	**97, 97%**	**97, 97%**	**98, 59%**	97, 81%	**97, 97%**
BPP	**98, 44%**	97, 81%	97, 66%	95, 47%	93, 59%	91, 87%
Gabor	**43, 44%**	38, 91%	34, 68%	**66, 41%**	63, 91%	61, 25%
JPEG coefficient	**95, 31%**	94, 69%	94, 84%	**95, 63%**	95, 31%	94, 69%
PHOG	97, 97%	**97, 97%**	96, 72%	98, 44%	**97, 97%**	97, 81%

```
 1
 2
 3  @relation FIT
 4
 5  ....
 6  @attribute 'Spatial Pyramid of Local Binary Pattern 748'
        numeric
 7  @attribute 'Spatial Pyramid of Local Binary Pattern 749'
        numeric
 8  @attribute 'Spatial Pyramid of Local Binary Pattern 750'
        numeric real
 9  @attribute 'Spatial Pyramid of Local Binary Pattern 751'
        numeric real @attribute 'Spatial Pyramid of Local Binary
        Pattern 752' numeric
10  @attribute 'Spatial Pyramid of Local Binary Pattern 753'
        numeric
11  @attribute 'Spatial Pyramid of Local Binary Pattern 754'
        numeric
12  @attribute 'Spatial Pyramid of Local Binary Pattern 755'
        numeric
13  @attribute class {ani, at, boland, bpm, ch, cheyer, chon,
        danieln, glickman, karyadi, ...}
14
15  @data
16  an2i_left_angry_open.jpg,5,8,5,0,9,1,0,1,15,1,0,0,1,,...
17  an2i_left_angry_sunglasses.jpg,5,8,6,0,8,1,1,0,15,1,0,...
18  an2i_left_happy_open.jpg,6,9,6,1,10,1,1,0,15,1,0,0,...
19  an2i_left_happy_sunglasses.jpg,6,9,6,1,9,1,1,...
20  an2i_left_neutral_open.jpg,6,9,6,1,10,1,1,1,15,...
21  an2i_left_neutral_sunglasses.jpg,6,9,6,1,9,1,0,1,...
22  an2i_left_sad_open.jpg,5,8,6,0,9,1,1,0,15,1,...
```

Fig. 2. Processed data

As shown in Table 1, the accuracy of the algorithms depends on the filter selection. The worst classification in both algorithms was obtained with the Gabor's filter, and the best results were obtained with the Edge histogram filter. The PM algorithm yielded the best result with the Edge histogram and Binary patterns pyramid (BPP) filters, and the EM algorithm with the Edge histogram and PHOG filter. For $k = 1$, the Binary patterns pyramid filter produced better results than the PM algorithm (by almost 3%), and the PHOG filter was more successful EM algorithm (approximately 0,5%). There are no large deviations between the algorithms, except for the Gabor's filter, where by increasing the number of neighbors the error is doubled by almost twice. Using Gabor's filter lead to ineffective classification results for the PM and EM with kNN.

Figures 3 and 4 show the ROC curves when using MPEG and PHOG filters, for both EM and PM algorithms, and for three values of $k = 1, 5, 10$. On x axis

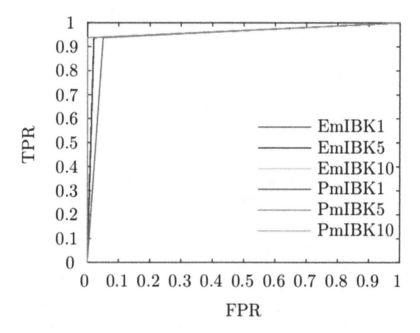

Fig. 3. ROC curve for MPEG filter for $k \in \{1, 2, 3\}$ and for PM and EM

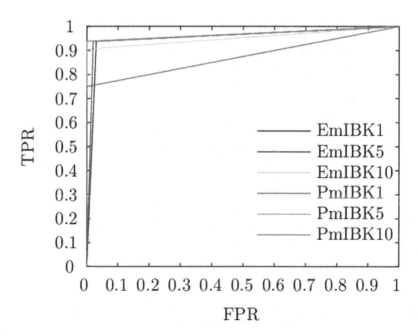

Fig. 4. ROC curve for PHOG filter for $k \in \{1, 2, 3\}$ and for PM and EM

Table 2. Comparison between the two algorithms

Filter	BPP		Edge histogram	
Variables	PM	EM	PM	EM
RMSE	**0,0395**	0,0666	0,0395	**0,0376**
MAE	**0,0047**	0,0076	0,0047	**0,0045**
RRSE	**18,1413%**	30,5454%	18,1413%	**17,2431%**
RAE	**4,9448%**	7,9647%	4,9448%	**4,7859%**

of the ROC curves we show the values of the False Positive Rate (FPR) and on the y axis we show the values of the True Positive Rate (TPR).

Table 2 shows a comparison between the two classification models for $k = 1$ and for two different filters, the BPP and the Edge histogram, for which both models perform successfully. The following statistical variables are compared: the root mean squared error (RMSE), the mean absolute error (MAE), the root relative squared error (RRSE) and the relative absolute error (RAE). With the BPP filter, the PM classification model has all the statistical values smaller than the EM classification model, therefore it follows that the PM classification model is more successful than the EM classification model. Conversely, the results of the Edge histogram filter show that all the statistical values of the EM classification model are smaller than those for PM classification model, hence it is considered as the better one.

5   Discussion and Conclusion

The paper is a first attempt of the authors to use the Poincaré metrics for classification. In particular we modified the well known kNN so that instead of the usual Euclidean metric for calculation of the distances between two points we applied the Poincaré metrics that is used for calculation of distances between points in hyperbolic spaces. We tested the modified kNN on a set of gray scale images that we obtained on the UCI Machine Learning Repository. The result was that the existing classification machine learning algorithm based on the Euclidean metric can be modified so that the Poincaré metric is used.

We have found that the choice of the filter that is used in the pre-processing phase influence the classification efficiency. In particular, the PM algorithm gave the best result with the Edge histogram filters and the Binary patterns pyramid. The EM algorithm provided the best results when used with the Edge histogram and PHOG filters. The comparison of both classification methods showed that the algorithm PM was better with the BPP filter (for $k = 1$ the classification was 98.44%). The same result was obtained with the Edge histogram filter, which was better than the BPP with the increase in the k number. The EM algorithm provided best results when used with the Edge histogram filter (for $k = 1$ the classification was 98.59%).

The future works include testing of the performance of the proposed classifier based on the Poincaré metric on color photos,and usage of other filters in the pre-processing phase. It could also be explored with other classification methods such as SVM, decision trees, or neural networks.

References

1. Chatzichristofis, S.A., Boutalis, Y.S.: FCTH: fuzzy color and texture histogram-a low level feature for accurate image retrieval. In: 2008 Ninth International Workshop on Image Analysis for Multimedia Interactive Services, pp. 191–196. IEEE (2008)
2. Climer, S., Bhatia, S.K.: Image database indexing using JPEG coefficients. Pattern Recogn. **35**(11), 2479–2488 (2002)
3. Han, J., Pei, J., Kamber, M.: Data Mining: Concepts and Techniques. Elsevier, Amsterdam (2011)
4. Hatem, H., Beiji, Z., Majeed, R.: A survey of feature base methods for human face detection. Int. J. Control Autom. **8**(5), 61–78 (2015)
5. Huang, H.M., Liu, H.S., Liu, G.P.: Face recognition using pyramid histogram of oriented gradients and SVM. AISS: Adv. Inf. Sci. Serv. Sci. **4**(18), 1–8 (2012)
6. Jafri, R., Arabnia, H.R.: A survey of face recognition techniques. Jips **5**(2), 41–68 (2009)
7. Jain, A.K., Bolle, R., Pankanti, S.: Biometrics: Personal Identification in Networked Society, vol. 479. Springer, Heidelberg (2006). https://doi.org/10.1007/978-0-387-32659-7
8. Krantz, S.G.: Complex Analysis: The Geometric Viewpoint, No. 23. MAA (2004)
9. Mohamed, A., Yampolskiy, R.: Recognizing artificial faces using wavelet based adapted median binary patterns. In: The Twenty-Sixth International FLAIRS Conference (2013)
10. Pietikäinen, M., Hadid, A., Zhao, G., Ahonen, T.: Local binary patterns for still images. Computer Vision Using Local Binary Patterns, vol. 40, pp. 13–47. Springer, London (2011). https://doi.org/10.1007/978-0-85729-748-8_2
11. Popova, A., Neshov, N.N.: Image similarity search approach based on the best features ranking. Egypt. Comput. Sci. J. **37**(1), 51–65 (2013)
12. Tan, P.N.: Introduction to Data Mining. Pearson Education India (2018)
13. Thorgeirsson, S.: Hyperbolic geometry: history, models, and axioms (2014)
14. Witten, I.H., Frank, E., Hall, M.A., Pal, C.J.: Data Mining: Practical Machine Learning Tools and Techniques. Morgan Kaufmann, Burlington (2016)
15. Zhang, D., Wong, A., Indrawan, M., Lu, G.: Content-based image retrieval using Gabor texture features. IEEE Trans. Pami **3656**, 1315 (2000)

Super-Resolution Reconstruction Using Deep Learning: Should We Go Deeper?

Daniel Kostrzewa[1,2]([✉]) [iD], Szymon Piechaczek[1], Krzysztof Hrynczenko[1],
Paweł Benecki[1,2] [iD], Jakub Nalepa[1,2] [iD], and Michal Kawulok[1,2] [iD]

[1] Future Processing, Bojkowska 37A, 44-100 Gliwice, Poland
{dkostrzewa,pbenecki,jnalepa,mkawulok}@future-processing.com
[2] Institute of Informatics, Silesian University of Technology,
Akademicka 16, 44-100 Gliwice, Poland
{daniel.kostrzewa,pawel.benecki,jakub.nalepa,michal.kawulok}@polsl.pl

Abstract. Super-resolution reconstruction (SRR) is aimed at increasing image spatial resolution from multiple images presenting the same scene or from a single image based on the learned relation between low and high resolution. Emergence of deep learning allowed for improving single-image SRR significantly in the last few years, and a variety of deep convolutional neural networks of different depth and complexity were proposed for this purpose. However, although there are usually some comparisons reported in the papers introducing new deep models for SRR, such experimental studies are somehow limited. First, the networks are often trained using different training data, and/or prepared in a different way. Second, the validation is performed for artificially-degraded images, which does not correspond to the real-world conditions. In this paper, we report the results of our extensive experimental study to compare several state-of-the-art SRR techniques which exploit deep neural networks. We train all the networks using the same training setup and validate them using several datasets of different nature, including real-life scenarios. This allows us to draw interesting conclusions that may be helpful for selecting the most appropriate deep architecture for a given SRR scenario, as well as for creating new SRR solutions.

Keywords: Super-resolution reconstruction · Image processing ·
Convolutional neural network · Deep learning

1 Introduction

Single-image super-resolution reconstruction (SRR) is a group of methods, whose main goal is to construct a high-resolution (HR) image on the basis of a single low-resolution (LR) input image [30,40]. There are a few main approaches to resolve this task: usage of predefined mathematical formula (e.g., bilinear, bicubic, and other interpolation methods), edge based methods [12,35,38], using a dictionary with pairs of matched LR and HR image fragments [6,13,14,19,23,41], heavy-tailed gradient distribution [32], sparsity property of large gradients [23], discrete and stationary wavelet decomposition [7,8],

© Springer Nature Switzerland AG 2019
S. Kozielski et al. (Eds.): BDAS 2019, CCIS 1018, pp. 204–216, 2019.
https://doi.org/10.1007/978-3-030-19093-4_16

and universal hidden Markov tree model [28]. However, most of recent works engage convolutional neural networks (CNNs) to transform LR image to HR output.

1.1 Related Work

Super-resolution CNN (SRCNN) is the first deep learning model whose authors claimed to surpass performance of other methods [9,10]. SRCNN is a single-image super-resolution reconstruction algorithm that optimizes an end-to-end mapping from LR to HR image. This model has been improved in two ways: by increasing depth of the network, and by introducing recursive learning. In very deep super-resolution CNN (VDSR) [21] and image restoration CNN (IRCNN) [42], additional convolutional layers have been added. Kim et al. proposed recursive learning for parameter sharing in their deeply-recursive convolutional networks (DRCNs) [22]. Afterwards, Tai et al. suggested the use of recursive blocks (deep recursive residual network—DRRN [36]) and memory blocks (Memnet [37]). All mentioned networks have the same drawback—they require input images that are of the same size as the desired output image. Therefore, an additional step is needed to interpolate the LR image to the expected size.

Fast super-resolution CNN (FSRCNN) [11], the successor to the SRCNN, solves aforementioned problem by taking the original LR image as an input, and using deconvolution layer to enlarge the image to the desired resolution. Moreover, the authors improved the previous design to accelerate the processing (up to 24 frames per second). In efficient sub-pixel CNN (ESPCN) [33], a new layer type was introduced to upscale the final LR feature maps into the reconstructed output. On this basis, SRResNet [27] and enhanced deep residual network (EDSR) were proposed [29], which exploit residual learning. In recent years, generative adversarial networks (GANs) gain popularity in image generation, and they outperform are deep learning techniques in SRR [27].

1.2 Contribution

We have witnessed a breakthrough in single-image SRR, underpinned with the use of deep neural networks. Architectures of different depth and complexity are employed for this purpose, and the widely shared opinion is that the deeper models are more capable of learning the relation between low and high resolution. However, the experimental results reported in the papers introducing new SRR techniques are often limited—the test sets are commonly composed of the images of the same kind, and LR images are obtained by downscaling and degrading HR images, which serve as a reference (grount-truth, GT) for evaluation.

In this paper, we report the results of our extensive experimental study to compare a number of different deep architectures that have been proposed for SRR. We consider the magnification factor of 2×, as we want to compare the networks taking into account their capability of reconstructing images based on the LR information. For larger magnification factors, the networks are, in fact, trying to "guess" the high-resolution appearance of the details that are not visible in the input image. Our main contribution lies in comparing the state-of-the-art deep SRR networks using the same training data and setup, for several test sets of different kind, namely: (i) artificially-degraded natural images, (ii) artificially-degraded satellite images, and (iii) real satellite data that are matched with HR images acquired using a different satellite of higher spatial resolution.

1.3   Paper Structure

The paper is structured as follows. Section 2 describes in details the CNNs investigated in this paper. In Sect. 3, we present the results of our experiments, which have been carried out to evaluate the implemented networks. Finally, Sect. 4 concludes the paper and shows the main goals of our ongoing research.

2   Convolutional Neural Networks for SRR

In this subsection, we discuss the implemented CNNs in detail. These networks have been experimentally validated in Sect. 3.

2.1   Super-Resolution Convolutional Neural Network

SRCNN is the first CNN that has been developed specifically for SRR (Fig. 1) [9,10]. The process of mapping LR to HR image can be separated into three steps:

1. **Patch extraction and representation**
   This part is done by the first convolutional layer. The *patch extraction* is the process of sliding kernel through the whole image with the overlaps. *Representation* is the resulting feature map which is a consequence of the aforementioned *patch extraction*.
2. **Non-linear mapping**
   The second convolving part non-linearly maps obtained feature map onto another one. This new set of features directly corresponds to the HR image.
3. **Reconstruction**
   The last part of the process gathers those HR representations and fuse them into an image which should be as similar as possible to ground-truth.

The SRCNN model is parametrized by a fairly small set of hyperparameters: the number of color channels, filter sizes in the first, second, and third layers (in this case, $k = 9$, $k = 1$, and $k = 5$, respectively), and the number of filters in all layers ($n = 64$, $n = 32$, and $n = 1$, respectively; note that the number of filters in the last layer corresponds to the number of color channels).

SRCNN requires a pre-magnification of the input image to the desired size. This design comes with one great advantage compared to other architectures, where the whole super-resolution process is done by the model itself. The advantage is that the network can be trained to restore images upscaled by any factor, including non-integer upscaling factors.

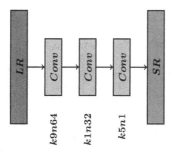

Fig. 1. A diagram of the SRCNN model

The experimental results reported in [9, 10] show that SRCNN achieves better numerical results than the state of the art at the time. It can be further tuned in terms of either time efficiency or performance by changing the previously mentioned hyperparameters.

2.2 Fast Super-Resolution Convolutional Neural Network

FSRCNN is built on the basis of SRCNN. There are two main high-level differences between these networks. The first one is that FSRCNN resizing process is performed by the part of the model itself. Thus, the preprocessing step needed by the SRCNN is eliminated. SRCNN requires that for different scales, therefore the network has to be trained from scratch.

FSRCNN (Fig. 2) is composed of five different parts, which resembles the SRCNN structure:

1. **Feature extraction**
   Feature extraction is made at the very beginning of the processing pipeline (first layer). Comparing to the SRCNN counterpart, its kernel size has been reduced due to the fact that the input LR image is of the original size (it is not interpolated).
2. **Shrinking**
   To improve processing speed, the second convolutional layer is appended with a kernel of size 1×1. Its purpose is to reduce the number of channels and as a result it lessens the number of parameters.
3. **Non-linear mapping**
   The non-linear mapping is performed by several convolutional layers with smaller kernels instead of one layer with a bigger filter of a greater size.
4. **Expanding**
   The features are "expanded" to correspond directly to the high-resolution image from which the final image is produced by the last part.
5. **Deconvolving**
   The last part of the model aggregates the features and performs the upscaling to produce the final high-resolution image.

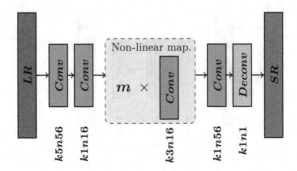

Fig. 2. A diagram of the FSRCNN model

The FSRCNN architecture improves computational speed without negatively impacting the performance. Both, the computational speed and the performance are highly adjustable with the number of non-linear mapping layers. Increasing that number results in growing complexity and mapping accuracy.

The results presented in [11] showed that the improved architecture is truly able to achieve real-time processing speed. Another effect of the changes is that the reconstruction quality also was improved. Multiple narrow layers (non-linear mapping layers) instead of one wider layer give better performance. The authors showed that including upscaling part as a deconvolution layer also positively affects the reconstruction quality. FSRCNN benefits from transfer-learning, once it is trained for one upscaling factor, only the last deconvolutional layer has to be retrained for other upscaling factor values.

2.3   Super-Resolution Residual Neural Network

Deeper networks have proven the ability to render performance impossible to obtain by the shallow networks—more layers allow for modeling mappings of a very high complexity. However, increasing the depth of the network is not as easy as adding more layers—the problems of convergence and degradation (of the network performance) can easily emerge. The vanishing/exploding gradient problems [3,15] can stop the network from converging—error gradient may bring too large or too small update for the network. Still, if the network manages to converge, its performance may deteriorate with the increase in depth [34]. The former problem has been addressed by normalized initialization [15,17,26,31] or batch normalization layers [20]; for the degradation problem, the proposed solutions encompass the highway networks [34] (utilizing information flow between layers) and residual networks [18] (with the input skipping layers).

The SRResNet [27] (Fig. 3) architecture belongs to the family of residual networks, and it was inspired by a network proposed by He et al. [18]. It deals with both vanishing/exploding gradient and degradation problems (by the use of intermediate normalization layers and residual connections). The *residual blocks* (RBs) are the groups of layers stacked together with the input of the block added

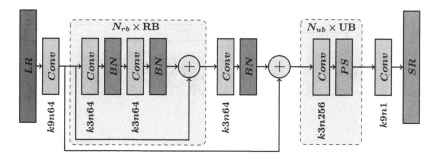

Fig. 3. A diagram of the SRResNet model

to the output of the final layer contained in this block. In SRResNet, each block encompasses two convolutional layers, each followed by a *batch normalization* (BN) layer that neutralizes the internal co-variate shift. The *upsampling blocks* (UBs) allow for image enlargement by *pixel shuffling* (PS) layers that increase the resolution of the features. The number of both RBs and UBs is variable— by increasing the number of RBs, the network may model a better mapping, whereas by changing the number of UBs, we may tune its scaling factor. However, by adding more blocks, the architecture of the network becomes increasingly complex, which makes it harder to train.

Throughout the whole network, Parametric ReLU (PReLU) is used as an activation function. Similarly to Leaky ReLU, it introduces a small slope for negative values, and PReLU enables the network to learn the optimal value for this slope. Both Leaky and Parametric versions of ReLU decrease the time needed for network to converge, and help with the dying ReLU problem [39].

2.4 Super-Resolution Generative Adversarial Network

GANs are rather complicated structures and may be described as two competing networks, the first one (generator) is trying to produce the image indistinguishable from the real (not generated) pictures, while the second (discriminator)— focuses on identifying these "fake" images. Once the discriminator starts to distinguish well between "real" and "fake" images, the generator network needs to produce increasingly better images. At the end of the training, discriminator should not be able to differentiate original images from those produced by the generator. In such a training scheme, the role of the discriminator may be perceived as an adaptive loss function, that moderates the training of the network by gradually increasing demand for the networks' performance.

Scenarios for the usage of GANs encompass image generation from noise, image inpainting and style transfer. In the context of single image super-resolution, GANs were firstly introduced by Ledig et al. [27]. In SRGAN (Fig. 4), both generator and discriminator (Fig. 5) are very deep networks; in fact SRRes-Net is used as a generative part of the network. To train the generator network, two loss functions are used and weighted—content loss is evaluated as an MSE

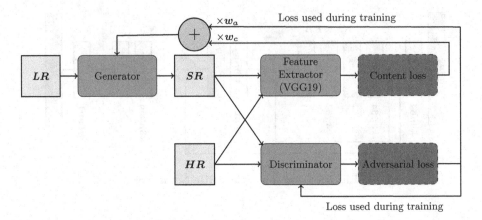

Fig. 4. A diagram of the SRGAN model

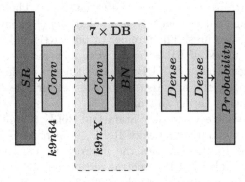

Fig. 5. A diagram of the discriminator model. Number of convolution filters (denoted by X) progressively increase, with $X \in \{64, 128, 256, 512\}$

between features of "real" and "fake" images (extracted with VGG19 network); adversarial loss favors the solutions that generate images unable to distinguish from the "real" ones. For discriminator, the loss proposed in [16] is used.

3 Experiments

All convolutional network models have been trained with DIV2K dataset [1]. The images were converted to 8-bit greyscale and downsampled $2\times$ using bicubic interpolation (in order to create LR patches). The set of prepared training and validation images is available online[1] under the license provided on the website [24].

Several popular benchmarks were selected for the experiments: **Set5** [5], **Set14** [41], and **BSD** [4]. In addition to these, we also created two datasets,

---

[1] https://doi.org/10.7910/DVN/DKSPJF.

based on B4MultiSR [25], consisting of satellite images. The first one, **Artificialy Degraded Satellite** dataset, is composed of images gathered during the Sentinel-2 mission. In this case, GT is the original image and LR is a bicubic downscaled counterpart. **Real Satellite** dataset is the most challenging scenario in our experiments—GTs are obtained by downscaling images from the Digital Globe WorldView-4 satellite (original ground sampling distance equals 30 cm/px), and LR are images from Sentinel-2.

To quantitatively compare the quality of the investigated models, we utilized several popular metrics: peak signal-to-noise ratio (PSNR), structural similarity index (SSIM), universal image quality index (UIQI), and visual information fidelity (VIF). Additionally, we exploit the following metrics: $PSNR_{HF}$ and KFS_{SIFT} [2]. In Table 1, we gathered all numeric results of conducted experiments, whereas Figs. 6, 7, and 8 render images for the visual comparison.

Table 1. The results of our experiments. We boldfaced the best results for each dataset, and the deep learning methods which render the results worse than Bicubic are annotated with the gray background.

Dataset name	Model	PSNR	SSIM	UIQI	VIF	$PSNR_{HF}$	KFS_{SIFT}
Set5	Bicubic	32.33	0.922	0.859	0.605	38.94	42.56
	SRCNN	34.96	0.944	0.871	0.672	45.40	44.67
	FSRCNN	**35.52**	**0.949**	**0.876**	**0.690**	46.70	44.73
	SRResNet	35.20	0.935	0.858	0.689	46.32	**44.97**
	SRGAN	33.86	0.931	0.828	0.637	**46.86**	44.57
Set14	Bicubic	28.75	0.857	0.781	0.500	34.34	42.46
	SRCNN	30.75	0.894	0.818	0.559	39.33	44.55
	FSRCNN	**31.11**	**0.900**	**0.825**	**0.574**	40.14	**44.69**
	SRResNet	30.29	0.889	0.813	0.558	39.21	44.62
	SRGAN	30.11	0.878	0.777	0.534	**40.49**	44.11
BSD	Bicubic	28.23	0.832	0.768	0.467	33.70	42.91
	SRCNN	29.88	0.876	0.814	0.518	38.45	45.06
	FSRCNN	**30.19**	**0.884**	**0.825**	**0.532**	**39.30**	**45.13**
	SRResNet	29.85	0.880	0.817	0.527	38.66	44.98
	SRGAN	29.88	0.876	0.814	0.518	38.45	45.06
Artificially Degraded Satellite	Bicubic	**30.22**	**0.899**	**0.873**	**0.517**	**41.97**	**43.54**
	SRCNN	26.84	0.829	0.800	0.440	34.76	41.85
	FSRCNN	26.11	0.812	0.783	0.425	33.43	41.49
	SRResNet	26.31	0.813	0.785	0.429	33.64	41.49
	SRGAN	24.60	0.727	0.696	0.363	30.43	39.92
Real Satellite	Bicubic	16.81	0.391	0.170	0.087	35.00	39.71
	SRCNN	16.64	0.345	0.138	0.075	35.68	40.55
	FSRCNN	16.72	0.343	0.139	0.076	35.81	40.34
	SRResNet	16.90	**0.394**	**0.179**	**0.090**	36.21	40.76
	SRGAN	**17.11**	0.391	0.173	0.088	**37.45**	**40.79**

For Set5 (Fig. 6), Set14 and BSD benchmarks, being most widely used datasets for SRR testing, FSRCNN renders best results both qualitatively and

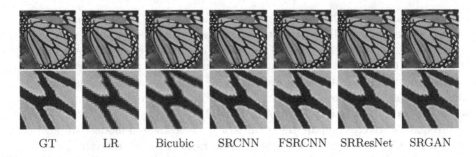

GT   LR   Bicubic   SRCNN   FSRCNN   SRResNet   SRGAN

Fig. 6. Example of the reconstructed image from the Set5 dataset. A part of the image is zoomed for clarity (second row).

quantitatively. On the other hand, deeper networks (SRResNet and SRGAN), introduced the halo effect around the edges, decreasing the numerical results as well. It is also worth mentioning that these benchmarks are composed of images in the `jpeg` format (which introduces artifacts)—this may degrade the results too (in fact, deeper networks seem to magnify such artifacts).

The results for the Artificially Degraded Satellite images "favor" the outcomes produced by the bicubic interpolation. However, the images obtained by the deep networks seem to be sharper than the high-resolution version of the image (Fig. 7). This may be the reason for the observed lower quantitative scores.

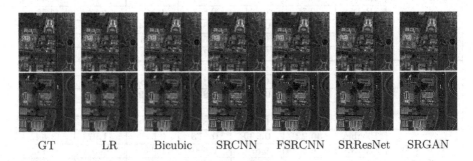

GT   LR   Bicubic   SRCNN   FSRCNN   SRResNet   SRGAN

Fig. 7. Example of the reconstructed image from the Artificially Degraded Satellite dataset. A part of the image is zoomed for clarity (second row).

Finally, for the images from the Real Satellite benchmark (Fig. 8), deeper networks obtained the highest scores. Still, in this case, compared images (GT and SR versions of the image) present the same area, however they do not share similar pixel values (e.g., due to the variable lighting conditions). This is the reason why some metrics (see e.g., PSNR) have such low values.

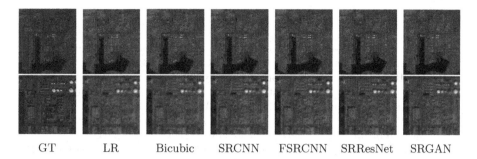

GT   LR   Bicubic   SRCNN   FSRCNN   SRResNet   SRGAN

Fig. 8. Example of the reconstructed image from the Real Satellite dataset. A part of the image is zoomed for clarity (second row).

4 Conclusions and Future Work

In this paper, we compared the performance of four different deep network architectures for single image super-resolution reconstruction. These networks were evaluated over five different benchmarks, including three standard ones (Set5, Set14, and BSD100) and two introduced in this work (Artificially Degraded Satellite and Real Satellite). The experiments showed that it is notably easier to reconstruct artificially degraded images (therefore, shallower networks can effectively cope with this task). In this case, deeper networks start to enhance the `jpeg` artifacts. On the other hand, reconstructing an image and comparing it to the one obtained by another sensor is much more difficult. As a result, deeper networks (with higher capacities) outperformed the others.

It is also worth mentioning, that deeper networks are much more complex structures, hence computational power needed to perform the training and prediction increase. In this article, we did not focus on efficiency of the deep networks, however our study shows that introducing more complexity to the model is not always worth it.

Currently, we are investigating the influence of the scale on the performance of deep networks (and how deeper networks deal with scales higher than 2). Also, we are focused on the comparison between the classical computer vision methods and deep networks for both single-image and multi-frame super-resolution.

Acknowledgements. This work was supported in part by the European Space Agency through the SuperDeep project. The authors were supported by Statutory Research funds of Institute of Informatics, Silesian University of Technology, Poland: BKM-556/RAU2/2018 (DK, PB, JN) and 02/020/BK_18/0128 (MK).

References

1. Agustsson, E., Timofte, R.: NTIRE 2017 challenge on single image super-resolution: dataset and study. In: The IEEE Conference on Computer Vision and Pattern Recognition (CVPR) Workshops, July 2017
2. Benecki, P., Kawulok, M., Kostrzewa, D., Skonieczny, L.: Evaluating super-resolution reconstruction of satellite images. Acta Astronaut. **153**, 15–25 (2018)
3. Bengio, Y., Simard, P., Frasconi, P.: Learning long-term dependencies with gradient descent is difficult. IEEE Trans. Neural Netw. **5**(2), 157–166 (1994)
4. Berkeley Segmentation Dataset: 01 December 2019. https://www2.eecs.berkeley.edu/Research/Projects/CS/vision/bsds/
5. Bevilacqua, M., Roumy, A., Guillemot, C., Alberi Morel, M.L.: Low-complexity single-image super-resolution based on nonnegative neighbor embedding. In: Proceedings of the British Machine Vision Conference, pp. 135.1–135.10. BMVA Press (2012)
6. Chang, H., Yeung, D.Y., Xiong, Y.: Super-resolution through neighbor embedding. In: Proceedings of the 2004 IEEE Computer Society Conference on Computer Vision and Pattern Recognition, 2004, CVPR 2004, pp. 275–282. IEEE (2004)
7. Chavez-Roman, H., Ponomaryov, V.: Super resolution image generation using wavelet domain interpolation with edge extraction via a sparse representation. IEEE Geosci. Remote Sens. Lett. **11**(10), 1777–1781 (2014)
8. Demirel, H., Anbarjafari, G.: Image resolution enhancement by using discrete and stationary wavelet decomposition. IEEE Trans. Image Process. **20**(5), 1458–1460 (2011)
9. Dong, C., Loy, C.C., He, K., Tang, X.: Learning a deep convolutional network for image super-resolution. In: Fleet, D., Pajdla, T., Schiele, B., Tuytelaars, T. (eds.) ECCV 2014. LNCS, vol. 8692, pp. 184–199. Springer, Cham (2014). https://doi.org/10.1007/978-3-319-10593-2_13
10. Dong, C., Loy, C.C., He, K., Tang, X.: Image super-resolution using deep convolutional networks. IEEE Trans. Pattern Anal. Mach. Intell. **38**(2), 295–307 (2016)
11. Dong, C., Loy, C.C., Tang, X.: Accelerating the super-resolution convolutional neural network. In: Leibe, B., Matas, J., Sebe, N., Welling, M. (eds.) ECCV 2016. LNCS, vol. 9906, pp. 391–407. Springer, Cham (2016). https://doi.org/10.1007/978-3-319-46475-6_25
12. Fattal, R.: Image upsampling via imposed edge statistics. ACM Trans. Graph. (TOG) **26**, 95 (2007)
13. Freeman, W.T., Jones, T.R., Pasztor, E.C.: Example-based super-resolution. IEEE Comput. Graph. Appl. **22**(2), 56–65 (2002)
14. Glasner, D., Bagon, S., Irani, M.: Super-resolution from a single image. In: 2009 IEEE 12th International Conference on Computer Vision, pp. 349–356. IEEE (2009)
15. Glorot, X., Bengio, Y.: Understanding the difficulty of training deep feedforward neural networks. In: Proceedings of the Thirteenth International Conference on Artificial Intelligence and Statistics, pp. 249–256 (2010)
16. Goodfellow, I., et al.: Generative adversarial nets. In: Advances in Neural Information Processing Systems, pp. 2672–2680 (2014)
17. He, K., Zhang, X., Ren, S., Sun, J.: Delving deep into rectifiers: surpassing human-level performance on imagenet classification. In: Proceedings of the IEEE International Conference on Computer Vision, pp. 1026–1034 (2015)

18. He, K., Zhang, X., Ren, S., Sun, J.: Deep residual learning for image recognition. In: Proceedings of the IEEE Conference on Computer Vision and Pattern Recognition, pp. 770–778 (2016)
19. Huang, J.B., Singh, A., Ahuja, N.: Single image super-resolution from transformed self-exemplars. In: Proceedings of the IEEE Conference on Computer Vision and Pattern Recognition, pp. 5197–5206 (2015)
20. Ioffe, S., Szegedy, C.: Batch normalization: accelerating deep network training by reducing internal covariate shift. arXiv preprint arXiv:1502.03167 (2015)
21. Kim, J., Kwon Lee, J., Mu Lee, K.: Accurate image super-resolution using very deep convolutional networks. In: Proceedings of the IEEE Conference on Computer Vision and Pattern Recognition, pp. 1646–1654 (2016)
22. Kim, J., Kwon Lee, J., Mu Lee, K.: Deeply-recursive convolutional network for image super-resolution. In: Proceedings of the IEEE Conference on Computer Vision and Pattern Recognition, pp. 1637–1645 (2016)
23. Kim, K.I., Kwon, Y.: Single-image super-resolution using sparse regression and natural image prior. IEEE Trans. Pattern Anal. Mach. Intell. **6**, 1127–1133 (2010)
24. Kostrzewa, D., Piechaczek, S., Hrynczenko, K., Benecki, P., Nalepa, J., Kawulok, M.: Replication data for: super-resolution reconstruction using deep learning: should we go deeper? (2019). https://doi.org/10.7910/DVN/DKSPJF
25. Kostrzewa, D., Skonieczny, Ł., Benecki, P., Kawulok, M.: B4MultiSR: a benchmark for multiple-image super-resolution reconstruction. Commun. Comput. Inf. Sci. **928**, 361–375 (2018)
26. LeCun, Y.A., Bottou, L., Orr, G.B., Müller, K.-R.: Efficient BackProp. In: Montavon, G., Orr, G.B., Müller, K.-R. (eds.) Neural Networks: Tricks of the Trade. LNCS, vol. 7700, pp. 9–48. Springer, Heidelberg (2012). https://doi.org/10.1007/978-3-642-35289-8_3
27. Ledig, C., et al.: Photo-realistic single image super-resolution using a generative adversarial network. In: 2017 IEEE Conference on Computer Vision and Pattern Recognition (CVPR), pp. 105–114. IEEE (2017)
28. Li, F., Jia, X., Fraser, D.: Universal HMT based super resolution for remote sensing images. In: 15th IEEE International Conference on Image Processing 2008, ICIP 2008, pp. 333–336. IEEE (2008)
29. Lim, B., Son, S., Kim, H., Nah, S., Lee, K.M.: Enhanced deep residual networks for single image super-resolution. In: The IEEE Conference on Computer Vision and Pattern Recognition (CVPR) Workshops, vol. 1, pp. 136–144 (2017)
30. Nasrollahi, K., Moeslund, T.B.: Super-resolution: a comprehensive survey. Mach. Vis. Appl. **25**(6), 1423–1468 (2014)
31. Saxe, A.M., McClelland, J.L., Ganguli, S.: Exact solutions to the nonlinear dynamics of learning in deep linear neural networks. arXiv preprint arXiv:1312.6120 (2013)
32. Shan, Q., Li, Z., Jia, J., Tang, C.K.: Fast image/video upsampling. ACM Trans. Graph. (TOG) **27**(5), 153 (2008)
33. Shi, W., et al.: Real-time single image and video super-resolution using an efficient sub-pixel convolutional neural network. In: Proceedings of the IEEE Conference on Computer Vision and Pattern Recognition, pp. 1874–1883 (2016)
34. Srivastava, R.K., Greff, K., Schmidhuber, J.: Highway networks. arXiv preprint arXiv:1505.00387 (2015)
35. Sun, J., Xu, Z., Shum, H.Y.: Image super-resolution using gradient profile prior. In: IEEE Conference on Computer Vision and Pattern Recognition 2008, CVPR 2008, pp. 1–8. IEEE (2008)

36. Tai, Y., Yang, J., Liu, X.: Image super-resolution via deep recursive residual network. In: 2017 IEEE Conference on Computer Vision and Pattern Recognition (CVPR), pp. 2790–2798. IEEE (2017)
37. Tai, Y., Yang, J., Liu, X., Xu, C.: MemNet: a persistent memory network for image restoration. In: Proceedings of the IEEE Conference on Computer Vision and Pattern Recognition, pp. 4539–4547 (2017)
38. Tai, Y.W., Liu, S., Brown, M.S., Lin, S.: Super resolution using edge prior and single image detail synthesis. In: 2010 IEEE Computer Society Conference on Computer Vision and Pattern Recognition, pp. 2400–2407. IEEE (2010)
39. Xu, B., Wang, N., Chen, T., Li, M.: Empirical evaluation of rectified activations in convolutional network. arXiv preprint arXiv:1505.00853 (2015)
40. Yue, L., Shen, H., Li, J., Yuan, Q., Zhang, H., Zhang, L.: Image super-resolution: the techniques, applications, and future. Signal Process. **128**, 389–408 (2016)
41. Zeyde, R., Elad, M., Protter, M.: On single image scale-up using sparse-representations. In: Boissonnat, J.D., et al. (eds.) Curves and Surfaces 2010. LNCS, vol. 6920, pp. 711–730. Springer, Heidelberg (2012). https://doi.org/10.1007/978-3-642-27413-8_47
42. Zhang, K., Zuo, W., Gu, S., Zhang, L.: Learning deep CNN denoiser prior for image restoration. In: IEEE Conference on Computer Vision and Pattern Recognition, vol. 2, pp. 3929–3938 (2017)

Application of Fixed Skipped Steps Discrete Wavelet Transform in JP3D Lossless Compression of Volumetric Medical Images

Roman Starosolski[✉]

Institute of Informatics, Silesian University of Technology,
Akademicka 16, 44-100 Gliwice, Poland
roman.starosolski@polsl.pl

Abstract. In this paper, we report preliminary results of applying a step skipping to the discrete wavelet transform (DWT) in lossless compression of volumetric medical images. In particular, we generalize the two-dimensional (2D) fixed variants of skipped steps DWT (SS-DWT), which earlier were found effective for certain 2D images, to a three-dimensional (3D) case and employ them in JP3D (JPEG 2000 standard extension for 3D data) compressor. For a set of medical volumetric images of modalities CT, MRI, and US, we find that, by adaptively selecting 3D fixed variants of SS-DWT, we may improve the JP3D bitrates in an extent competitive to much more complex modifications of DWT and JPEG 2000.

Keywords: Medical imaging · Image processing ·
Lossless image compression · Medical image compression ·
Volumetric image compression · Image compression standards ·
JPEG 2000 · JP3D · DWT · RDLS · Step skipping · Fixed SS-DWT

1 Introduction

The reversible variant of the discrete wavelet transform (DWT) is used in the lossless JPEG 2000 compression to decompose an image into subbands of different characteristics, that are then independently entropy coded [16,33]. In [27] we noticed, that the lifting steps (LS) [7,32] employed by DWT may propagate noise between subbands and worsen the compression effects. To limit the noise propagation we proposed the reversible denoising and lifting step (RDLS), which is built based on LS. It prevents the noise propagation by exploiting denoising filters while retaining other desirable effects and properties of LS (like the perfect reversibility). We found that RDLS improves the compression effects of algorithms exploiting DWT [27] and color space transforms [25]. In [28] we showed, for a color space transform consisting only of LSs, that by applying RDLS with a special denoising filter we may practically skip selected LSs of the transform

© Springer Nature Switzerland AG 2019
S. Kozielski et al. (Eds.): BDAS 2019, CCIS 1018, pp. 217–230, 2019.
https://doi.org/10.1007/978-3-030-19093-4_17

or the entire transform. Results obtained in [27] suggested, that a partial skipping of DWT may be worthwhile. Since DWT consists not only of LSs but also of the non-lifting reorder step, in [29] we proposed the skipped steps DWT (SS-DWT) obtained from DWT by skipping selected steps of its computation. To construct SS-DWT in an image-adaptive way, we employed a heuristic and, based on its outcomes, we defined two simple fixed SS-DWT variants FIX1 and FIX2. SS-DWT resulted in compression ratio (bitrate) improvements similar to the RDLS-modified DWT (RDLS-DWT), but obtained at a smaller cost. The fixed variants are especially interesting from a practical standpoint because they allow obtaining high bitrate improvements at a very small cost and are compliant to the JPEG 2000 part 2 standard [13,30].

So far, the step skipping was applied in compression of two-dimensional (2D) images only. In this paper, we report the preliminary results of applying it to the 3-dimensional DWT (3D-DWT) in lossless compression of volumetric medical images. In particular we generalize the FIX1 and FIX2 variants to a 3D case and apply them in JP3D, i.e., volumetric extension of JPEG 2000 standard [4,14].

The task of improving the lossless JP3D bitrates of volumetric medical images is not simple and not many attempts are reported in the literature. Two interesting approaches were evaluated in [4]. In that study, the direction-adaptive DWT (DA-DWT), earlier used for 2D data [5,8,9], was applied to volumetric medical images. Also the other approach, the block-based intra-band prediction of DWT transformed subbands (JP3D+BP), was an adaptation of the method presented in [24] for 2D images. These approaches improve average bitrates of JP3D for medical volumes by less than 1% at a cost of a high increase in the compression process complexity. We will compare our results with findings reported in [4] and, for this purpose, we will use the same set of test data. We note, that video coding algorithms, like the state-of-the-art High Efficiency Video Coding (HEVC) standard [15,35], may be applied to medical volumes treated as sequences of volume slices, however, the average lossless HEVC bitrate for the aforementioned set is inferior to the JP3D bitrate [4].

We also note, that RDLS and step skipping may technically be applied to lossy compression, however, to our best knowledge, no such attempt has been undertaken. An application of lossy compression to medical images is still disputable, guidelines issued by various professional bodies recommend different lossy compression ratios, and in some cases lossy compression of such images is forbidden by law [6,19]. For some modalities the compression algorithm and lossy compression ratio utilized may have a negative impact on the results of automatic analyses of the decompressed images (e.g., in biomedical applications exploiting texture operators [18,22]).

The remainder of this paper is organized as follows. In subsections of the next Section, we first briefly describe 2D- and 3D-DWT (Sect. 2.1) and the modifications of DWT investigated earlier: RDLS-DWT, SS-DWT, and fixed variants FIX1 and FIX2 of 2D-SS-DWT (Sect. 2.2). Then, in Sect. 2.3, we propose 3D variants of FIX1 and FIX2. Section 2.4 contains the experimental procedure including the description of the test images and the used implementations. The experimental results are presented and discussed in Sect. 3. Section 4 summarizes the findings.

2   Materials and Methods

2.1   Lifting-Based Discrete Wavelet Transform

DWT employed in image compression algorithms decomposes an image into subbands of different characteristics (they represent image details of different orientations and sizes). A subband is easier to encode efficiently than the original image because its well-defined characteristics allow for better statistical modeling. The decomposition into subbands has many additional advantages for the lossless and lossy coding, e.g., it allows for various kinds of progressive coding. For brevity, as in the previous works [29, 30], we describe here only the lifting-based reversible DWT with Cohen Daubechies Faveau (5,3) wavelet filter, reduced to essentials. Among others, it is exploited in lossless JPEG 2000 compression of 2D and volumetric images. For further details, as well as for more general characteristics of various variants of DWT, their application in JPEG 2000, and the JPEG 2000 standard, the reader is referred to [1, 4, 13, 14, 16, 33].

The one-dimensional DWT (1D-DWT) transforms in-place a discrete signal $S = s_0 s_1 s_2 \ldots s_{l-1}$ of finite length l into two subbands:

- a low-pass filtered signal L, that represents the low-frequency features of S;
- a high-pass filtered signal H containing high-frequency features that, along with the L signal, allows the perfect reconstruction of the original signal.

The transformation of S is done in 3 steps. First, in the prediction step, we perform the high-pass filtering of odd samples (hereafter, the parity of the sample is determined by its location and not its value) by applying to each of them the LS presented in below Eq. 1:

$$s_x \leftarrow s_x - \lfloor (s_{x-1} + s_{x+1})/2 \rfloor. \tag{1}$$

In each LS a single signal sample is modified by adding to it a linear combination of other samples (in a general case the sum may be negated). A transform performed as a sequence of LSs has advantageous properties: it may be computed in-place and it is easily and perfectly invertible.

Another LS (Eq. 2) is then, during the update step, applied to each even sample:

$$s_x \leftarrow s_x + \lfloor (s_{x-1} + s_{x+1} + 2)/4 \rfloor. \tag{2}$$

Finally, in the reorder step, we reposition even samples to the lower half of the original signal, preserving their ordering (sample s_x is moved to $s_{x/2}$), and odd samples are moved to the upper half. We obtain separate subbands L and H, respectively. The reorder step is not an LS.

The 2D-DWT of an image is obtained by first applying 1D-DWT to each image column, which results in L and H subbands of the image (Fig. 1A–B).

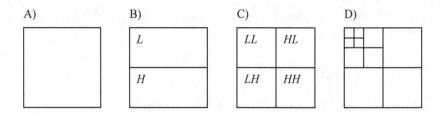

Fig. 1. 1-level 2D-DWT (A–C) and 3-level 2D-DWT (D)

Then, by applying 1D-DWT to each row, we obtain the 1-level DWT consisting of LL and HL subbands (transformed from L subband) and LH and HH subbands (from H subband)—see Fig. 1C.

The 3D-DWT of a volumetric image is obtained by first applying 1D-DWT in the axial direction, which produces two (volumetric) subbands L and H (Fig. 2A–B). Next, we proceed like in the 2D-DWT. That is, we apply 1D-DWT to the volume in the vertical direction (Fig. 2C), which results in the LL and HL subbands (transformed from the L subband) and LH and HH subbands (from the H subband). Finally we apply 1D-DWT horizontally obtaining the 1-level 3D-DWT of the volume (Fig. 2D), that consists of 8 subbands: LLL and HLL (transformed from the LL subband), LHL and HHL (from HL), LLH and HLH (from LH), and LHH and HHH (from HH). In Fig. 2B–D each subband is labeled with a number (in round brackets) of the group of steps (GS) that has the most recently been used to filter all the samples contained in the subband. For instance, GS 4 in Fig. 2C corresponds to all the prediction LSs employed when transforming the H subband into LH and HH. We will use GSs in defining 3D fixed variants of SS-DWT.

The higher-level DWT, that provides multiresolution image representation, is obtained by Mallat decomposition [20]. The $t+1$-level transform is obtained by applying the 1-level transform to the low-pass subband of the t-level transform, i.e., to the LL in the cases of 2D-DWT (Fig. 1D) and LLL for 3D-DWT (Fig. 2E).

2.2   Reversible Denoising and Lifting Steps and Step Skipping

The idea of skipping selected steps of a transform originates from RDLS. An unwanted side effect of LS is that the sample being modified by LS (filtered by LS in the case of DWT) gets contaminated by noise from other samples. In the case of DWT, the noise gets propagated between subbands. JPEG 2000 encodes the DWT subbands independently; the noise propagation increases the amount of information that has to be encoded and thus worsens the compression effects. RDLS is a modification of LS (and consequently of the lifting-based transform) that integrates LS with denoising filters in order to avoid noise propagation while preserving other properties of LS (and of the lifting-based transform). For instance, in RDLS-DWT the prediction (Eq. 1) and update (Eq. 2) steps are

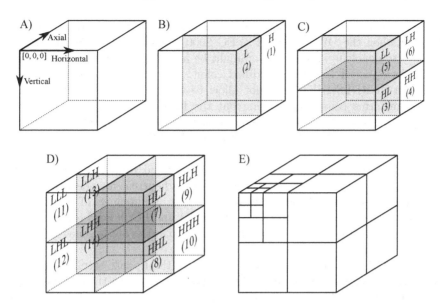

Fig. 2. 1-level 3D-DWT (A–D) and 3-level 3D-DWT (E); coordinate system presented in panel (A), in round brackets presented are GS numbers

replaced by RDLSs constructed based on them, i.e., by:

$$s_x \leftarrow s_x - \lfloor (s_{x-1}^d + s_{x+1}^d)/2 \rfloor \text{ and} \tag{3}$$

$$s_x \leftarrow s_x + \lfloor (s_{x-1}^d + s_{x+1}^d + 2)/4 \rfloor, \tag{4}$$

respectively, where s_i^d denotes the denoised sample s_i.

RDLS was successfully applied to DWT [27] and to reversible color space transforms—first [25] to a simple RDgDb [26] and then [28] to more complex color space transforms: LDgEb [26], RCT [16], and YCoCg-R [21]. Both in the cases of reversible color space transforms (for JPEG-LS [12,34], JPEG 2000, and JPEG XR [10,17]) and the DWT exploited in lossless JPEG 2000, the application of RDLS resulted in practically useful improvements of image compression ratios of certain types of images. For example, in the latter case the lossless compression ratios of non-photographic images were improved by about 14% [27].

RDLS has very interesting properties. Despite exploiting the inherently irreversible denoising, it is perfectly reversible. The RDLS-modified transform is more general than the original one. Among others, by employing special denoising filters we may obtain, as a special case of the RDLS-modified transform, the unmodified transform. Such a special filter case denoted None, for which $s_i^d = s_i$, was proposed and investigated in [27]. Another special filter, the Null filter proposed in [28] (for which $s_i^d = 0$), allows to practically skip the step. For further details and properties of the RDLS approach we refer the reader to [27,28].

In [27] we found that the noise filtering in RDLS-DWT was the most effective in improving lossless JPEG 2000 bitrates when applied during computing of some RDLS-DWT subbands only and since in some cases the best bitrates were obtained when the DWT stage of JPEG 2000 was skipped, we suspected that similarly to denoising, the optimum might be in-between skipping and applying DWT. As opposed to RDLS-modified color space transforms, employing the Null filter in RDLS-DWT does not allow to skip the entire transform. Color space transforms consist of nothing but the lifting steps, whereas DWT and RDLS-DWT, besides the lifting prediction and update steps (that may be turned into $s_x \leftarrow s_x$ by employing the Null filter in RDLS-DWT), perform the sample reordering steps.

Therefore in [29] we proposed SS-DWT obtained from DWT by skipping selected steps of its computation. We employed a heuristic for selecting steps to be skipped in an image-adaptive way and, based on outcomes of the heuristic, defined two simple fixed SS-DWT variants FIX1 and FIX2. For brevity, we refer the reader to [29] for a detailed description and properties of the general SS-DWT case and describe below only the fixed variants.

In the FIX1 variant of SS-DWT we skip all the update steps. In FIX2 we skip all the update steps (as in FIX1) and additionally skip the prediction step for the HH subband as well as the reorder step for HH and LH. Noteworthy, FIX2 results in a decomposition of an image into fewer subbands than the regular DWT or FIX1 because the HH and LH subbands are not created from the H subband that remains unchanged after the transform—see Fig. 3. As we noticed in [29] and verified experimentally in [30], the FIX1 and FIX2 variants of SS-DWT are compliant with the JPEG 2000 part 2 by exploiting standard extensions defined in annexes H (FIX1) or F and H (FIX2) [13].

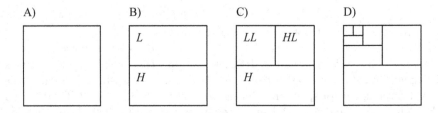

Fig. 3. 1-level FIX2 variant of SS-DWT (A–C) and 3-level FIX2 (D)

The most interesting results, from a practical standpoint, were obtained by applying entropy estimation of JPEG 2000 coding effects for selecting among the fixed SS-DWT variants, the unmodified DWT, and the skipping of the DWT stage of JPEG 2000. The average bitrate improvement due to selecting among the above-mentioned fixed variants was similar to that of RDLS-DWT (roughly 14% for non-photographic images), but it was obtained at a significantly smaller cost; the overall compression time was only 3% greater than that of the unmodified JPEG 2000. We remark, that so far the research on RDLS has been carried out using very simple denoising filters only (we used simple RCRS filters [11] like the median filter). We suspect, that RDLS effects could be improved by the use of more sophisticated filters, e.g., based on detector precision characteristics or operating in the transform domain [2,3]. In [29] we also found, that by combining SS-DWT and RDLS-DWT even greater bitrate improvements (of up to about 17.5% for non-photographic images) might be obtained at a significantly increased cost of heuristic-based selecting, based on the actual bitrate instead of an estimated one, of the steps to be skipped and the denoising filters. However, we do not expect so high bitrate improvements in the case of volumetric medical images, because the characteristics of these images resemble characteristics of photographic images, for which the JPEG 2000 bitrate improvement due to applying RDLS-DWT and SS-DWT was below 1%.

2.3   Fixed Variants of Three-Dimensional Skipped Steps Discrete Wavelet Transform

2D-DWT may be obtained as a special case of 3D-DWT when the volume size in the axial direction is 1. Thus, 3D variants of fixed SS-DWT, should be defined in such a way, that for the above volume they become their 2D counterparts (FIX1 and FIX2) and in general follow the same principles (when to skip the step) as their 2D counterparts. A three-dimensional variant of FIX1 (3D-FIX1), naturally, is obtained from 3D-DWT by skipping all the update steps. In FIX2 we additionally skipped prediction of the HH subband (i.e., of samples that had already been subjected to prediction), and reordering of samples from HH and LH (i.e., we skipped reordering of samples if for each of the signal samples to be reordered during 1D-DWT the prediction or update has been skipped). Fortunately, this may be extended to the 3D case. In 3D-FIX2 variant of 3D-SS-DWT (Fig. 4) we skip

- all the update steps,
- prediction for HH and reorder for HH and LH (the H subband of the volume does not get decomposed into LH and HH, once created, the H subband remains unchanged after the transform), and
- prediction for HHL and reorder for HHL and LHL (the HL subband of the volume does not get decomposed into LHL and HHL and remains unchanged after the transform).

Some of the volumes we used for evaluation in this paper use the same resolution in all directions, whereas for others a different resolution is used in the axial direction. Another case, when the axial direction should be treated differently than others, happens when a volume data represents a time series of 2D images. Above suggests, that the pixel correlation in the axial direction may be different than in other directions. Therefore we define also fixed variants of 3D-SS-DWT that either always perform the 1D-DWT in the axial direction (variants denoted 3D-FIX1p and 3D-FIX2p) or always skip such transform (denoted 3D-FIX1s and 3D-FIX2s) and to each volume slice apply 2D FIX1 or FIX2, respectively. In Table 1, for each 3D fixed SS-DWT variant, we list which GSs of 3D-DWT (recall Fig. 2B–D) are skipped by this variant; when applying 1D-DWT, if both prediction and update steps are skipped, then the reorder step is skipped as well.

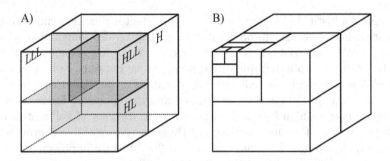

Fig. 4. 1-level 3D-FIX2 (A) and 3-level 3D-FIX2 (B)

Table 1. The investigated variants of 3D fixed SS-DWT

GS #	LS type	Sub-band	Fixed 3D-SS-DWT variant					
			3D-FIX1	3D-FIX2	3D-FIX1p	3D-FIX2p	3D-FIX1s	3D-FIX2s
1	pred.	H	+	+	+	+	-	-
2	upd.	L	-	-	+	+	-	-
3	pred.	HL	+	+	+	+	+	+
4	pred.	HH	+	-	+	+	+	+
5	upd.	LL	-	-	-	-	-	-
6	upd.	LH	-	-	-	-	-	-
7	pred.	HLL	+	+	+	+	+	+
8	pred.	HHL	+	-	+	-	+	-
9	pred.	HLH	+	-	+	+	+	+
10	pred.	HHH	+	-	+	-	+	-
11	upd.	LLL	-	-	-	-	-	-
12	upd.	LHL	-	-	-	-	-	-
13	upd.	LLH	-	-	-	-	-	-
14	upd.	LHH	-	-	-	-	-	-

GS and subband – see Fig. 2B–D, + – perform the LS, - – skip the LS.

Variants 3D-FIX2, 3D-FIX2p, 3D-FIX1s, and 3D-FIX2s result in decomposition of a volumetric image into fever subbands, than 3D-DWT. Only some of the 3D fixed SS-DWT variants are compliant to the JPEG 2000 standard and may be obtained without altering the standard implementation. 3D-FIX1 may be obtained by exploiting JPEG 2000 part 2 Annex H (i.e., the arbitrary wavelet transform kernels) supported by JP3D (although not supported by the implementation we used in the research reported herein). Unfortunately, the JPEG 2000 part 2 Annex F that would allow obtaining 3D-FIX2 is not supported by JP3D. Out of the remaining 3D fixed SS-DWT variants, only the 3D-FIX1s may be obtained in compliance with the standard by using JPEG 2000 part 2 extension defined in Annex H and 0-level decomposition in the axial direction (allowed by JP3D).

2.4   Procedure

We investigated the effects of 3D fixed SS-DWT variants on compression ratios of the JP3D compressor. For SS-DWT variants we employed the RDLS-SS-DWT version 0.9[1]—our implementation of SS-DWT (and of RDLS-DWT). RDLS-SS-DWT is available as a patch to the IRIS-JP3D version 1.1.1[2]—a JPEG 2000 part 10 [4,14] reference software, developed by Tim Bruylants from Vrije Universiteit Brussel (VUB) and the Interdisciplinary Institute for BroadBand Technology (IBBT). In our implementation, the SS-DWT variants were obtained by modifying the DWT stage of JP3D; the transformed subbands were compressed by further JP3D stages as if the regular 3D-DWT was applied. For instance, 3D-FIX2 skips 1D-DWT in vertical direction for the HL subband not decomposing it into LHL and HHL (see Fig. 4A and Fig. 2D). In our implementation, the JP3D statistical modeling for arithmetic coding of HL symbols was divided into two stages between which the model was reinitialized. This could slightly worsen the obtained bitrates—only slightly because, as compared to the 2D image compression [30], the subbands of volumetric images are much larger and the impact on bitrate, of the unnecessary re-adaptation of the model, is correspondingly smaller.

In experiments, we used the set of medical volumetric images, that was earlier used in [4] for evaluation of other modifications of 3D-DWT and was made available to us thanks to the courtesy of Tim Bruylants. The set contains 11 images of various modalities:

- 6 computed tomography scans (CT1...CT6) of 12-bit depth and sizes (width × height × depth, in pixels) from $512 \times 512 \times 44$ to $512 \times 512 \times 672$,
- 3 magnetic resonance imaging scans (MRI1...MRI3) of 12-bit depth and sizes from $256 \times 256 \times 100$ to $432 \times 432 \times 250$, and
- 2 ultrasound volumes (US1 and US2) of 8-bit depth and sizes $500 \times 244 \times 201$ and $352 \times 242 \times 136$, respectively.

Images are described in detail in [4].

The compression ratio or bitrate r, expressed in bits per pixel (bpp), is calculated using the total size in Bytes of the compressed volumetric image including the compressed file format header. The bitrate is directly proportional to the compressed file size, hence smaller bitrate means better compression result. The effects of 3D fixed SS-DWT variants on JP3D bitrate were analyzed based on bitrate changes with respect to the reference bitrate obtained using unmodified 3D-DWT. The bitrate change Δr was expressed in percentage of the reference bitrate.

---

[1] http://sun.aei.polsl.pl/~rstaros/rdls-ss-dwt/.

[2] http://www.irissoftware.be/.

Besides comparing bitrates obtained using the 3D fixed SS-DWT variants presented in Table 1 to the bitrate of unmodified 3D-DWT, we also tested effects of skipping the transform stage (NO-DWT). For DWT and SS-DWT, we used the 3-level decomposition (0-level for NO-DWT). Except for setting the transform variant and the decomposition level, we invoked the compressor with default parameters.

3   Results and Discussion

In Table 2 we present JP3D bitrates (r) for volumetric images from our test-set and bitrate changes (Δr) obtained by replacing 3D-DWT with 3D fixed SS-DWT variants listed in Table 1 or by skipping the DWT stage. For the latter case, we see that skipping the DWT stage results in a significant bitrate worsening for all images, the average bitrate increase is about 57%. These results are consistent with the earlier research on 2D-SS-DWT [29,30], where although omitting DWT allowed for improving the bitrates of some of the images, but almost exclusively they were non-photographic images (e.g., computer-generated, composed from others including natural ones, or screen content images [23]). In the above research, for evaluating the SS-DWT effects on single-component 2D image compression, the green components of images from a large CT2 set introduced in [31] were used. For photographic images (typical natural continuous-tone images) that have characteristics resembling medical volumes, by skipping the DWT stage an improvement was obtained for 1 out of 499 such images. Also the SS-DWT variants, in which 1D-DWT in the axial direction is skipped for the entire volumetric image (3D-FIX1s and 3D-FIX2s), worsen bitrates for all images, although to a lesser extent (by 11.06% and 12.47%, respectively). We generally see that DWT is an appropriate transform for the preparation of volumetric medical image data for further stages of compression.

However, skipping some transform steps only or even skipping 1D-DWT in a certain direction, but only for a part of the volumetric image being transformed, may result in bitrate improvements. Figure 5 shows the cases for which a bitrate improvement was obtained by using one of the proposed variants of 3D fixed SS-DWT. The best results were obtained for 3D-FIX1p—this variant improves bitrates of 4 images (CT4...CT6 and MR1) on average by 1.23%. Note, that this variant applied to all images worsens the bitrate by 0.09% on average. Therefore, in order to obtain an improvement of the JP3D bitrate with the use of 3D fixed SS-DWT variants, the effects of applying the transform variant to a given volume should be checked (or estimated) and SS-DWT should be applied only if it results in a bitrate improvement. By using 3D-FIX1p in such image-adaptive way, we get an average bitrate improvement for the whole set of images by 0.45%.

The remaining variants (3D-FIX1, 3D-FIX2, and 3D-FIX2p) allow improving bitrates of fewer images (each time being an image, whose bitrates are improved by 3D-FIX1p). For some images, choosing instead of 3D-FIX1p a different 3D fixed SS-DWT variant gives greater bitrate savings. By adaptively choosing the best variant for the image (for images from the exploited test set, the choice

Table 2. JP3D bitrate changes due to applying various 3D fixed SS-DWT variants

Image	3D-DWT r (bpp)	NO-DWT Δr	3D-FIX1 Δr	3D-FIX2 Δr	3D-FIX1p Δr	3D-FIX2p Δr	3D-FIX1s Δr	3D-FIX2s Δr
CT1	4.911	53.48%	0.48%	2.65%	0.92%	1.73%	16.66%	17.82%
CT2	7.632	44.69%	1.57%	2.35%	1.17%	2.06%	9.34%	10.29%
CT3	5.437	48.50%	0.47%	3.43%	0.13%	1.58%	10.33%	12.25%
CT4	3.844	68.55%	−0.10%	5.81%	−1.56%	−0.28%	10.11%	11.36%
CT5	2.822	89.20%	−2.12%	−1.00%	−1.80%	−1.43%	17.79%	18.55%
CT6	5.029	41.70%	0.67%	4.80%	−0.64%	0.91%	5.56%	7.11%
MRI1	3.503	72.02%	−1.97%	−2.12%	−0.91%	−0.95%	15.82%	15.37%
MRI2	4.091	129.13%	0.23%	0.73%	0.77%	1.78%	31.65%	33.05%
MRI3	6.588	37.92%	4.09%	2.29%	2.52%	1.82%	7.57%	6.83%
US1	4.840	25.52%	1.01%	5.66%	0.38%	4.22%	5.25%	9.25%
US2	5.233	17.88%	1.43%	6.60%	0.01%	3.72%	1.85%	5.60%
Average	4.903	57.14%	0.52%	2.84%	0.09%	1.38%	11.06%	12.47%

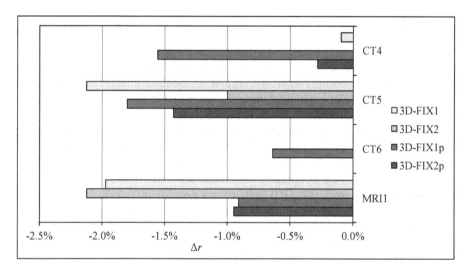

Fig. 5. Bitrate improvements obtained using 3D fixed SS-DWT variants

can be limited to 3D-FIX1, 3D-FIX2, and 3D-FIX1p) the average improvement for images, for which SS-DWT is effective, increases from 1.21% to 1.61%. The average improvement for the whole set of images increases to 0.59%.

Looking at the modalities of images we see, that the fixed SS-DWT variants are not effective for ultrasound volumes and in the case of other modalities they are effective for some images. However, the numbers of images in individual modalities are too small to draw significant conclusions with respect to the medical volumetric image modality.

The obtained bitrate improvement appears small from a practical point of view, but the result is not bad compared to other methods of improving the DWT effects for JPEG 2000 compression of continuous-tone images. In [30], using adaptively selected fixed SS-DWT variants for the 499 aforementioned 2D photographic images, an average improvement of 0.62% was obtained for the baseline JPEG 2000; exploiting the extensions of part 2 of JPEG 2000 standard resulted in an improvement of 0.68%. In [29] we showed that the improvement achieved using the general SS-DWT case is greater than for the fixed variants and can be further increased by combining SS-DWT with RDLS-DWT (for the aforementioned 2D images, the improvement was 0.92%). Then, by using only the general SS-DWT case, but with an improved heuristic of the adaptive selection of steps to be skipped, an improvement of 0.98% was obtained. In [4], for the same set of volumetric medical images that is used in this paper, by using much more complex modifications of 3D variants of DWT and JPEG 2000 (i.e., the DA-DWT and JP3D+BP mentioned in Sect. 1), the bitrates were improved only little more, than by using the adaptively selected 3D fixed SS-DWT variants. JP3D+BP was more effective than DA-DWT and resulted in a bitrate improvement of 0.87%.

4   Conclusion

In this paper, we report preliminary results of applying the step skipping to 3D-DWT in lossless compression of volumetric medical images. In particular, we generalized the FIX1 and FIX2 variants of SS-DWT to the 3D case (obtaining 6 fixed variants 3D-FIX1, 3D-FIX2, 3D-FIX1p, 3D-FIX2p, 3D-FIX1s, and 3D-FIX2s) and employed them in the JP3D compressor. We evaluated these variants on a set of 11 medical volumetric images (modalities CT, MRI, and US) and found, that 3D-FIX1, 3D-FIX2, 3D-FIX1p, and 3D-FIX2p allowed for bitrate improvements. By adaptively selecting a fixed 3D variant of SS-DWT for each image, we improved the JP3D bitrate on average by 0.59%. This result is not impressive form a practical standpoint, but it is competitive to much more complex modifications of DWT and JPEG 2000. The results we obtained for 3D data along with earlier results for 2D images suggest, that further JP3D bitrate improvements are possible by using an adaptively constructed general case SS-DWT, probably also combined with RDLS, which is currently being investigated.

Acknowledgment. This work was supported by the 02/020/BK_18/0128 grant from the Institute of Informatics, Silesian University of Technology.

References

1. Addison, P.S.: The Illustrated Wavelet Transform Handbook: Introductory Theory and Applications in Science, Engineering, Medicine and Finance. CRC Press, Boca Raton (2017)
2. Bernas, T., Starosolski, R., Robinson, J.P., Rajwa, B.: Application of detector precision characteristics and histogram packing for compression of biological fluorescence micrographs. Comput. Methods Prog. Biomed. **108**(2), 511–523 (2012). https://doi.org/10.1016/j.cmpb.2011.03.012

3. Bernas, T., Starosolski, R., Wójcicki, R.: Application of detector precision charac-teristics for the denoising of biological micrographs in the wavelet domain. Biomed. Sig. Process. Control **19**, 1–13 (2015). https://doi.org/10.1016/j.bspc.2015.02.010

4. Bruylants, T., Munteanu, A., Schelkens, P.: Wavelet based volumetric medical image compression. Sig. Process. Image Commun. **31**, 112–133 (2015). https://doi.org/10.1016/j.image.2014.12.007

5. Chang, C., Girod, B.: Direction-adaptive discrete wavelet transform for image com-pression. IEEE Trans. Image Process. **16**(5), 1289–1302 (2007). https://doi.org/10.1109/TIP.2007.894242

6. Clunie, D.: What is different about medical image compression? IEEE Commun. Soc. MMTC E-Lett. **6**(7), 31–37 (2011)

7. Daubechies, I., Sweldens, W.: Factoring wavelet transforms into lifting steps. J. Fourier Anal. Appl. **4**(3), 247–269 (1998). https://doi.org/10.1007/BF02476026

8. Ding, W., Wu, F., Wu, X., Li, S., Li, H.: Adaptive directional lifting-based wavelet transform for image coding. IEEE Trans. Image Process. **16**(2), 416–427 (2007). https://doi.org/10.1109/TIP.2006.888341

9. Dong, W., Shi, G., Xu, J.: Adaptive nonseparable interpolation for image com-pression with directional wavelet transform. IEEE Sig. Process. Lett. **15**, 233–236 (2008). https://doi.org/10.1109/LSP.2007.914929

10. Dufaux, F., Sullivan, G.J., Ebrahimi, T.: The JPEG XR image coding standard. IEEE Sig. Process. Mag. **26**(6), 195–204 (2009). https://doi.org/10.1109/MSP.2009.934187

11. Hardie, R.C., Barner, K.E.: Rank conditioned rank selection filters for signal restoration. IEEE Trans. Image Process. **3**(2), 192–206 (1994). https://doi.org/10.1109/83.277900

12. ISO/IEC, ITU-T: Information technology - lossless and near-lossless compression of continuous-tone still images - Baseline (2000). ISO/IEC International Standard 14495-1 and ITU-T Recommendation T.87

13. ISO/IEC, ITU-T: Information technology - JPEG 2000 image coding system: extensions (2004). ISO/IEC International Standard 15444-2 and ITU-T Recom-mendation T.801

14. ISO/IEC, ITU-T: Information technology - JPEG 2000 image coding system: extensions for three-dimensional data (2011). ISO/IEC International Standard 15444-10 and ITU-T Recommendation T.809

15. ISO/IEC, ITU-T: Information technology - high efficiency coding and media deliv-ery in heterogeneous environments - Part 2: high efficiency video coding (2015). ISO/IEC International Standard 23008-2 and ITU-T Recommendation H.265

16. ISO/IEC, ITU-T: Information technology - JPEG 2000 image coding system: core coding system (2016). ISO/IEC International Standard 15444–1 and ITU-T Rec-ommendation T.800

17. ISO/IEC, ITU-T: Information technology - JPEG XR image coding system - image coding specification (2016). ISO/IEC International Standard 29199-2 and ITU-T Recommendation T.832

18. Janhom, A., van der Stelt, P., Sanderink, G.: A comparison of two compression algorithms and the detection of caries. Dentomaxillofacial Radiol. **31**(4), 257–263 (2002). https://doi.org/10.1038/sj.dmfr.4600704

19. Liu, F., Hernandez-Cabronero, M., Sanchez, V., Marcellin, M., Bilgin, A.: The current role of image compression standards in medical imaging. Information **8**(4), 131 (2017). https://doi.org/10.3390/info8040131

20. Mallat, S.: A theory for multiresolution signal decomposition: the wavelet representation. IEEE Trans. Pattern Anal. Mach. Intell. **11**, 674–693 (1998). https://doi.org/10.1109/34.192463

21. Malvar, H.S., Sullivan, G.J., Srinivasan, S.: Lifting-based reversible color transformations for image compression. In: Applications of Digital Image Processing XXXI, Proceedings of SPIE, vol. 7073, p. 707307 (2008). https://doi.org/10.1117/12.797091

22. Obuchowicz, R., Nurzynska, K., Obuchowicz, B., Urbanik, A., Piorkowski, A.: Caries detection enhancement using texture feature maps of intraoral radiographs. Oral Radiol. (2018). https://doi.org/10.1007/s11282-018-0354-8

23. Peng, W.H., et al.: Overview of screen content video coding: technologies, standards, and beyond. IEEE J. Emerg. Sel. Top. Circ. Syst. **6**(4), 393–408 (2016). https://doi.org/10.1109/JETCAS.2016.2608971

24. Sanchez, V., Abugharbieh, R., Nasiopoulos, P.: Symmetry-based scalable lossless compression of 3D medical image data. IEEE Trans. Med. Imaging **28**(7), 1062–1072 (2009). https://doi.org/10.1109/TMI.2009.2012899

25. Starosolski, R.: Reversible denoising and lifting based color component transformation for lossless image compression. arXiv:1508.06106 [cs.MM] (2016). http://arxiv.org/abs/1508.06106

26. Starosolski, R.: New simple and efficient color space transformations for lossless image compression. J. Visual Commun. Image Represent. **25**(5), 1056–1063 (2014). https://doi.org/10.1016/j.jvcir.2014.03.003

27. Starosolski, R.: Application of reversible denoising and lifting steps to DWT in lossless JPEG 2000 for improved bitrates. Sig. Process. Image Commun. **39**(A), 249–263 (2015). https://doi.org/10.1016/j.image.2015.09.013

28. Starosolski, R.: Application of reversible denoising and lifting steps with step skipping to color space transforms for improved lossless compression. J. Electron. Imaging **25**(4), 043025 (2016). https://doi.org/10.1117/1.JEI.25.4.043025

29. Starosolski, R.: Skipping selected steps of DWT computation in lossless JPEG 2000 for improved bitrates. PLOS ONE **11**(12), e0168704 (2016). https://doi.org/10.1371/journal.pone.0168704

30. Starosolski, R.: A practical application of skipped steps DWT in JPEG 2000 Part 2-compliant compressor. In: Kozielski, S., Mrozek, D., Kasprowski, P., Małysiak-Mrozek, B., Kostrzewa, D. (eds.) BDAS 2018. CCIS, vol. 928, pp. 334–348. Springer, Cham (2018). https://doi.org/10.1007/978-3-319-99987-6_26

31. Strutz, T.: Multiplierless reversible colour transforms and their automatic selection for image data compression. IEEE Trans. Circ. Syst. Video Technol. **23**(7), 1249–1259 (2013). https://doi.org/10.1109/TCSVT.2013.2242612

32. Sweldens, W.: The lifting scheme: a custom-design construction of biorthogonal wavelets. Appl. Comput. Harmon. Anal. **3**, 186–200 (1996). https://doi.org/10.1006/acha.1996.0015

33. Taubman, D.S., Marcellin, M.W.: JPEG2000 Image Compression Fundamentals, Standards and Practice. Springer, New York (2004). https://doi.org/10.1007/978-1-4615-0799-4

34. Weinberger, M.J., Seroussi, G., Sapiro, G.: The LOCO-I lossless image compression algorithm: principles and standardization into JPEG-LS. IEEE Trans. Image Process. **9**(8), 1309–1324 (2000). https://doi.org/10.1109/83.855427

35. Xu, J., Joshi, R., Cohen, R.A.: Overview of the emerging HEVC screen content coding extension. IEEE Trans. Circuits Syst. Video Technol. **26**(1), 50–62 (2016). https://doi.org/10.1109/TCSVT.2015.2478706

Bioinformatics and Biomedical Data Analysis

A Novel Approach for Fast Protein Structure Comparison and Heuristic Structure Database Searching Based on Residue EigenRank Scores

Florian Heinke[1,2(✉)], Lars Hempel[1], and Dirk Labudde[1,3]

[1] Bioinformatics Group, University of Applied Sciences Mittweida,
Technikumplatz 17, 09648 Mittweida, Germany
heinke@hs-mittweida.de
[2] TU Bergakademie Freiberg, Akademiestrasse 6, 09599 Freiberg, Germany
[3] Fraunhofer Cybersecurity, Darmstadt, Germany
http://www.hs-mittweida.de/bigm

Abstract. With the rapid growth of public protein structure databases, computational techniques for storing as well as comparing proteins in an efficient manner are still in demand. Proteins play a major role in virtually all processes in life, and comparing their three-dimensional structures is essential to understanding the functional and evolutionary relationships between them.

In this study, a novel approach to compute three-dimensional protein structure alignments by means of so-called EigenRank score profiles is proposed. These scores are obtained by utilizing the LeaderRank algorithm—a vertex centrality indexing scheme originally introduced to infer the opinion leading role of individual actors in social networks. The obtained EigenRank representation of a given structure is not just highly-specific, but can also be used to compute profile alignments from which three-dimensional structure alignments can be rapidly deduced. This technique thus could provide a tool to rapidly scan entire databases containing thousands of structures.

Keywords: Protein structure alignment · LeaderRank · EigenRank

1 Introduction

In the process of understanding the molecular machinery facilitating life on its very basis, research is often revolving around the smallest functional molecular entity: the protein. Although being only combined from the small set of 20 canonical amino acids, proteins yield a plethora of functional roles determined by three-dimensional structures characteristic for each individual protein [14,20].

To gain knowledge about a protein's functional characteristics, analyses on the molecular structure level often involve the comparison of two or more homologous structures. Homology, in a broader sense, refers to functional similarity

© Springer Nature Switzerland AG 2019
S. Kozielski et al. (Eds.): BDAS 2019, CCIS 1018, pp. 233–247, 2019.
https://doi.org/10.1007/978-3-030-19093-4_18

that originated from a common ancestor protein from which the proteins in question evolved from. In the process, conserved functionality is apparent by their three-dimensional conformation. However, their amino acid sequences can be highly dissimilar. Comparing protein structures is thus one of the key tasks of structural bioinformatics. In this context, searching databases for significant similarities has to be carried out efficiently at high biological sensitivity [9].

Although numerous methods have been developed for this task, such as FAT-CAT [21], CE [15] and TM-align [23], searching similarities between structures efficiently remains computational demanding, due to the complexity of the problem of finding pairs of matching three-dimensional atom coordinates and deducing the optimal spatial superimposition. This problem is further exacerbated by multi-domain and polymeric architectures commonly present in protein structures, as well as, rather trivially, the large number of structures involved in such studies. For example, at the time of writing the Protein Data Bank (PDB) holds about 140,000 protein structures, of which approximately 80,000 comprise more than one peptide chain [4].

In this paper, a novel technique to align protein structures is presented. In principle, a protein structure is represented as a residue-residue interaction network. By applying multiple interaction distance cutoffs a structure-specific set of adjacency matrices is obtained from the network.[1] In Fig. 1 an example illustrating this idea is given. By employing the LeaderRank algorithm [8] to each adjacency matrix, structure-specific EigenRank scores for each residue are computed. It is demonstrated that sequences of EigenRank scores—so-called EigenRank profiles—can be aligned by means of a dynamic programming approach. Obtained EigenRank profile alignments are then used to deduce pairs of spatially matching residues from which the structural alignment can be inferred efficiently. Thus, the presented method circumvents the computational bottleneck of finding a solution to the coordinate superimposition problem in the original three-dimensional coordinate space. Instead, the three-dimensional structure space is reduced to one-dimensional EigenRank profiles. Computational speed-up in structure alignment calculation is achieved by two factors. Firstly, EigenRank profile alignments can be computed rapidly and identifying pairs of matching residues from them is straightforward. Secondly, the sets of matching residues between both structures are of the same size. This aspect greatly simplifies the problem of deducing an optimal superimposing coordinate rotation and translation. It thus can be argued that achieved computational speed-up could provide the basis for a fast structure database search heuristic.

The paper is structured as follows: before giving detailed information on how EigenRank profiles can be obtained from a protein structure, we first give an overview of the LeaderRank algorithm. Next, the dynamic programming algorithm for computing EigenRank profile alignments is elucidated. We subsequently demonstrate how structure alignments can be deduced from said

---

[1] The set of adjacency matrices can actually be used to derive one single matrix containing binned residue-residue distances. The underlying structure of the protein can be reconstructed from that matrix.

alignments. Furthermore, a strategy to benchmark biological sensitivity of the algorithm is presented. Finally, we demonstrate and discuss its performance in comparison with commonly utilized structure alignment methods as well as sequence-based homology/fold recognition techniques.

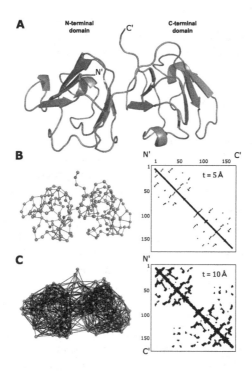

Fig. 1. Abstracting a given protein structure as a network of interacting residues, and computing adjacency matrices accordingly, forms the basis to derive a structure-specific EigenRank score profile. (A) Protein structure of the γ-B crystallin tandem domains (PDB-Id: 1amm) as cartoon representation. (B) Residue-residue interaction network and adjacency matrix computed for 1amm. C_α coordinates are used as reference points. Here, the residue-residue distance cutoff t, at which interactions are assumed, is 5 Å. (C) Interaction network and adjacency matrix at $t = 10$ Å. Note, that the spatially distant domains are clearly visible in both adjacency matrices, as each of them is indicated as a cluster of adjacent matrix elements.

2   Methods

In this section the process of computing EigenRank score profiles and comparing two given protein structures based on such is elucidated. In summary, the proposed technique involves the calculation of a residue-residue distance matrix of a given protein structure in the first step, followed by deducing adjacency matrices based on varying distance cutoffs t in the range of 5 to 15 Å. In the next step, LeaderRank score profiles are computed for each adjacency matrix, and the EigenRank score profile is obtained by deducing the first principle component from these LeaderRank profiles. Given the EigenRank profiles of two structures, a dynamic programming approach is utilized next to identify the best matching sequence between EigenRank profile scores in order to retrieve corresponding atom coordinates in both structures. These coordinates form the basis for structure alignment calculation. In Fig. 2 the process flow is illustrated as a visual guidance through these next subsections, wherein all of the involved steps are elucidated in detail. In addition, means for benchmarking performance are presented in Subsect. 2.5.

2.1   The LeaderRank Algorithm

The LeaderRank algorithm [7,8] has been originally proposed as a means to infer the opinion leading role of individuals in social networks and, thus, to identify potential opinion leaders [18]. The network is represented as a graph, whereas members and their interactions are represented as nodes and directed edges, respectively. The LeaderRank algorithm provides a node centrality measure which aims to describe the emergence of opinions and the flow of information by heuristically approximating a probabilistic infection model (for a recent comprehensive overview on this subject see [3]). In this model, the centrality of a node—herein equivalent to the corresponding member's role in information spreading—is equal to the number of nodes it is able to infect. This infection model is similar to a probabilistic breadth-first graph traversal and represents our intuitive understanding of information spreading very well. However, this model is too computationally intensive to be of practical use. This is mainly due the vast number of simulation runs to be conducted for each of the nodes, in order to approximate its infection potential. Furthermore, network-specific infection probabilities have to be meaningfully chosen or deduced from simulation.

In this heuristic approximation, the obtained LeaderRank score s_i of a given node v_i is equivalent to its infection potential. LeaderRank scores for all nodes are obtained by iteratively applying an update function $s_i^{\tau+1} = f(v_i, s_i^\tau)$ to all nodes until an adequate convergence criterion is reached. The function reads as follows:

$$s_i^{\tau+1} = \sum_{j=1}^{N+1} \frac{A_{ji}}{deg^{out}(v_j)} s_j^\tau, \tag{1}$$

where N is the number of nodes in the network, A corresponds to the adjacency matrix, and $deg^{out}(v_j)$ gives the number of outgoing edges of node v_j. In addition, a ground node v_g is introduced to the network, which is bidirectionally connected to all nodes and ensures convergence and graph connectivity. The calculation is initialized at τ_0 by setting $s_i^{\tau_0} = 1$ for all N nodes and $s_g^{\tau_0} = 0$ for the ground node. Upon reaching convergence after τ_c iterations, the final LeaderRank scores are obtained by distributing the LeaderRank score of the ground node evenly to all other nodes:

$$s_i = s_i^{\tau_c} + \frac{s_g^{\tau_c}}{N}. \tag{2}$$

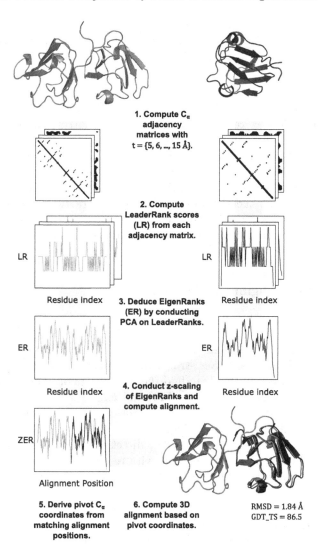

Fig. 2. Process flow of the proposed technique. In this example, the tandem structure of two γ-B crystallin domains is considered on the left side (PDB-Id: 1amm). On the right, a single γ-B crystallin structure is given (PDB-Id: 1dsl). The structural alignment obtained by EigenRank scores is illustrated in the bottom right. Both structures show good structural agreement.

2.2 EigenRank Calculation from Protein Structure Data

EigenRank scores for a given protein structure are obtained from LeaderRank score profiles computed from residue-residue interaction matrices. The latter are calculated at various interaction cutoff distances t. In this context, each interaction matrix corresponds to an adjacency matrix that can be input to the LeaderRank algorithm. In this study the $\mathbf{C}_{\alpha,i}$ coordinates of a given residue i are considered as spatial points of reference. Residue-residue adjacency A_{ji} between residues i and j is assumed, if $d(\mathbf{C}_{\alpha,j}, \mathbf{C}_{\alpha,i}) < t$. Calculations were made with t ranging from 5 to 15 Å at 1 Å increments. In the next step, all of the resulting eleven LeaderRank sets $S^t = \{s_1^t, s_2^t, \ldots, s_N^t\}$, to which we refer as LeaderRank profiles in the following text, provide an $N \times 11$ matrix that is decomposed by using PCA. The EigenRank profile $ER = \{e_1, \ldots, e_N\}$ corresponds to the scores of the first principal component along the protein sequence. Tested on the COPS query structure set (see Sect. 2.5), obtained projections on the first principal component yield 80% of relative variance on average ($sd = 3.4\%$), thus it can be argued that these scores yield highly characteristic representations of the three-dimensional residue-residue interaction landscapes and, hence, the given protein structures themselves. All EigenRank calculations in this study were made using the Bio3D and centiserve R-packages.

2.3   EigenRank Alignment Calculation

EigenRank profiles can be aligned by means of dynamic programming; an approach similar to the dynamic programming solution of the longest common subsequence problem (LCSP). Similarly to the LCSP, where the solution can be understood as editing two given strings by finding (mis-)matching characters and introducing gaps, the goal is to find the sequence of optimal matches in two EigenRank score profiles, denoted by ER^q and ER^p.

The process of EigenRank profile alignment calculation is conducted similar to the classic Needleman-Wunsch algorithm commonly used in sequence bioinformatics [10]. Note that, in order to allow direct comparison of two EigenRank profiles, both of these have to be z-scaled with respect to their arithmetic means and standard deviations. Scaled EigenRank profiles are referred to as ZER in the following text.

The algorithm proposed here involves the iterative calculation of an optimization matrix $F_{(m+1)\times(n+1)}$, where m and n are the number of residues in structures q and p, respectively. Additionally, the traceback matrix T allows to reconstruct the final alignment based on the individual steps in calculating the matrix elements $F_{i,j}$. Herein, ε acts as a positive constant to penalize the introduction of gaps. The elements in T ('up', 'left' and 'diag') indicate whether a gap is introduced to ZER^p or ZER^q, or if a match of scaled EigenRank scores ZER_{i+1}^q and ZER_{j+1}^p is accepted. The pseudo-code of the algorithm is given in Algorithm listing 1. The index offset of 1 is due to penalizing terminal gaps at the beginning of both ZER profiles.

Algorithm 1. ZER Profile Alignment

1: **function** ALIGN(ZER^q, ZER^p)
2:     $m \leftarrow |ZER^q|, n \leftarrow |ZER^p|$
3:     initialize optimization matrix $F_{(m+1)\times(n+1)}$
4:     initialize edit traceback matrix $T_{(m+1)\times(n+1)}$
5:     fill $\forall i \in [1, ..., m+1] : F_{i,1} \leftarrow -\varepsilon \cdot (i-1)$
6:     fill $\forall j \in [1, ..., n+1] : F_{1,j} \leftarrow -\varepsilon \cdot (j-1)$
7:     fill $\forall i \in [2, ..., m+1] : T_{i,1} \leftarrow$ 'up'
8:     fill $\forall j \in [2, ..., n+1] : T_{1,j} \leftarrow$ 'left'
9:     $F_{1,1} \leftarrow 0$
10:     $T_{1,1} \leftarrow$ 'stop'
11:     **for** $i \leftarrow 2, ..., m+1$ **do**
12:         **for** $j \leftarrow 2, ..., n+1$ **do**
13:             $\delta_{ij} \leftarrow 1 - |e^q_{i-1} - e^p_{j-1}|$
14:             $u \leftarrow F_{i-1,j} - \varepsilon$
15:             $l \leftarrow F_{i,j-1} - \varepsilon$
16:             $d \leftarrow F_{i-1,j-1} + \delta_{ij}$
17:             $c \leftarrow \max(u, l, d)$
18:             $F_{i,j} \leftarrow c$
19:             $T_{i,j} \leftarrow \begin{cases} \text{'up'}, & u = c \\ \text{'left'}, & l = c \\ \text{'diag'}, & \text{else} \end{cases}$
20:     infer alignment path P from traceback on T with $T_{m+1,n+1}$ as initial element
21:     introduce gaps to ZER^q and ZER^p according to P
22:     return aligned scaled EigenRank profiles ZER^q and ZER^p.

2.4   From EigenRank Alignments to Structure Alignments

Given an EigenRank profile alignment, it can be hypothesized that ungapped positions in the alignment report pairs of residues that are also of spatial correspondence in both underlying structures. Thus, from the m and n residues of the first and second structure, the alignment yields a subset of l residues that can be meaningfully aligned on the structure level. However, it has to be addressed that not all matches are necessarily of relevance to the structural alignment calculation. Such false matches can be simply caused by inaccuracies in the ZER alignment and by the strict design of the ZER alignment algorithm (f. e. using a constant gap penalty cost without taking the functional role of underlying residues into consideration). Hence, referencing all ungapped ZER alignment positions can not only lead to false assumptions of spatial correspondence between residues, but also to a completely incorrect structural alignment. Henceforth, a filter is proposed that simply assumes that a pair of aligned residues is to be neglected in deriving the structural alignment if their absolute ZER score difference is greater than a preset threshold. Benchmarking showed that a preset threshold of 0.50 yields best performance (see Sect. 3 for further information).

Upon filtering and selecting ungapped positions from the ZER alignment, the corresponding subset of l C_α coordinates is retrieved from structure p and structure q. With a well-defined set of l three-dimensional coordinates to be aligned, the problem of deriving the optimal structural alignment is straightforward by finding the best rotation and translation of one set of coordinates onto the other. This can be efficiently realized by means of the Kabsch algorithm [6].

In order to quantify structural similarity based on the resulting set of l aligned C_α coordinates, one can utilize common similarity measures, such as the root mean square deviation (RMSD) or the total global distance test score (GDT_TS) [22]. By both measures, the Euclidean distances between aligned coordinates of structure q (the set of C_α coordinates $\{\mathbf{C}^q_{\alpha,1}, \mathbf{C}^q_{\alpha,2}, \ldots, \mathbf{C}^q_{\alpha,l}\}$) and structure p, denoted by $\{\mathbf{C}^p_{\alpha,1}, \mathbf{C}^p_{\alpha,2}, \ldots, \mathbf{C}^p_{\alpha,l}\}$ are used. The RMSD, as shown in Eq. 3, is based on the arithmetic mean of spatial distances.

$$\text{RMSD} = \sqrt{\frac{1}{l} \sum_{u=1}^{l} [d(\mathbf{C}^q_{\alpha,u}, \mathbf{C}^p_{\alpha,u})]^2}. \tag{3}$$

Although the RMSD is commonly applied to quantify structural alignment quality, one should be careful as its descriptive power can be overstated. As a straightforward arithmetic mean it can be sensitive to outliers and could thus give false implications on alignment quality. Hence, additional measures, such as the GDT_TS, should be provided. The GDT_TS is the average relative fraction of Euclidean distances within preset distance bins, and is defined as:

$$\text{GDT\_TS} = \frac{1}{4} \sum_{t \in \{1,2,4,8 \text{ Å}\}} \frac{100 \cdot H_t}{l}, \tag{4}$$

where H_t gives the absolute number of aligned C_α coordinate pairs $(\mathbf{C}^q_{\alpha,u}, \mathbf{C}^p_{\alpha,u})$ with an Euclidean distance $d(\mathbf{C}^q_{\alpha,u}, \mathbf{C}^p_{\alpha,u}) < t$, and t corresponding to the preset distance bin. The GDT_TS is defined in $[0, 100]$, whereas a score of 100 indicates very high structural similarity.

Both measures were used independently to quantify performance using the COPS benchmark. In Fig. 2 the structural alignment of two γ-B crystallin structures (PDB-Ids 1amm and 1dsl) is shown. Here, the smaller structure 1dsl, shown on the right hand side of the figure, is structurally highly similar to the C-terminal domain of 1amm. This fit is also clearly visible in the ZER profile alignment, where the ZER profile of 1dsl is almost identical to the C-terminal ZER profile region in 1amm. As reported, the RMSD of the structure alignment is 1.84 Å. The GDT_TS of 86.5 underlines the good structural agreement.

2.5   The COPS Benchmark

The COPS benchmark [5, 19] provides a standardized dataset to analyze homology detection and fold recognition techniques. All structures in the dataset are single functional domains, thus consisting of only one peptide chain per structure. The dataset is comprised of two subsets: a query subset and a target subset (referred to as database in the original study as well as in the following text). The query subset comprise 176 proteins. For each of the queries, the database holds six homologous proteins, resulting to 1,056 target proteins in total. In general, the average sequence identity (19.9%, sd = 5.2%) and average similarity (32.3%, sd = 9.0%) computed from all query-target alignments are notably low, providing a sound basis to measure the performance of sequence-based homology detection methods.

As to be elucidated in the Results and Discussion section, common structure alignment techniques show superior performance on average compared to sequence-based techniques, which of course is to be expected. Yet, quite strikingly, even these tested structure alignment techniques showed minor inaccuracies in the benchmark. Thus, although originally intended to evaluate the performance of sequence-based homology detection and fold recognition methods, this benchmark is also suitable to assess structure alignment performance.

Benchmarking Strategy and Performance Measures. For benchmarking, Frank et al. [5] proposed a hide-and-seek strategy. This strategy states that any structure- or sequence-based similarity search method shows perfect performance if it is able to correctly identify all six homologue targets in the database for each query structure. In a broader sense, if said method provides a measure for similarity (or dissimilarity) scoring, an accordingly sorted score list for each query should hold the six homologue targets at the top six ranks. By means of common receiver operating characteristic measures deduced from all lists, inaccuracies and the overall performance can be quantified. In the original work [5], specificity and recall are proposed as measures for ROC analysis.

In summary, benchmarking a given similarity searching method involves the following steps:

1. For each query, the similarity to each target structure is scored and a ranked list is obtained.
2. From each of the ranked lists and each of the six top ranks $k \in \{1, 2, \ldots, 6\}$ the following figures are derived:
   - true-positives (TP_k): the number of homologs with ranks $\leq k$,
   - false-positives (FP_k): the number of non-homologs with ranks $\leq k$,
   - false-negatives (FN_k): the number of homologs with ranks $> k$.
3. TP_k, FP_k and FN_k are summed for all 176 queries and six ranks k.

With all figures obtained, recall Rec_k and specificity Spec_k can be computed for each top-six rank k according to Eqs. 5 and 6.

$$\mathrm{Rec}_k = \frac{\sum TP_k}{\sum TP_k + \sum FN_k}. \tag{5}$$

$$\mathrm{Spec}_k = 1 - FPR_k. \tag{6}$$

Here, FPR_k is the rank-specific false-positive rate, which corresponds to:

$$FPR_k = \frac{\sum FP_k}{1056}. \tag{7}$$

The derived six recall and specificity coordinates can be plotted, and performance be qualitatively assessed between utilized similarity detection methods. Additionally, in order to provide a quantitative measure, the deviation-from-gold-standard (DGS) is proposed in this study. The gold standard corresponds to perfect benchmark performance, trivially holding a specificity Spec_k^{gs} of 1.0 independent of k and $\mathrm{Rec}_k^{gs} = k/6$. The DGS of a given method corresponds to the average Euclidean distance of its ($\mathrm{Spec}_k, \mathrm{Rec}_k$) coordinates to the gold standard:

$$DGS = \sqrt{\frac{1}{6}\sum_k (\mathrm{Spec}_k^{gs} - \mathrm{Spec}_k)^2 + (\mathrm{Rec}_k^{gs} - \mathrm{Rec}_k)^2}. \tag{8}$$

Assessing and Comparing Benchmark Performance of EigenRank-Based Structure Alignments. Originally, the COPS benchmark only considered sequence-based homology detection and fold recognition techniques [5]. These include classic sequence database search methods, namely SSearch [11,16] (an implementation of the rigorous Smith-Waterman sequence alignment algorithm), FASTA [11] and BLAST [1], as well as the more sensitive PSI-BLAST [2]. In addition, multiple sequence alignment- and Hidden Markov model-based methods were considered as well (COMPASS [13] respectively HHsearch [17]).

Furthermore, the performance of commonly utilized structure alignment algorithms in the COPS benchmark was assessed in this study. This thus allows to compare the proposed EigenRank-based method to sequence-based search methods on the one hand and state-of-the-art structure alignment techniques on the other hand. Additionally assessed structure alignment techniques are FATCAT [21], CE [15], TM-align [23] and a sequence-guided structure alignment method [12]. The latter, referred to as SW-SuperPos. in text and figures, derives structure alignments based on sequence alignments computed by the Smith-Waterman algorithm [16]. In the case of FATCAT, CE and SW-SuperPos., the implementations provided by the BioJava library [12] were utilized. In addition, FATCAT and CE each provide two distinctive algorithmic prerequisites that affect their underlying alignment strategies. These two strategies are the so-called flexible and rigid mode in case of FATCAT, and the default mode and the circular permutation extension mode in case of CE. The performances of both algorithms and their respective strategies were individually tested in the COPS benchmark.

3   Results and Discussion

Performance statistics for HHsearch, COMPASS, PSI-BLAST, SSearch, BLAST and FASTA were retrieved from the COPS benchmark database [5]. Performance statistics for all named structural alignment algorithms, including the proposed ZER-based method, were computed as described in Subsect. 2.5. DGS statistics were derived for all techniques. With respect to the ZER-based structural alignment method, multiple ΔZER thresholds were considered in an effort to quantify their effect on alignment quality and resulting benchmark performance. Tested thresholds were 0.05, 0.25, 0.50 and 1.00. Furthermore, quality was assessed for structural alignments obtained without ZER filtering.

In Fig. 3 a graphical overview on obtained performance statistics is given. In general, gold standard performance yields specificity-recall coordinates along the recall axis (see Subsect. 2.5) in this representation. Thus, the closer a method's respective specificity-recall curve is located at the recall axis, the better its overall biological sensitivity. This deviation is averaged by the DGS measure. Note that only for a selection of methods the performance statistics are given in Fig. 3 in order to provide visual clarity. Neglected methods showed either little performance discrepancy to reported methods, or their performance is well-represented by their DGS values. A complete overview of all utilized methods, including their alignment modes and considered similarity measures (if applicable) as well as resulting DGS values, is given in Table 1.

As shown in Table 1 and Fig. 3, BLAST, FASTA and SSearch show inferior performance with respect to the more advanced sequence-based search methods PSI-BLAST, COMPASS and HHsearch as well as to all tested structure alignment methods, including the proposed ZER-based structure alignment method. Interestingly, even applied structure alignment methods showed inaccuracies in the benchmark, with FATCAT (rigid) yielded performance slightly inferior to the best method, HHsearch.

ZER-based structure alignment quality is sensitive to applied filter thresholds. Additionally, overall benchmark performance is further impacted by the considered structural similarity measure. Alignments assessed by means of the RMSD are prone to be less accurately evaluated on average, leading to generally higher DGS values compared to applying the GDT_TS scoring scheme. It can be observed that GDT_TS scoring yields an overall performance comparable to commonly utilized structural alignment techniques in the best case, where a ΔZER filter threshold of 0.50 was utilized. Other filter thresholds lead to a drop in performance, at which it is comparable to COMPASS and PSI-BLAST. If the filter threshold is chosen too strict (ΔZER < 0.05), the ZER-based structural alignment method yielded inferior performance compared to most methods, except SSearch, BLAST and FASTA, and performed comparably to SW-SuperPos. This loss in performance can be attributed to the aspect, that at ΔZER < 0.05 only a small set of matching residues are retrieved from any ZER alignment on average. This leads to an insufficient number of coordinates to be aligned and results to suboptimal structure alignments, even in cases where query-homolog pairs are considered. Especially, scoring obtained structure alignments by means of the outlier-affected RMSD, assessed structural similarity between query structures and their homologs become less distinguishable from non-homologs.

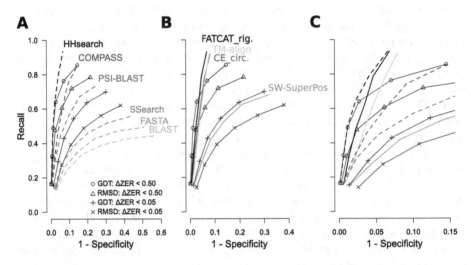

Fig. 3. Specificity-recall curves of protein search methods, evaluated by means of the COPS benchmark. Curve deviations from the recall axis indicate reduced biological sensitivity and decreased benchmark performance. (A) Curves of sequence-based search methods, as originally reported in [5], compared to the proposed EigenRank-based structural alignment method. ΔZER filter thresholds impact the set of C_α coordinates used to compute structural alignments and, in consequence, alignment quality. (B) Comparison to common structural alignment methods. (C) Comparison to sequence- and structure-based methods. Note the 1-Specificity range of 0 to 0.15 in order to discuss detailed performance discrepancies. Color and line type schemes are in correspondence to method labels given in (A) and (B).

When considering the individual ranks instead of overall performance, the ZER-based structural alignment method performed slightly better at ranks 1 to 4 compared to FATCAT. In this range, its performance was on par with HHsearch. However, as performance decreases as more false-positives were reported on ranks 5 and 6, sensitivity lowered to a level at which it was comparable to COMPASS.

Table 1. Considered methods in the COPS benchmark, sorted by the deviation-from-gold-standard statistic (DGS). The similarity measures considered to obtain ranked lists in the benchmark are given if applicable/available. Methodological categories (Seq.: sequence-based; Struc.: structure-based) are given for each method. Note that SW-SuperPos is a sequence-guided structure alignment technique, thus belonging to both categories.

Algorithm	Similarity measure	Parameter/ΔZER filter	DGS	Seq.	Struc.
HHsearch [17]	n.a.		.047	x	
FATCAT [21]	p-value	rigid	.054		x
CE [15]	CE score		.055		x
CE [15]	CE score	circ. permutations	.066		x
TM-align [23]	TM-score		.067		x
FATCAT [21]	p-value	flexible	.085		x
EigenRank	GDT	ΔZER < 0.50	.094		x
EigenRank	GDT	no filter	.096		x
EigenRank	GDT	ΔZER < 1.00	.096		x
EigenRank	GDT	ΔZER < 0.25	.104		x
COMPASS [13]	n.a.		.109	x	
EigenRank	RMSD	ΔZER < 0.50	.146		x
EigenRank	RMSD	no filter	.149		x
EigenRank	RMSD	ΔZER < 1.00	.149		x
EigenRank	RMSD	ΔZER < 0.25	.155		x
PSI-BLAST [2]	e-value		.187	x	
EigenRank	GTD	ΔZER < 0.05	.224		x
SW-SuperPos. [12]	p-value		.241	x	x
EigenRank	RMSD	ΔZER < 0.05	.298		x
SSearch [11,16]	n.a.		.342	x	
FASTA [11]	n.a.		.397	x	
BLAST [1]	E-value		.443	x	

4   Conclusions and Future Work

Although the proposed technique in its current form is less sensitive on average compared to other structural alignment methods, its potential in using it as a fast heuristic search method is apparent. Instead of trying to find a solution to align m to n three-dimensional coordinates, its circumvents this bottleneck by selecting l residue coordinate pairs by means of aligning one-dimensional profiles, which can be conducted rather quickly. The set of thereby obtained l coordinates can be efficiently aligned utilizing the Kabsch algorithm [6].

The benchmark showed that sensitivity of the proposed method is affected by the used ZER filter threshold. Future work thus involves the development of a unified scheme that generates an ensemble of structure alignments based on multiple ZER filter thresholds from which an averaged assessment is made. In addition, it has to be addressed that the computation of an EigenRank profile can be a computational bottleneck, as multiple LeaderRank profiles need to be computed beforehand. Memory and time demands in computing a LeaderRank profile are highly dependent on the number of residues in the protein and its structural complexity (the graph topologies derived at different residue-residue interaction cutoffs). Thus, to scan a query structure against a database containing thousands of structures becomes computationally intensive, and other structural alignment methods are of greater practical usability. This issue can be however overcome by pre-computing a database of ZER profiles, whereas the proposed technique can outperform common techniques with respect to time demands. Furthermore, further computational speed-up can be achieved by utilizing word-indexing techniques, as used by rapid sequence search heuristics such as BLAST and PSI-BLAST. Database indexes can thus allow to narrow the number of ZER profiles to scan against, whereas time and computational demands are further reduced.

References

1. Altschul, S.F., Gish, W., Miller, W., Myers, E.W., Lipman, D.J.: Basic local alignment search tool. J. Mol. Biol. **215**, 403–410 (1990). https://doi.org/10.1016/S0022-2836(05)80360-2
2. Altschul, S.F., et al.: Gapped BLAST and PSI-BLAST: a new generation of protein database search programs. Nucleic Acids Res. **25**, 3389–3402 (1997)
3. Bamakan, S.M.H., Nurgaliev, I., Qu, Q.: Opinion leader detection: a methodological review. Expert Syst. Appl. **115**, 200–222 (2019). https://doi.org/10.1016/j.eswa.2018.07.069. http://www.sciencedirect.com/science/article/pii/S0957417418304950
4. wwPDB consortium: Protein Data Bank: the single global archive for 3D macromolecular structure data. Nucleic Acids Research, October 2018. https://doi.org/10.1093/nar/gky949
5. Frank, K., Gruber, M., Sippl, M.J.: COPS benchmark: interactive analysis of database search methods. Bioinformatics **26**, 574–575 (2010). https://doi.org/10.1093/bioinformatics/btp712. (Oxford, England)
6. Kabsch, W.: A solution for the best rotation to relate two sets of vectors. Acta Crystallogr. Sect. A **32**(5), 922–923 (1976). https://doi.org/10.1107/S0567739476001873
7. Li, Q., Zhou, T., Lü, L., Chen, D.: Identifying influential spreaders by weighted LeaderRank. Phys. A Stat. Mech. Appl. **404**(Supplement C), 47–55 (2014). https://doi.org/10.1016/j.physa.2014.02.041
8. Lü, L., Zhang, Y.C., Yeung, C.H., Zhou, T.: Leaders in social networks, the delicious case. PloS One **6**(6), e21202 (2011). https://doi.org/10.1371/journal.pone.0021202
9. Mrozek, D.: Scalable Big Data Analytics for Protein Bioinformatics. CB, vol. 28. Springer, Cham (2018). https://doi.org/10.1007/978-3-319-98839-9
10. Needleman, S.B., Wunsch, C.D.: A general method applicable to the search for similarities in the amino acid sequence of two proteins. J. Mol. Biol. **48**, 443–453 (1970)

11. Pearson, W.R., Lipman, D.J.: Improved tools for biological sequence comparison. Proc. Nat. Acad. Sci. U.S.A. **85**, 2444–2448 (1988)
12. Prlic, A., et al.: BioJava: an open-source framework for bioinformatics in 2012. Bioinformatics **28**, 2693–2695 (2012). https://doi.org/10.1093/bioinformatics/bts494. (Oxford, England)
13. Sadreyev, R.I., Grishin, N.V.: Accurate statistical model of comparison between multiple sequence alignments. Nucleic Acids Res. **36**, 2240–2248 (2008). https://doi.org/10.1093/nar/gkn065
14. Schulz, G.E., Schirmer, R.H.: Principles of Protein Structure, 5th edn. Springer, New York (1984)
15. Shindyalov, I.N., Bourne, P.E.: Protein structure alignment by incremental combinatorial extension (CE) of the optimal path. Protein Eng. **11**, 739–747 (1998)
16. Smith, T., Waterman, M.: Identification of common molecular subsequences. J. Mol. Biol. **147**(1), 195–197 (1981). https://doi.org/10.1016/0022-2836(81)90087-5. http://www.sciencedirect.com/science/article/pii/0022283681900875
17. Soeding, J.: Protein homology detection by HMM-HMM comparison. Bioinformatics **21**(7), 951–960 (2005). https://doi.org/10.1093/bioinformatics/bti125
18. Spranger, M., Becker, S., Heinke, F., Siewerts, H., Labudde, D.: The infiltration game: artificial immune system for the exploitation of crime relevant information in social networks. In: Proceedings of Seventh International Conference on Advances in Information Management and Mining (IMMM), pp. 24–27. IARIA. ThinkMind Library (2017)
19. Suhrer, S.J., Wiederstein, M., Gruber, M., Sippl, M.J.: COPS - A novel workbench for explorations in fold space. Nucleic Acids Res. **37**, W539–W544 (2009). https://doi.org/10.1093/nar/gkp411
20. Surade, S., Blundell, T.L.: Structural biology and drug discovery of difficult targets: the limits of ligandability. Chem. Biol. **19**(1), 42–50 (2012). https://doi.org/10.1016/j.chembiol.2011.12.013
21. Ye, Y., Godzik, A.: Flexible structure alignment by chaining aligned fragment pairs allowing twists. Bioinformatics **19**(Suppl 2), ii246–ii255 (2003). (Oxford, England)
22. Zemla, A.: LGA: a method for finding 3D similarities in protein structures. Nucleic Acids Res. **31**(13), 3370–3374 (2003). https://doi.org/10.1093/nar/gkg571
23. Zhang, Y., Skolnick, J.: TM-align: a protein structure alignment algorithm based on the TM-score. Nucleic Acids Res, **33**, 2302–2309 (2005). https://doi.org/10.1093/nar/gki524

The Role of Feature Selection in Text Mining in the Process of Discovering Missing Clinical Annotations – Case Study

Aleksander Płaczek[1,2(✉)], Alicja Płuciennik[1,3], Mirosław Pach[1,2], Michał Jarząb[4], and Dariusz Mrozek[2]

[1] Research and Development Department, WASKO S.A.,
ul. Berbeckiego 6, 44-100 Gliwice, Poland
a.placzek@wasko.pl
[2] Institute of Informatics, Silesian University of Technology,
ul. Akademicka 16, 44-100 Gliwice, Poland
dariusz.mrozek@polsl.pl
[3] Institute of Automatic Control, Silesian University of Technology,
ul. Akademicka 16, 44-100 Gliwice, Poland
[4] Maria Skłodowska-Curie Memorial Cancer Center and Institute of Oncology,
Gliwice Branch, Gliwice, Poland

Abstract. Vocabulary used by the doctors to describe the results of medical procedures changes alongside with the new standards. Text data, which is immediately understandable by the medical professional, is difficult to use in mass scale analysis. Extraction of data relevant to the given case, e.g. Bethesda class, means taking on the challenge of normalizing the freeform text and all the grammatical forms associated with it. This is particularly difficult in the Polish language where words change their form significantly according to their function in the sentence. We found common black-box methods for text mining inaccurate for this purpose. Here we described a word-frequency-based method for annotation of text data for Bethesda class extraction. We compared them with an algorithm based on a decision tree C4.5. We showed how important is the choice of the method and range of features to avoid conflicting classification. Proposed algorithms allowed to avoid the rule-base limitations.

Keywords: Text mining · Feature selection ·
Unstructured medical text · Inverse document frequency ·
Text tiding

1 Introduction

The rapidly rising incidence of thyroid cancer [19] causes that physicians have to treat more patients during the same working hours. Patients' cases become

S. Kozielski et al. (Eds.): BDAS 2019, CCIS 1018, pp. 248–262, 2019.
https://doi.org/10.1007/978-3-030-19093-4_19

progressively complicated, the co-occurrence of diseases appears more and more often. Diagnostics and Treatment of Thyroid Carcinoma guideline [8] show how complicated the proper diagnosis of patient's disease based on the symptoms and limited evidence is. Meanwhile, the physicians very rarely use the knowledge they have access to in their medical systems. In most cases, the results of patients' examination are stored there as plain text which is sufficient for doctors' day-to-day work but is inefficient for large-scale data analysis for populations. Mainly this prevents them from using the data.

While taking part in the MILESTONE project, which aims to create new tools for supporting decisions on local treatment of breast, thyroid and prostate cancers, to achieve the reduction in therapy aggressiveness, we were motivated to use retrospective data collected over two years to develop solutions support-ing medical inference. Usually, to prove the necessity of surgery the clinician orders a Fine Needle Aspiration Biopsy (FNAB) [5,19]. It is a diagnostic pro-cedure performed in order to investigate nodules by a pathologist. The sample is stained and then examined under a microscope, sometimes it can be also analyzed chemically. The results of the examination should be in a form of FNA report where physicians should find (a) information related to the nodule location and its features enabling its identification, (b) information concerning FNA representativeness - both qualitative and quantitative, (c) description of cytological examination of each nodule assessed, and finally diagnostic conclu-sion that classifies FNA findings as one of six Bethesda classes [1]. The class attribute results from the previous three subsections, where the quality of par-ticular findings depends on pathologist experience and knowledge in the field of oncology and the additional clinical information stored in treatment records. Even in Guidelines of Polish National Societies prepared by Polish Group for Endocrine Tumours some cytologic criteria are mentioned as subjective, others have their reliability and limitations. Such ambiguous features are presented in Fig. 1. The Bethesda system (TBS) is used to uniform terminology occurring in FNAB reports and categorize samples into one of six categories. Each category has an implied risk of malignancy [4].

Feature	Follicular lesion of undetermined significance	Suspicious for a follicular neoplasm
Hypercellular aspirate (subjective)	Rather yes	Yes
Prominent population of small arrangement (groups, nests, rosets)	Yes	Yes
Sheets of follicular cells	Might be seen	No or single
Colloid in background	Might be seen	No or trace
Foamy macrophages	Might be present	No or single
Anisocytosis/anisokaryosis	No or a little	No
Lymphocytes/plasmatic cells	No or single	No

Fig. 1. Example of differentiating characteristics of phenomena. Column no. 2 - Bethesda III, Column no. 3 - Bethesda IV. Source: [8], page 42.

We have initially analyzed the depersonalized data in the form of results of thyroid biopsies collected from 2007 till 2017 extracted and transformed from databases of Cancer Center And Institute of Oncology - a partner in the MILE-STONE project. Analysis of the data mined in the data sources showed us what kind of unstructured data we have to deal with. In fact, the data had no pre-defined model in the sense of structure, but the content was mainly consistent. Records very seldom meet the criteria for being analyzed according to fixed rules all physicians adhere to. Descriptions are written using natural language which does not require extra vocabulary and dictionary knowledge. We found records only from the last two years, where natural language description is supplemented by TBS classification. Before we started using them as a training set we have made some assumptions: we do not know anything about the recommended and unacceptable terminology, authors of the description and their common vocab-ulary from the past year, their experience and additional information they had access to. In other words, we have made no assumptions about the context of the written text.

Our motivation was to develop a reliable data extraction and classification mechanism which can be used with retrospective data collected in data sources in many organizational health care units. We assume that our solution should be ready to also analyze new descriptions prepared not only by the staff belonging to the team of pathologists whose records were analyzed during the learning stage. That is why the goal of our work is to check and develop methods based on statistical inference where we take into consideration medical ambiguities in the assessment of biological materials, different vocabulary and how far the implementation of TBS has progressed.

This work is structured as follows. Section 2 presents the related works. Section 3 describes actions which had to be taken to prepare the unified data model. It is followed by Sect. 4 where we presented our approach and quality assessment of developed methods. In Sect. 5 we made the analysis of achieved experimental results. Section 6 presents our final conclusions.

2   Related Works

The text-mining technique is quite common for processing electronic health records [7]. Recently the Systematized NOMenclature of MEDicine - Clinical Terms (SNOMED CT) has become a huge convenience. This approach allows to successfully apply natural language processing [13], but SNOMED CT is not in common use in Hospital Information Systems reports yet. The common approaches for classification of health records include the SVM classifiers and Naive Bayes methods [9]. The Neural Networks were also used for classification for thyroid disease diagnosis [16]. All these methods require a set of features, and taking into account the complexity and flexibility of Polish language, we decided to search the methods for feature selection from the description of examination. Despite the fact that many language-processing systems were designed during last decades, the analysis of Polish texts is still problematic. Using the aforementioned methods it is difficult to extract and analyze the importance of particular features. Thus, we have found this solutions not accurate enough for our purpose.

The Bethesda System is generally a new approach in Polish pathology [20], despite the fact that guidelines suggest to use it for FNAB of thyroid since 2007. This standard is only a recommendation. Therefore we expected the large variety and variability-over-time of vocabulary used in the descriptions for FNAB results in our data sets. For that reason we had to be prepared for handling conflict cases. We wanted to compare discovered rules to the clinical set of rules (presented in Table 1). Thus we had searched for methods where it would be possible to export the rules in a human-readable form (to be validated by an expert). For our purpose, we checked a simple method for overt feature selection from a text (avoiding black box methods), which is also commonly used in medical data analysis - the binary decision trees. One of the popular algorithms, C4.5, was successfully applied to the diagnosis of coronary heart disease [22] and other decision support systems [11]. The algorithm creates a binary tree where rules are made based on the presence or absence of a feature. The pruning process provides the generalization of decision rules. However, in this paper we would like to pay attention to the role of a feature selection stage, which can affect not only the performance but also later classification accuracy and potential overfitting. The overfitting is highly undesirable to infer from evolving medical terminology.

3 Materials and Methods

Our goal was to prepare transformed data sets which would be necessary to create a Bayesian Network to infer unobserved variables as the long-term goal - in our case the probability of malignancy. Bayesian Network is a probabilistic graphical model (a type of statistical model) that represents a set of variables and their conditional dependencies via a directed acyclic graph (DAG) [6]. To properly build the network we need to provide learning parameters which use the data gathered historically. The problem occurs when data is missing or incomplete. One of the most important features supporting medical decision is Bethesda class mentioned above. For most of our retrospective data set the Bethesda class was not pointed but it can be inferred from description. The worst case is when during the analyzed period significant changes have occurred in the way of describing the variable. For example, in 2016 after long-term follow-up of patients with the previously suspected malignant lesion, the new subgroup of cases were recognized with classification changed to benign [17].

For our experiments, we used the BioTest environment [14]. To enhance the ability to work with dynamic tables we used special components of Galaxy Server allowing users to do interactive data processing from within Galaxy. We integrated Galaxy with the R environment gaining access to all files and database connections prepared previously for integration with relational medical databases. Our motivation to use BioTest was also the fact that is was installed on the infrastructure of the Ziemowit computer cluster (www.ziemowit. hpc.polsl.pl) in the Laboratory of Bioinformatics and Computational Biology. In the future, parallel calculations are intended, because the features can be extracted independently.

3.1   Data Sources

The process of de-identification was performed by the third party to protect the patient's personal data. Only fully deidentified diagnoses were analyzed without access to full patient data. Also, the medical staff's data were anonymized. The data sources contained information about more than 2500 patients, 2700 diagnostic procedures (FNAB) and more than 3200 examined thyroid nodules. The procedures were performed from 2016 till 2017, by different doctors. We had no information about their experience. Starting from 2016 the TBS was introduced partially, but examinations were also performed by an external specialist and third parties. Thus, some nodules have cytological classes assigned. At the same time, others are only described in natural language.

The comparison between nodules where TBS was identified and not is presented in Fig. 2.

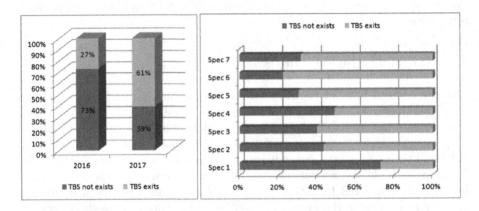

Fig. 2. (a) TBS classified/not classified nodules in 2016 and 2017; (b) ratio of cases containing classification label (TBS) to an indeterminate ones for each doctor (specialist) in 2017

More than 25 specialists were engaged in examination every year, but we checked only those who performed more than 50 examinations per year. We wanted to consider doctors with the right level of skills to properly set the Bethesda class. Especially that the Bethesda system was introduced just two years earlier and some of them had no enough experience in using it.

The results showed us that even in more experienced group there are no exceptions. Each specialist had cases where no TBS class was set. In our opinion, it shows how complicated procedure is the specimen's examination and how hard is to properly assess the findings. That is why we expected a lot of ambiguity and dependency in the data.

We divided our data into two sets: the first one with training data which contains input and output pair (explicitly indicated value of TBS class), the other one with test data without the assigned labels. In this paper, we are focused mainly on training sets. For the needs of this study, no distinction was made between patients on whom the biopsy was performed to help with medical diagnosis and patients with suspicion of tumor recurrence.

As was mentioned above, specialists try to keep consistent information content when they write down their observations. Reports consist of macroscopic and microscope descriptions. We extracted the data from the latter using the R environment and performed some transformations on them as described below. In the result, we had prepared texts consisting of canonical forms of previously used words without words considered to be "stop words".

3.2 Source Data Extraction and Destination Data Model Preparation

The process of text transformation to meaningful features is widely presented in publications [2, 21]. Among all tasks, the key ones are: text clean-up, normalization, and unification. Below are some issues which had to be resolved before using such data to discover missing annotations, especially in medicine. According to best practice, we first tried to standardize the form of extracted sections. First of all, we extracted explicitly written down TBS classes using regular expressions to use them as labels. They were the part of unstructured text. Next, we excluded these labels and all words directly related to them such as 'category', 'bethesda', 'class', 'group' etc. Then, the remaining text of FNAB report was split into tokens using the tidytext R package [18] and used in supervised analysis. Each record in the training set was split into tokens using the tidytext R package [18].

Tidy Data Sets. In our approach, we used the term frequency-inverse document frequency (TF-IDF [10,15]) to reflect how extracted features were important. We defined each FNAB report as a page written by a particular specialist in a summary document related to one of six TBS classes - on the basis of previously extracted labels, six documents in total. Each page consisted of between one and dozens of paragraphs. The number of paragraphs depended on identified nodules during the diagnostic procedure and the results of the examination. We used the tokenizers package to split each line of text in the original documents into tokens. We defined a tidy text data set as a table with one-token-per-row for each paragraph. In our approach, we used words (without knowledge of the context of usage), 2-grams and 3-grams (meaningful sentences) as tokens. We chose

3-grams as a final level of tokenization on the basis of analysis of the clinical set of rules prepared by specialists. After reading the stemmed records from the database we prepared working data sets using each variant of n-grams to identify the set of all possible features using TF-IDF to reflect how these features were important to each type of TBS class. We calculated factors for each identified feature in the group for future term-weighting when more than one classification decision is possible.

Stemming. In our project, we used Morfologik stemming library [12]. Because the FNAB report is focused on the description of characteristics of phenomena we did not pay much attention to part of speech of extracted words. We were aware of possible loss of accuracy, but as was mentioned before, we operated on phrases without the knowledge of the context and therefore it would not be possible to discriminate between words which have different meanings depending on the part of speech. Such example is the word 'podejrzenie' (suspicion) which can be tagged as (a) gerund (b) substantive and have different lemma depending on the part of speech. To avoid complications and ambiguity we always chose the shorter form. Another issue we had to deal with was a negation. Searching for the common canonical form we had to be aware that some negative form has the same lemma as positive ones. Again the example is the word 'niepodejrzana' (not suspicious) which is negation but has the same lemma as 'podejrzenie' (suspicion). In fact, the final meaning is reverse and may lead to various conclusions. Fortunately, each word in the dictionary has its own attributes. So we checked word's annotations to find out which word was marked as negation. If the verification was positive we added a negative particle at the beginning of the selected lemma.

Special Transformation Due to Medical Context. Medical texts are specific and their meaning depends on used vocabulary. One of the problems is that specialists write down the same information using three languages interchangeably. They can describe the same findings in Polish, English or Latin language. So, we had to find similar concepts (words, phrases or even sentences) and standardize them before undergoing transformations. Such examples are presented below:

```
Rak brodawkowaty = Papillary thyroid cancer = Carcinoma papillare

Rak brodawkowaty podtyp p{\k{e}}cherzykowaty = Papillary thyroid
cancer follicular variant = Carcinoma papillare typus follicularis
```

Before we started analyzing each FNAB report in the training set we translated all English and Latin names of diseases to adequate polish ones. Medical texts are characterized by the usage of these three languages interchangeably. We used the Guidelines of Polish National Societies Diagnostics and Treatment of Thyroid Carcinoma to define the correct translations.

We also had to deal with negative/positive sense of the sentence. Some specialists wrote down information about the features they had not found in their report, which further complicated the analysis. We assumed words 'nie' (not) and 'bez' (without) as a denial of the relevance of the next words and we ignored the detection of the key feature if it occurred very next to negation word - not far then 4 words after it (determined on the basis of clinical set of rules analysis).

We needed to achieve informative, discriminating and independent characteristic of a phenomenon being observed [3].

3.3 Clinical Set of Rules

The challenge of this work was to choose the best way to discover one of the Bethesda classes based only on descriptions derived from medical databases. So, first, we decided to ask clinicians (experienced specialists with 18 years of experience, having a leading influence on the diagnosis and treatment of patients) to provide their ideal set of rules. Such mapping is presented in Table 1. It is consistent with the standards. Such contents of FNAB reports should allow to automatically assign the proper Bethesda category. We have made an experiment to check how many records from the training set would be classified according to the proposed terminology. Some extra transformations were needed to allow text comparison - those will be described in the following subsections. The results of such an experiment can be observed in Fig. 3. The results showed that despite the fact, that updated guidelines provide recommended, accepted and restricted terminologies to help specialist to assign one of the classes to the specimen, many texts deviate from the standards. We would like to stress the fact that training set contains specimen examined during the last two years, our future work will be an assessment of the descriptions from a 10-years period when such guidelines were absent or unused. We achieved 50% accuracy when tried to reclassify training set with provided clinical rules - for 1600 records we found at least one key feature which described the Bethesda group (details presented in Table 1). For these records, we resulted in 95% properly assigned Bethesda classes (accuracy). The group II is the one were most descriptions follow the recommendations. On the other hand, pathologists had to deal with the most problematic issues in groups VI and V, where they had to decide whether it was cancer (94–96% probability of malignancy) or only they suspected them to be (45–60%). The worse accuracy of classification in these groups probably resulted from more specifically written diagnoses and longer texts. That is why we decided to use one of the most popular algorithm C4.5, but we observed that most of misclassified probes resulted from conflicting data. Rejection of them would require the step of preprocessing (removing inadequate labels with descriptions) what could be done only by manual review. We decided to propose a new approach without the need of preprocessing described above. Our implementation is based on features selection and text mining techniques.

Table 1. Recommended terminology and phrases

Bethesda category	Vocabulary
I	'non-diagnostic biopsy' or 'inadequate cellularity'
II	'benign nodule'
III	'follicular lesion of undetermined significance'
IV	'suspicious for a follicular neoplasm' or 'suspicious for a oxyphilic neoplasm'
V	'suspicious for malignancy' or 'suspicion for papillary thyroid carcinoma' or 'suspicion for medullary thyroid carcinoma'
VI	'papillary thyroid carcinoma' or 'anaplastic thyroid cancer' or 'medullary thyroid carcinoma'

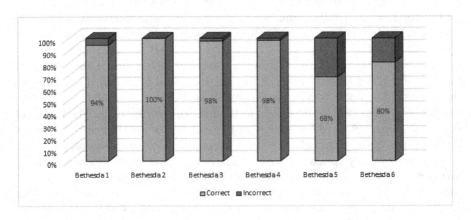

Fig. 3. Bethesda classification accuracy for clinical set of rules

4   Proposed Approach

A central question in text mining and natural language processing is how to quantify what a document is about? We used numerical statistic called term frequency-inverse document frequency (TF-IDF calculated) to find which words or sentences (later N-grams) best describe each group (Bethesda class) and at the same time are not used very often in reports of other ones (details in Sect. 3.2). We assumed they would identify key features for each group. At the beginning, we rejected N-grams with frequency in the group less than mean value for all the N-grams in the same group (without stop words). Then, to avoid the huge number of N-grams found in the training set, we set a parameter which is responsible for selecting top N values (10, 20) for each group. The number of key features in each group can be more than N if there are ties. Such preprocessing, in our opinion, minimizes the possibility of over-fitting. The process of identification of the set of features best describing each category of Bethesda and next using it for classification is presented in a form of RetroNGSC algorithm. The pseudocode is presented below (Algorithm 1).

Algorithm 1. NGram Selection Classifier for retrospective data (RetroNGSC)

1: Fetch records in canonical forms, without stop words
   Features identification and mining

2: **for** N-gram in 3,2,1 **do**
3:      Prepare tokens, group data according to Bethesda
4:      Standardize between English, Polish and Latin grams
5:      Calculate statistic indicators for tokens
6:      $KeyFeatures \leftarrow$ Select top(n) features based on TF-IDF
7: **end for**
8: **for** every record **do**
9:      $Tokens \leftarrow$ Tokenize(report description)
10:     $Findings \leftarrow$ Search($Tokens$ in $KeyFeatures$)
11:     $RecordFeatures \leftarrow$ Match($Findings$)
12: **end for**
    Classification and conflict resolution

13: $CResults \leftarrow$ ClassifyRecords($RecordFeatures$)
    ▷ Verification if all found features describe the same class
14: $NonUnique \leftarrow$ SelectAmbiguousRecords($CResults$)
15: **for** each record in $NonUnique$ **do**
16:     $BestFeature \leftarrow$ MaxImportanceForClass($RecordFeatures$)
17:     **if** no two equally important features **then**
18:         $BethesdaClass \leftarrow$ AssignClass($BestFeature$)
19:     **else**
20:         **for** N-gram in 3,2,1 **do**
21:             $Tokens \leftarrow$ Tokenize(report description,N-Gram)
22:             $BestFeature \leftarrow$ MaxImportanceForClass($Tokens$,Max(TF-IDF))
23:             **if** Count(BestFeature) ≥ 2 **then**
24:                 Continue
25:             **else**
26:                 Break
27:             **end if**
28:         **end for**
29:         $BethesdaClass \leftarrow$ Assign($BestFeature$)
30:     **end if**
31: **end for**

From a short analysis of the clinical set of rules we concluded that in many cases, trigrams (3-gram continuous sequences of words) brings the sufficient context to properly classify the specimen on the basis of natural language description. That's why we started from trigrams but continued with 2-grams and words - the first version of our algorithm. When at least one key feature was found, the record was rejected from the next step. This results in smaller data set: only 14%

of records had to be analyzed using 2-grams and only 2% using words. The rest was classified using 3-grams. But we noticed that this was a feature dependent solution and might influence the cross-validation process. We implemented a second version, which tried to match all records to three-word features and treated them as one main set of features. Such approach increased the number of features, but at the same time, we could eliminate the feature-dependent issues. It resulted in the increase of accuracy, because when 3-grams key features were similar to each other in different groups, the 2-grams features differentiated the cases. We called it as a conflict resolution.

4.1   Quality Assessment

In this study, we compared a popular C4.5 algorithm for data mining with proposed algorithms. To analyze the quality of classification for used methods we performed 10-fold cross-validation. We divided our dataset into 10 subsets. For each iteration we selected one subset as testing and the rest as training set. Then we identified feature set for this training set and applied it on a test set. For both methods we have used only a feature set which has been defined during training iteration of the algorithm. For each iteration as an indicator of the quality of studied algorithms, we took the percent of assigned Bethesda categories consistent with the ones assigned by the doctor. The other applied measure of quality

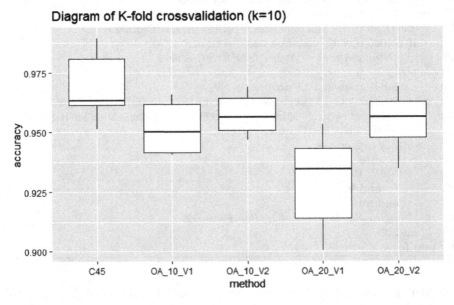

Fig. 4. Comparison between C4.5 and RetroNGSC (k-fold accuracy). C45 - algorithm based on decision tree, OA-10-V1 - RetroNGSC version 1 for 10 meaningful features, OA-10-V2 - RetroNGSC version 2 for 10 meaningful features, OA-20-V1 - RetroNGSC version 1 for 20 meaningful features, OA-20-V2 - RetroNGSC version 2 for 20 meaningful features.

was the number of records where the algorithm did not assign any category. This measure was also applied for assessment of algorithms for cases where the Bethesda class was not given in the text of microscopic examination. The results are presented in Fig. 4. Lower accuracy (less then one percentage point difference in mean values) of RetroNGSC resulted from proposed corrections to medical decisions (see Results section). C4.5 in some cases didn't follow the guidelines, probably because of description inconsistency with reference classification.

5   Results and Discussion

All three algorithms achieved similar accuracy for training data. In RetroNGSC, increasing the number of key features allowed more records to be classified generally, without decreasing the accuracy. All have some problems with the classification of specific FNAB reports, what is presented in Figs. 5 and 6, but when it comes to RetroNGSC it's decisions could be explained and justified after studying the original text.

No.	Type	The content of the FNAB report (canonical form) / C4.5 rule / Our-Algorithm (selected top 3 features)	Decision
1	FNAB	tyreocyty polimorficzny jądro cech metaplazja oksyfilnej płat grupa rozproszyć brak koloid obraz cytologiczny zmiana łagodny	Bethesda 3
	C4.5	zmiana ; brak ; jądro & *rak ; nowotwór ; blisko ; carcinoma ; aspirat ; oksyfilnym ; histopatologiczny ; nowotworu; nieokreślić ; pobiopsyjnymi ; stosować ; niediagnostyczny ; krotny ; powtórny*	Bethesda 3
	Our-Alg.	[1] feature: "zmiana łagodny" -> Bethesda 2 (13%) [2] feature: "zmiana" -> Bethesda 3(5%) [3] feature: "tyreocyty" -> Bethesda 3 (1,5%)	Bethesda 2
2	FNAB	materiał bogatokomórkowy polimorficzny tyreocyty metaplazja oksyfilną liczny płat grupa rozproszyć tło krew brak koloid podejrzeć nowotwór pęcherzykowy typ oksyfilny obraz wymagać różnicować rak rdzeniastym tarczyca	Bethesda 5
	C4.5	rak ; podejrzeć ; pęcherzykowy; bogatokomórkowy & *aspirata ; wskazać ; ekscentrycznie ; brodawkowaty*	Bethesda 5
	Our-Alg.	[1] feature:"podejrzeć nowotwór pęcherzykowy"-> Bethesda 4(22%) [2] feature: "podejrzeć" -> Bethesda 5 (7%) [3] feature: "rak" -> Bethesda 5 (5%)	Bethesda 4

Fig. 5. Comparison between C4.5 and RetroNGSC (mismatching) (Color figure online)

The table presented in Fig. 5 shows the examples of FNAB reports which were classified by RetroNGSC differently (in opposite to C4.5). C4.5 found words required by the rules and checked against the ones which are restricted (marked in red). It followed the same clinical decision. RetroNGSC checked the TF-IDF indicator for each feature, taking into account the N-grams and decided that the most important features are related to different class - group II in No. 1 (benign nodule) and group IV in No. 2 (suspicion of follicular thyroid cancer). We checked the descriptions manually and noticed that the results of RetroNGSC are consistent with the standards (guidelines). So, we found them interesting to discuss with clinicians and saved them for further mining e.g. the second record in Fig. 5 contains only the recommendation for an extra test for medullary thyroid cancer, but not assert suspicion of malignant lesion.

No.	Type	The content of the FNAB report (canonical form) / C4.5 rule / Our-Algorithm (selected top 3 features)	Decision
1	FNAB	uzyskać aspirant ubogo komórkowy duży domieszka krew widoczny nieliczny grupa tyreocytów budzić podejrzeć złośliwość	Bethesda 5
	C4.5	widoczny & rak ; nowotwór ; histopatologiczny ; blisko ; nieokreślić ; pobiopsyjnymi ; stosować ; zmiana ; brodawkowaty ; cytoplazma ; płat ; zmiana łagodna ; część ; syderofagi ; grudka ; histiocyty ; limfocyty ; olbrzym ; rozproszeniu ; torbiel ; węzeł	Bethesda 2
	Our-Alg.	[1] feature:"budzić podejrzeć" -> Bethesda 5(5,6%) [2] feature: "budzić podejrzeć" -> Bethesda 4 (5,3%) [3] feature: „podejrzeć złośliwość" -> Bethesda 5 (4,7%)	Bethesda 5
2	FNAB	komórka kwasochłonny cytoplazma częściowo wydłużyć jądro liczny bruzda inkluzja jądrowe płat grupa obraz brodawkowaty tarczyca	Bethesda 6
	C4.5	łagodny ; atypia ; pasm	Bethesda 2
	Our-Alg.	[1] feature: "brodawkowaty tarczyca" - > Bethesda 6 (7,8%) [2] feature: "brodawkowaty tarczyca" -> Bethesda 5 (4,2%) [3] feature: "bruzda inkluzja"-> Bethesda 6 (4%)	Bethesda 6

Fig. 6. Comparison between C4.5 (mismatching) and RetroNGSC

The table presented in Fig. 6 shows example reports which were classified by RetroNGSC according to a doctor (in opposite to C4.5). The rules found by C4.5 contained conflicting variables. RetroNGSC on the basis of the importance of N-grams decided that the most important features are related to the same class - group V in No. 1 (malignancy suspicion) and group VI in No. 2 (papillary thyroid cancer). Furthermore, none of the less important features assigned to case No. 1 was not related to group II. Since our results are consistent with the guidelines again, we found them interesting to discuss with clinicians and saved it for further mining.

6   Conclusions and Future Work

Increasing interest in automatic text classification is the result of growing complexity and quantity of medical cases. The final report does not always contain Bethesda classification which is one of most important features taken into account during medical consilium. That's why health care applications should include algorithms supporting clinicians to make decisions on the basis of natural language description. As the result of the study, we found the C4.5 algorithm better as a tool for prediction which can suggest the Bethesda category to a clinician when a new examined specimen comes. It's because it uses static rules which are more dependent on vocabulary. Besides, to become up to date it requires repeating the learning stage to construct new rules from new data (maybe some new words will become key ones). We found it worse as a tool for analyzing and suggesting mistakes in retrospective researches (historical data).

C4.5 using the learned rules can only classify the data, without reporting possible mistakes resulting from deviations from the recommendation, as was shown in the previous section. In our study, when using C4.5, we had problems with the occurrence of conflicting classification values for identical input values - it was very hard to find and resolve them quickly.

In turn, our both versions of the algorithm presented better results in the scope of suggesting corrections without losing in the accuracy. They can update their key feature set during classification, because they use the whole data set to find key ones. They can also produce selected feature in meaningful forms, what is practically impossible when working with C4.5.

RetroNGSC had insignificantly worse accuracy in comparison with C4.5, but our research showed that in case of mismatch of description for classification (e.g. clinical mistakes, low text specificity) C4.5 could achieve bad results. Because, it analyzed data without the clinical context. Our approach built such a context using N-grams.

We are aware that we can improve the accuracy much more when we set N to maximum, but it will result in poor performance. In the future, we plan to check how words rotation can influence accuracy. But the final test will be comparison classified retrospective data with risk of malignancy [4] based on historical data.

Acknowledgments. This work was supported by The National Center for Research and Development project MILESTONE under the program STRATEGMED (contract No. STRATEGMED2/267398/4/NCBR/2015). Full protocol of study was approved by ethics committee. This work was partially supported by the Polish Ministry of Science and Higher Education as part of the Implementation Doctorate program at the Silesian University of Technology, Gliwice, Poland (contract No. 10/DW/2017/01/1).

References

1. Al Dawish, M.A., et al.: Bethesda system for reporting thyroid cytopathology: a three-year study at a tertiary care referral center in Saudi Arabia. World J. Clin. Oncol. **8**(2), 151–157 (2017)
2. Allahyari, M., et al.: A brief survey of text mining: classification, clustering and extraction techniques. arXiv preprint arXiv:1707.02919 (2017)
3. Bishop, C.: Pattern Recognition and Machine Learning. Information Science and Statistics. Springer, New York (2006)
4. Cibas, E.S., Ali, S.Z.: The 2017 Bethesda system for reporting thyroid cytopathology. Thyroid **27**(11), 1341–1346 (2017)
5. Gharib, H.: Fine-needle aspiration biopsy of thyroid nodules: advantages, limitations, and effect. Mayo Clin. Proc. **69**(1), 44–49 (1994)
6. Guo, Z., Gao, X., Di, R.: Learning Bayesian network parameters with domain knowledge and insufficient data, vol. 73, pp. 93–104 (2017)
7. Iavindrasana, J., Cohen, G., Depeursinge, A., Müler, H., Meyer, R., Geissbuhler, A.: Clinical data mining: a review. Yearb. Med. Inform. **18**(1), 121–133 (2009)
8. Jarząb, B., et al.: Guidelines of Polish national societies diagnostics and treatment of thyroid carcinoma. 2018 update. Endokrynologia Polska **69**(1), 34–74 (2018)

9. Kocbek, S., et al.: Text mining electronic hospital records to automatically classify admissions against disease: measuring the impact of linking data sources. J. Biomed. Inform. **64**, 158–167 (2016)
10. Kwon, O.S., Kim, J., Choi, K.H., Ryu, Y., Park, J.E.: Trends in deqi research: a text mining and network analysis. Integr. Med. Res. **7**(3), 231–237 (2018)
11. Lamy, J.B., Ellini, A., Ebrahiminia, V., Zucker, J.D., Falcoff, H., Venot, A.: Use of the C4.5 machine learning algorithm to test a clinical guideline-based decision support system. Stud. Health Technol. Inform. **136**, 223–228 (2008)
12. Miłkowski, M.: Morfologik: LanguageTool 2.5. http://morfologik.blogspot.com/2014/03/languagetool-25.html
13. Nguyen, A.N., et al.: Symbolic rule-based classification of lung cancer stages from free-text pathology reports. J. Am. Med. Inform. Assoc. **17**(4), 440–445 (2010)
14. Psiuk-Maksymowicz, K., et al.: A holistic approach to testing biomedical hypotheses and analysis of biomedical data. In: Kozielski, S., Mrozek, D., Kasprowski, P., Małysiak-Mrozek, B., Kostrzewa, D. (eds.) BDAS 2015–2016. CCIS, vol. 613, pp. 449–462. Springer, Cham (2016). https://doi.org/10.1007/978-3-319-34099-9_34
15. Qaiser, S., Ali, R.: Text mining: use of TF-IDF to examine the relevance of words to documents. Int. J. Comput. Appl. **181**(1), 25–29 (2018)
16. Razia, S., Rao, M.R.N.: Machine learning techniques for thyroid disease diagnosis - a review. Indian J. Sci. Technol. **9**(28), 1–9 (2016)
17. Seethala, R.R., et al.: Noninvasive follicular thyroid neoplasm with papillary-like nuclear features: a review for pathologists, **31**(1), 39–55. https://doi.org/10.1038/modpathol.2017.130
18. Silge, J., Robinson, D.: tidytext: text mining and analysis using tidy data principles in R. https://doi.org/10.21105/joss.00037
19. Song, J.S.A., Hart, R.D.: Fine-needle aspiration biopsy of thyroid nodules. Can. Fam. Phys. **64**(2), 127–128 (2018)
20. Stanek-Widera, A., Biskup-Frużyńska, M., Zembala-Nożyńska, E., Śnietura, M., Lange, D.: The diagnosis of cancer in thyroid fine needle aspiration biopsy. Surgery, repeat biopsy or specimen consultation? Pol. J. Pathol. **67**(1), 19–23 (2016)
21. Szwed, P.: Enhancing concept extraction from Polish texts with rule management. In: Kozielski, S., Mrozek, D., Kasprowski, P., Małysiak-Mrozek, B., Kostrzewa, D. (eds.) BDAS 2015–2016. CCIS, vol. 613, pp. 341–356. Springer, Cham (2016). https://doi.org/10.1007/978-3-319-34099-9_27
22. Wiharto, W., Kusnanto, H., Herianto, H.: Interpretation of clinical data based on C4.5 algorithm for the diagnosis of coronary heart disease. Healthc. Inform. Res. **22**(3), 186–195 (2016)

Fuzzy Join as a Preparation Step for the Analysis of Training Data

Anna Wachowicz$^{(\boxtimes)}$ and Dariusz Mrozek$^{(\boxtimes)}$

Institute of Informatics, Silesian University of Technology,
ul. Akademicka 16, 44-100 Gliwice, Poland
anna.m.wachowicz@gmail.com, dariusz.mrozek@polsl.pl

Abstract. Analysis of training data has become an inseparable part of sports preparation not only for professional athletes but also for sports enthusiasts and sports amateurs. Nowadays, smart wearables and IoT devices allow monitoring of various parameters of our physiology and activity. The intensity and effectiveness of the activity and values of some physiology parameters may depend on weather conditions in particular days. Therefore, for efficient analysis of training data, it is important to align training data to weather sensor data. In this paper, we show how this process can be performed with the use of the fuzzy join technique, which allows to combine data points shifted in time.

Keywords: Fuzzy sets · Fuzzy logic · Data analysis · Smart devices · Human activity monitoring · IoT · Sensors

1 Introduction

Physical activity, along with proper nutrition, is the most important issue of health-care. Some of the sports disciplines require advanced equipment while the other - almost anything. For years, only professional athletes could monitor their training parameters with wearable devices as it was too expensive and unapproachable for the most of amateurs. Along with the evolution of IoT devices and various types of wearables more and more people can control workout parameters. The easiest way is to use one of the numerous fitness mobile applications available on all types of smartphones. As they monitor usually raw training features – distance, time or location, more and more sports enthusiasts prefer to use smart bands and watches. Depending on the type of wearable device user has the ability to control not only the raw features but also the parameters such as heart rate, average speed, number of burnt calories, the level of body hydration or even the recognition of practiced workout. More and more often the amateurs decide to practice with others, under the control of a coach. Together, they motivate each other to perform their best in safe and healthy way. Also in social media and online forums the users of wearables form large communities. They exchange experience, give advice and upload their workout

© Springer Nature Switzerland AG 2019
S. Kozielski et al. (Eds.): BDAS 2019, CCIS 1018, pp. 263–273, 2019.
https://doi.org/10.1007/978-3-030-19093-4_20

results. Finally, having a review of previous activities, it is possible to analyze the data on a larger scale and draw some useful conclusions.

One of the most popular issues raised among members of online communities is how the weather affects the training performance. The impact of the atmospheric conditions on the quality and intensity of the workout is considered in many scientific papers. The analysis itself may be challenging but the weather and activity data need to be properly combined together. Online weather services provide access to measurements but they determine the interaction with the user. Moreover, the weather data are presented hourly while physical activities may start at any time during the day. This implies the need for using intelligent data join operations so that the times of workout and weather measurements are best fitted.

The purpose of this paper is to present a fuzzy join technique as a preparation step for data analysis. The technique was implemented in the system that analyses the impact of weather conditions on the training parameters. The training data are gathered with the use of smartwatch used for monitoring the running training. With the use of the fuzzy join algorithm, data are combined with the most accurate weather measurements retrieved from an online weather service. The output is then loaded to the Microsoft Azure cloud data center where the analysis can be conducted. In the paper, we show the duration of particular operations related to data pre-processing performed by the fuzzy join module responsible for flexible combining of data points.

2   Related Works

As wearable devices are widely used for monitoring different types of activities, many research was published in this field. Mukhopadhyay [14] presents a review on various types of wearable devices and inner sensors used for observing human activities. In his work, the author also describes the architecture of monitoring systems and networks. Mostly, the devices consist of the processor, various sensors, a display where the collected data may be presented to the user and, if the device has an option of wireless transmission, the transceiver that enables to send data to a paired application. Since the design issues for wearable devices are also important, to meet all users' requirements smart wearables have to be light and low-energy consuming [16]. Devices should be able to send data to the paired application, so users can access the history of performed activity, like training, or be informed that there is something wrong with a monitored person, regardless of whether it is a sports amateur or an elderly that wears the smart band.

Similarly, in [8] Lara and Labrador present the areas of utilization the human activity recognition (HAR) systems. The authors mention the types of activities recognized by *state-of-the-art* HAR systems which may cover areas such as daily activities, fitness, military, ambulation or transportation. They also describe the data flow in training and testing such systems, focus on the architecture of the systems, design issues and methods that may be used for activity recognition.

The feature extraction step is widely explained, with the differentiation of the techniques used for measured attributes that are grouped into three sets: *time-domain, frequency-domain, and others*. The most popular methods for the time-domain group are mean, standard deviation, variance, mean absolute deviation and correlation between axes, whereas the techniques for the frequency-domain set are Fourier Transform and Discrete Cosine Transform.

Health-care monitoring systems using IoT technologies become more and more popular. The platform proposed in [18] uses data gathered from electrocardiograph and accelerator to analyze the posture and fatigue in a smartphone or in the Cloud. The paper presents an overview of IoT technologies and the architecture of the system and data processing steps performed in the Horton Works Data Platform (HDP). The article [15] describes the wearable and implantable sensors for distributed mobile computing. It also presents the difficulties and complications that may happen while using wearables. Similarly to the system that we have built and present in Sect. 4, the system presented in [15] aims to analyze the impact of weather conditions on the running training parameters.

The assumption that weather conditions may influence the training effectiveness is not only intuitive but also confirmed by published research. In [4,17] and [5], the authors study the impact of the temperature and other atmospheric measurements on the performance of marathon runners. All of these articles describe the optimal temperatures for the best running performance among men and women which are between $10\,°C-15\,°C$. One of the most recent studies that investigate the relationship between various physical exercises (including the running) and weather conditions are presented in [3]. The author analyzes seasonal upper-body strength resistance and running endurance performance, and studies if there are any relationships between the efficiency of these activities and weather conditions. Comparing to previous papers, the running distance is 5 km and the research shows that participants of the conducted experiment gained better results in summer and spring, which are hotter seasons. In our paper, we do not investigate these relationships (although, it is possible in the system that we have developed), but focus on the fuzzy join as a data preparation step before the main analysis begins.

The term *fuzzy join* is widely used in scientific literature but may have different meanings and applications. Many of published papers, including [1,2,7,19], use the term while combining data sets on the basis of flexible character data matching and string similarity with the use of various distance functions, like the Hamming distance. Meanwhile, articles [6] by Khorasani et al. and [9] by Małysiak-Mrozek et al. show how to flexibly combine big data sets with the use of fuzzy join operation by applying the fuzzy sets-based techniques on the numerical attributes. In our paper, we show how we utilize this idea on the numerical values of the time attribute while combining sensor data from a wearable device (a smartwatch) with meteorological data from weather sensors.

3 Fuzzy Sets

The classical set theory states that membership of an object to a set is bivalent - the object belongs to the set or does not belong to it. In comparison with classical set theory, the theory of fuzzy sets assumes that the membership of an object $\mu(x)$ to a set may be represented with countless values within the unit interval [0,1] [20]. Assuming that X is the universe of points (objects) and x is an element of X, the fuzzy set A in X is defined as follows:

$$A = \{(x, \mu_A(x)|x \in X)\}, \tag{1}$$

where $\mu_A(x)$ is *membership (characteristic) function* of the element x to set A and takes a value from 0 to 1.

Graphically, the *membership function* is usually represented as a triangular or trapezoidal function. Triangular function is used when there is only one situation such as the value of membership is equal to 1. This type of characteristic function is defined within parameters a, b and c, where $a \leq b \leq c$ as given by:

$$\mu_A(x; a, b, c) = \begin{cases} 0, & x \leq a \\ \frac{x-a}{b-a}, & a < x \leq b \\ \frac{c-x}{c-b}, & b < x \leq c \\ 0, & c < x \end{cases} \tag{2}$$

The triangular membership function is illustrated in Fig. 1.

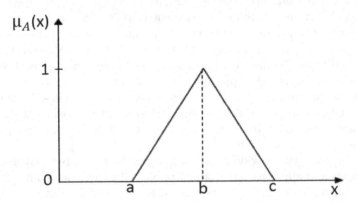

Fig. 1. Triangular membership function.

The trapezoidal characteristic function, in comparison to triangular one, is described within four parameters a, b, c and d, where $a \leq b \leq c \leq d$ and is defined as follows:

$$\mu_A(x; a, b, c, d) = \begin{cases} 0, & x \leq a \\ \frac{x-a}{b-a}, & a < x \leq b \\ 1, & b < x \leq c \\ \frac{d-x}{d-c}, & c < x \leq d \\ 0, & d < x \end{cases} \tag{3}$$

The shape of trapezoidal membership function is illustrated in Fig. 2.

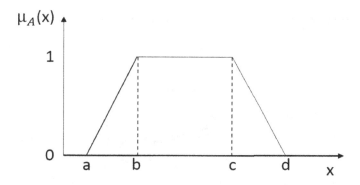

Fig. 2. Trapezoidal membership function.

4   Architecture of the System

Fuzzy sets defined by triangular membership functions were used in joining training data with weather parameters from meteorology sensors, which was a data preparation phase preceding data analysis on the Spark cluster in the Azure cloud. The architecture of the whole solution is shown in Fig. 3. During physical activity (we analyzed the effectiveness of running) training parameters were gathered by various sensors with the use of smartwatch. The data were saved into *.tcx* formatted files. *.tcx* is the acronym for Training Center XML introduced by Garmin Company and enables to exchange GPS tracks as an activity with parameters such as heart rate, running cadence, bicycle cadence, calories apart from raw GPS points. Independently, with the use of meteorology sensors, weather station collects atmospheric conditions. By using appropriate API (Application Programming Interface), the Dark Sky Web service provides data as JSON objects that consist of attribute-value pairs. Both training and meteorological data are transferred to a smartwatch application and weather service databases. Then, the data are transferred to the Cloud-based analytical system, which prepares the data for further analysis.

The Fuzzy Join module is responsible for data preparation, supplementation and combining before sending the data for further analysis. This phase consists of merging data collected by wearable sensors and atmospheric conditions measurements provided by an online weather Web service (available through appropriate API). The output data are saved into *.csv (comma-separated) values* files. This format enables to store tabular data in plain text. The full algorithm of the fuzzy join will be presented in Sect. 5 of this article.

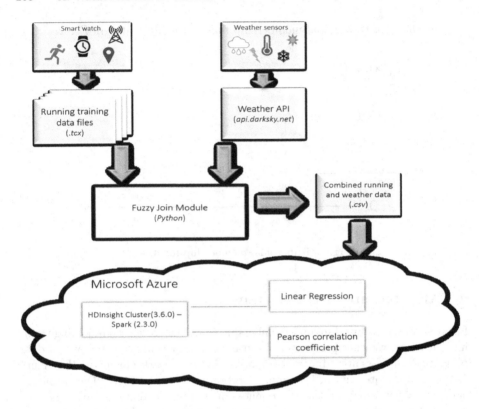

Fig. 3. Architecture of the Cloud-based system for training data analysis.

Due to large volumes of the training data that can be analyzed and wide scaling capabilities, the analysis phase is performed with use of the Apache Spark engine in the HDInsight cluster in the Microsoft Azure cloud platform [11,12]. Nowadays, cloud solutions are becoming increasingly popular. Microsoft Azure is one of the biggest public cloud platforms [10,13]. It offers many services that may be customized by the user. Azure is scalable and performs well with large volumes of data. Apache Spark, on the other hand, is a unified analysis platform that enables efficient processing of various types and amounts of data. As it can be used with Java, Python, R and SQL shell so creating applications does not require the knowledge of unknown programming languages. It has built-in libraries, including machine learning MLib. Due to its wide range of usage possibilities and popularity among developers, it is well documented. All of the above plus the ability to use it within Microsoft Azure Cloud made us choose this framework for data analysis. Our application uses Linear Regression and Pearson's correlation coefficient (both from MLib library) for evaluation of the impact of weather conditions on the quality and efficiency during the training (running/jogging). The training efficiency is measured by average running speed and the number of calories burnt during exercises.

The training data that we collected concerned running on an average distance of 10 km. The runners were amateurs with different level of running experience, various running habits, and frequency. Among them, there were men and women, at the age between 20 and 55.

5   Fuzzy Umbrella

The Fuzzy Join module processes training data stored in *.tcx* files and weather conditions that are accessed through the Dark Sky Web service API. The training data are gathered with the use of a wearable device (a smartwatch) and contains information about training parameters, such as the time stamp and coordinates of the beginning of the training, its duration, distance, average heart rate, calories burnt during the training, and many others. Weather conditions are retrieved by specifying the date, time, and coordinates of the training. They are collected by the application installed on the smartwatch as *.json* objects. This information contains hour-by-hour daily measurements of air temperature (real and apparent), the percentage level of air humidity, dew point, wind speed, air pressure, and others. The aim of the fuzzy join is to find the most accurate weather conditions based on the time the training begins.

The fuzzy join algorithm used in this module calculates the value of a membership degree for each full hour of weather measurement. The fuzzy join algorithm consists of the following steps (Fig. 4):

1. Convert the time of weather conditions measurement (e.g., t_{m_2} in Fig. 4) and the beginning of training (t_t) into seconds (e.g., 10 AM is converted to 36000 s).
2. Find times of meteorological measurements neighbouring to full hour of t_{m_2} (e.g., t_{m_1}, t_{m_3}) and convert them into seconds.
3. Compute the differences between the weather conditions measurement hours (t_{m_1}, t_{m_2}, t_{m_3}) and the beginning of the training (t_t) (values in seconds). Take the absolute values of the results.
4. Compute the ratio of the value from point 3 to the number of seconds in one hour (3600 s).
5. For each hour of weather conditions measurement the value of membership function is calculated as the difference of 1 and value calculated above (point 4). If the calculated value is not between 0 and 1 it has to be rejected, as it is out of the range of the membership function.
6. Take the maximum of all values of the calculated membership degree.

The use of fuzzy join allows to find the best matching of the weather conditions to the training as they are chosen by the nearest hour of measurement. The concept of the fuzzy umbrella is presented in Fig. 4. On the OX axis, there are hours of weather conditions measurements from sensors in meteorological stations. On the OY axis, there are placed values of membership function computed for each hour of weather measurement.

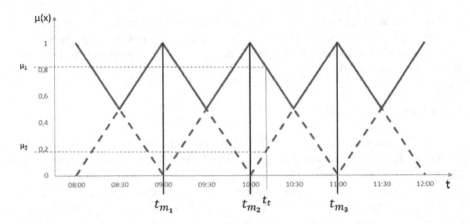

Fig. 4. The idea of fuzzy umbrella for joining weather and training data by time.

6   Experimental Results

We tested the performance of the fuzzy join operation in the environment presented in Fig. 3. The Fuzzy Join module combines data before sending the data
to the Cloud for further analysis and resides on an IoT field gateway. The performance tests were conducted on PC station with 8 GB RAM and processor
Intel(R) Core(TM) i7-3537U CPU @ 2.00 GHz, controlled by the 64-bit Windows operating system. To obtain the most reliable results, during experiments
no other applications were running on the machine.

We tested the performance of the fuzzy join on the data set consisting of 850
files (.tcx) with data from real training (running/jogging). During our tests, for
each out of the 850 data files we measured: the time of reading and processing the
data file (presented in Table 1 under the term *Load file*), the URL request execution time (meteorology data reading), the duration of computing the membership
function value, the time of saving object that contains the combined training and
weather data into the .csv file, and finally, the overall duration of all the operations performed on the particular .tcx file. The results of performance tests
are presented in Table 1. To provide an easy comparison of the duration of each
of the performed processes the times are given in milliseconds. Apart from the
operations presented in Table 1, there are also instructions that prepare objects,
such as the access key and the URL address for the weather service, the variables
and tables to store data or the output files. As these operations are not much
time-consuming they are not listed in the Table 1.

The results of the experiments show that the operation of Fuzzy Join takes
the least time in comparison to other operations. On average, it takes less than
0.20 ms which is 0.0015% of overall time spent on processing a particular data
file. The maximum time of fuzzy join process is more than four times greater
than the minimum value but it is probably due to the non-isolated environment

of conducted tests. The mean and median times are similar and the standard deviation value is comparatively small which implies that the results are reliable.

Table 1. Duration of particular operations on one set of training data.

Time (ms)	Overall	Load file	URL request	Fuzzy Join	Save object
Mean	1213.841	1058.480	50.753	0.181	5.357
Median	671.875	625.000	46.875	0.177	4.620
Std dev.	1134.257	874.849	17.188	0.046	5.664
Min.	171.875	78.125	15.625	0.070	4.113
Max.	7515.625	5953.125	296.875	0.400	102.468

The loading file operation consists of opening the *.tcx* training data file, iterating on the data inside to gather information about training parameters, such as the date and the time of the training, coordinates of the beginning of the training, distance and duration of the training, average heart rate, the number of calories, etc. Some of the data need to be calculated – for example, the average running pace is not given in a file, so it has to be computed based on the time and the distance of training. As the inspected files have various sizes, the time of the analysis also differs. The file size depends on the number of measurements and is related to the model of smart watch and the settings customized by the user. The operation of loading file is the longest operation performed in the system and takes on average 80% of the overall time. Using the online weather service to access atmospheric conditions explains an enormous difference between the minimum and the maximum execution time of URL requests. This step depends on the network traffic and its efficiency. The value of standard deviation is not as small as in fuzzy join operation, which means that the particular results are varying from each other.

In Table 1 we also presented the duration of saving combined training and weather data to an output file performed by the Fuzzy Join module. The time of this operation is on average about 5 milliseconds, which is about 0.44% of the overall time of one file processing. Although the maximum time attracts attention due to its high value, the values of other measures point that it is rather a singular outlying result than a frequent issue.

7   Conclusion

In the paper, we showed the idea of using the fuzzy join operation for combining training data with weather parameters. The fuzzy join operation is a preparation step for further data analysis.

The preparation step consists of reading files with training data, gathering the important information from the data files, sending URL requests to weather Web services, combining the atmospheric and training parameters together and saving the results to output files. Although the operation of the fuzzy join takes the least time out of all processes, it is very important as it supplements the existing training data with additional information that may shed new light on the analyzed data. The results of the performed experiments show that the algorithm used to perform the fuzzy join as a data preparation step for data analysis is time-efficient and that the execution time of the operation is negligible.

In comparison to the other works presented in Sect. 2 this paper describes simple but fast algorithm for flexible combining of training data from wearable sensors with the most close weather parameters. The algorithm operates on numerical representation of time stamps, similar to the solution presented in [9] that operates on numbers in Big Data cloud environments, and in contrast to works [1,2,19] that operate on strings. Moreover, likewise it was presented in works by Yamato [18], Revathi Pulichintha Harshitha et al. [15], Małysiak-Mrozek et al. [9] the whole solution is built upon the cloud infrastructure, which ensures scalability while analyzing the data. However, the fuzzy join is performed on the Edge, which frees the Cloud data center from the necessity to pre-process the data before the analysis.

Acknowledgements. This work was supported by Microsoft Research within Microsoft Azure for Research Award grant, pro-quality grant for highly scored publications or issued patents of the Rector of the Silesian University of Technology, Gliwice, Poland (grant No 02/020/RGJ19/0167), and partially, by Statutory Research funds of Institute of Informatics, Silesian University of Technology, Gliwice, Poland (grant No BK/204/RAU2/2019).

References

1. Afrati, F.N., Sarma, A.D., Menestrina, D., Parameswaran, A., Ullman, J.D.: Fuzzy joins using MapReduce. In: 2012 IEEE 28th International Conference on Data Engineering, pp. 498–509 (04 2012). https://doi.org/10.1109/ICDE.2012.66

2. Deng, D., Li, G., Hao, S., Wang, J., Feng, J.: MassJoin: a MapReduce - based method for scalable string similarity joins. In: 2014 IEEE 30th International Conference on Data Engineering, pp. 340–351 (03 2014)

3. Dhahbi, W.: Seasonal weather conditions affect training program efficiency and physical performance among special forces trainees: a long-term follow-up study. PLoS ONE **13**(10) (10 2018). https://doi.org/10.1371/journal.pone.0206088

4. El Helou, N., Tafflet, M., Berthelot, G., Tolaini, J., Marc, A., Guillaume, M., Hausswirth, C., Toussaint, J.F.: Impact of environmental parameters on marathon running performance. PLoS ONE (05 2012). https://doi.org/10.1371/journal.pone.0037407

5. Ely, M.R., Cheuvront, S.N., Roberts, W.O., Montain, S.J.: Impact of weather on marathon-running performance. Med. Sci. Sports Exerc. **39**(3), 487–493 (2007)

6. Khorasani, E.S., Cremeens, M., Zhao, Z.: Implementation of scalable fuzzy relational operations in MapReduce. Soft Comput. **22**(9), 3061–3075 (2018). https://doi.org/10.1007/s00500-017-2561-3

7. Kimmett, B., Srinivasan, V., Thomo, A.: Fuzzy joins in MapReduce: an experimental study. In: Proceedings of VLDB Endow, pp. 1514–1517 (08 2015). https://doi.org/10.14778/2824032.2824049

8. Lara, O.D., Labrador, M.A.: A survey on human activity recognition using wearable sensors. IEEE Commun. Surv. Tutorials **15**(3), 1192–1209 (2013)
9. Małysiak-Mrozek, B., Lipińska, A., Mrozek, D.: Fuzzy join for flexible combining big data lakes in cyber-physical systems. IEEE Access **6**, 69545–69558 (2018). https://doi.org/10.1109/ACCESS.2018.2879829
10. Małysiak-Mrozek, B., Stabla, M., Mrozek, D.: Soft and declarative fishing of information in big data lake. IEEE Trans. Fuzzy Syst. **26**(5), 2732–2747 (2018). https://doi.org/10.1109/TFUZZ.2018.2812157
11. Małysiak-Mrozek, B., Baron, T., Mrozek, D.: Spark-IDPP: high-throughput and scalable prediction of intrinsically disordered protein regions with Spark clusters on the Cloud. Cluster Comput., November 2018. https://doi.org/10.1007/s10586-018-2857-9
12. Mrozek, D., Daniłowicz, P., Małysiak-Mrozek, B.: HDInsight4PSi: boosting performance of 3D protein structure similarity searching with HDInsight clusters in Microsoft Azure cloud. Inf. Sci. (2016). https://doi.org/10.1016/j.ins.2016.02.029
13. Mrozek, Dariusz: Scalable Big Data Analytics for Protein Bioinformatics. CB, vol. 28. Springer, Cham (2018). https://doi.org/10.1007/978-3-319-98839-9
14. Mukhopadhyay, S.C.: Wearable sensors for human activity monitoring : a review. IEEE Sensors J. **15**(3), 1321–1327 (2015)
15. Revathi Pulichintha Harshitha, S., Narramneni, P., Raghavee, N.S.: Body sensor using internet of things (IoT). ARPN J. Eng. Appl. Sci. **13**(8) (2018)
16. Toh, W.Y., Tan, Y.K., Koh, W.S., Siek, L.: Autonomous wearable sensor nodes with flexible energy harvesting. IEEE Sensors J. **14**, 2299–2306 (2014)
17. Vihma, T.: Effects of weather on the performance of marathon runners. Int. J. Biometeorol. **54**(3), 297–306 (2010). https://doi.org/10.1007/s00484-009-0280-x
18. Yamato, Y.: Proposal of vital data analysis platform using wearable sensor. In: Proceedings of the 5th IIAE International Conference on Industrial Application Engineering (2017)
19. Yan, C., Zhao, X., Zhang, Q., Huang, Y.: Efficient string similarity join in multi-core and distributed systems. PLoS ONE **12**(3), 1–16 (2017)
20. Zadeh, L.A.: Fuzzy sets. Inf. Control **8**, 338–353 (1965)

Industrial Applications

On the Interdependence of Technical Indicators and Trading Rules Based on FOREX EUR/USD Quotations

Bartłomiej Kotyra[✉] and Andrzej Krajka[✉]

Institute of Informatics, Maria Curie-Skłodowska University,
Pl. Marii Curie-Skłodowskiej 1, Lublin, Poland
bartlomiej.kotyra@gmail.com, akrajka@gmail.com

Abstract. The general aim of this paper is to investigate the interdependence within the wide set (2657) of technical analysis indicators and trading rules based on daily FOREX quotations from 01.01.2004 to 20.09.2018. For the purpose of this paper, we have limited our study to *EUR/USD* quotations only. The most frequently used methods for FOREX behavior modeling are regression, neural networks, ARIMA, GARCH and exponential smoothing (cf. [1,3,6]). They are used to predict or validate inputs. Inputs interdependence may cause the following problems:

- The error term is obviously not normally distributed (for regression it is heteroscedastic). Therefore we lose the main tool for the model validation.
- R-squared becomes a useless measure.
- The obtained model is problematic for forecasting purposes. We would normally like to forecast the probability of a certain set of independent variables to create a certain output - the FOREX observations.
- Regression and neural network methods use the inverses of some matrices. When we use two identical variables as input, the matrix is singular. When we use some dependent variables, the matrix is "almost" singular. It leads to model instability (assuming that computations are possible at all).
- It is meaningless to evaluate which of the two identical inputs is more significant.

Therefore the independence of inputs is a crucial problem for FOREX market investigation. It may be done directly, as shown in this paper, or by means of PCA techniques, where the inputs are mapped into the small set of independent variables. Unfortunately, in the second case, the economical meaning is lost.

The obtained results may be treated as the base for building FOREX market models, which is one of our future goals.

Keywords: FOREX · Market indicators · Technical analysis

© Springer Nature Switzerland AG 2019
S. Kozielski et al. (Eds.): BDAS 2019, CCIS 1018, pp. 277–290, 2019.
https://doi.org/10.1007/978-3-030-19093-4_21

1   Introduction

The prediction of market behavior is one of the most crucial problems for traders. There are a lot of methods and procedures which can be used in attempt to achieve financial success. We can assume that many of such procedures are not known to the public. On the other hand, there is a rich bibliography of scientific investigation on market behavior. The most frequently used prediction methods include regression, ARIMA, GARCH, exponential smoothing and neural networks. These can be based on historical market quotations, as well as some indicators and technical trading rules. This kind of data can be used as the input for prediction models. There is a lot of software (for example Meta-Trader platform) allowing users to construct and examine such indicators. On the other hand, prediction methods are often used inconsistently with their basic assumptions. In regression methods, for example, it is assumed that inputs are independent. This assumption is present in neural networks models, too. In order to achieve this kind of independence, the Principal Component Analysis (PCA) is often applied. An interesting generalization of this method - $(2D)^2PCA$ - is described and discussed in [8]. However, since the PCA methods lead to input data transformation, we lose some of its original financial meaning. In this paper we investigate the interdependencies within the rich set of market indicators and trading rules.

The investigation of data independence is generally a problematic task. Although it is easy to compute Person's, Spearmann's or Kendall's correlation, we should not assume that uncorrelated random variables are truly independent. This assumption could be valid for gaussian attributes, but FOREX EUR/USD quotations are not gaussian (Shapiro-Wilk's test $p < 2.2 * 10^{-16}$ for OPEN, LOW, HIGH and CLOSE, Hurst ratio is between 0.391 and 0.749). The dependence should be investigated using the χ^2 test (noncontinuous observations) with binding observations to some intervals (windows). However, the assumption that each window should contain at least 5 observations is often difficult to satisfy. Furthermore, it is known that three pairwise independent random variables are not necessarily independent. Thus the χ^2 test should be repeated for large combinations of attributes.

1.1   Input Data, Indicators and Their Classifications

The input data comprises a large set of FOREX EUR/USD quotations ($T = 3841$ observations from 01.01.2004 to 20.09.2018) named $DATE_t, OPEN_t, HIGH_t,$ $LOW_t, CLOSE_t, 1 \leq t \leq T$.

On the basis of FOREX EUR/USD quotations we computed a large set of 2657 technical indicators and trading rules (cf. Murphy [5] and Schwager [7]). We decided to classify the content of our data set in two ways. The first classification (referred to as $Cl(A)$ classification) was based on the technical method used. Here we distinguished the classes described in Table 1.

Table 1. Classification *Cl(A)* of indicators and rules

Name	Abbr.	Number of series	Short description
BOLLINGER	BB	266	Bollinger Bands
CCI	CCI	10	Commodity Channel Index
DEMA	DEM	337	Double Exponential Moving Average
DI	DI	82	Directional Indicator
DMI	DMI	27	Directional Movement Index
EMA	EMA	289	Exponential Moving Average
HL-INDEX	HLI	192	Conditions based on highest and lowest prices (HI - highest index, LI - lowest index)
MOMENTUM	MTM	29	Momentum Indicator
PRICE	PRI	48	Conditions based directly on recent price values
PRICE_CHANNEL	PCH	24	Conditions related to price channels
RANGE	RAN	99	Simple volatility measure based on bar ranges
RSI	RSI	328	Relative Strength Index
SMA	SMA	5	Simple (Arithmetic) Moving Average
STS	STS	453	Stochastic Oscillator
TEMA	TEM	337	Triple Exponential Moving Average
TRUE_RANGE	TR	99	True Range - widely used volatility measure
VHF	VHF	27	Vertical Horizontal Filter
WMA	WMA	5	(Linearly) Weighted Moving Average

For each indicator we implemented a few different versions. The choice of parameters (period, smoothing, deviation multiplier) for these versions is based on the values commonly used by trades (cf. [5,7]).

Bollinger Bands are a widely used tool for characterizing price volatility. Their formula is based on moving averages and standard deviation, calculated using recent price values. Two bands (referred to as an "upper band" and a "lower band") are constructed around a moving average, expressing a price volatility range. There are many possible ways to use this indicator, but the most common rules are related to price crossing one of the bands. In our research, we constructed multiple versions of Bollinger Bands, utilizing two types of moving averages (EMA and SMA) with different periods and deviation multipliers. We implemented trading rules based on the most common interpretations.

Commodity Channel Index (CCI) is one of the most popular oscillators used by traders and market analysts. It is often used to identify probable trend reversal points. The formula is based on a simple moving average and mean absolute deviation:

$$CCI_t = \frac{1}{0.015} \frac{p_t - SMA(p_t)}{MD(p_t)},$$

where $p_t = \frac{HIGH_t + LOW_t + CLOSE_t}{3}$. We implemented the CCI oscillator along with relatively simple rules of its interpretation. We used a set of different periods to construct diverse versions of the indicator.

Double Exponential Moving Average (DEMA) is an indicator aiming at improving the characteristics of standard moving averages. The idea is based on transforming an exponential moving average in order to reduce its time lag:

$$DEMA_t = 2 * EMA_t - EMA(EMA_t).$$

We implemented multiple versions of the indicator, utilizing different period lengths. Our set of trading rules consists mostly of simple conditions, comparing values of current price and different versions of the indicator.

Directional Indicator set contains Positive Directional Indicator (DI$^+$) and Negative Directional Indicator (DI$^-$). These two are often used together to measure the upward and downward trends. In our research, we used both indicators with a few different periods. We implemented trading rules based on the relation between DI$^+$ and DI$^-$ values.

Directional Movement Index (DMI) is closely related to the idea of Directional Indicator (they are usually implemented together as a single tool). Values of Positive Directional Indicator and Negative Directional Indicator are used here to measure the strength of the trend, without informing about its direction.

$$DMI_t = \frac{|(DI_t^+) - (DI_t^-)|}{(DI_t^+) + (DI_t^-)} * 100.$$

We constructed a few versions of this indicator, using different period lengths. Our set includes simple rules based on both raw DMI and its smoothed version (referred to as ADX - Average Directional Index).

Exponential Moving Average (EMA) is one of the most commonly used type of moving averages. Its exponential smoothing treats the most recent input values as the most important ones.

$$EMA_t = \begin{cases} Y_1 & t = 1 \\ \alpha * Y_t + (1 - \alpha) * EMA_{t-1} & t > 1 \end{cases},$$

where $\{Y_t, 1 \le t \le T\}$ is one from $\{OPEN_t, HIGH_t, CLOSE_t, LOW_t, 1 \le t \le T\}$. We implemented multiple versions of this moving average, utilizing different period lengths and input series. Our set of rules includes a lot of commonly used ways to apply this tool.

Highest/Lowest Index set contains rules based on measuring the highest and lowest prices from a given period. We consider both the value and placement of the highest and lowest price.

Momentum (MTM) is a relatively simple indicator used to determine direction and strength of a price trend. It compares the current price to price values from the past (k days ago).

$$MTM_t(k) = CLOSE_t - CLOSE_{t-k}, 1 \le k < t \le T.$$

We added a few versions of this indicator to our set, using different period lengths. We implemented a set of rules based on popular interpretations of this tool.

Price set contains simple rules and conditions based on recent price values. We observed how the price behaved over the last few days.

Price Channels are based on measuring the highest and the lowest points of a price movement. This technique is used to determine an area where the price is expected to remain. We used a few different approaches to this technique. Our set contains rules based on the most common ways to apply them.

Range category contains rules related to a simple volatility measure. Conditions are based on the average range of price bars from the recent days. We used a few different periods to diversify this set.

Relative Strength Index (RSI) is an oscillator commonly used to recognize overbought and oversold states. It measures recent price movements to estimate the current condition of the market. We implemented a few versions of this indicator (using different period lengths and optional smoothing) and a set of trading rules based on these. We covered the most common ways to interpret this kind of tool.

Simple Moving Average (SMA) is probably the most recognizable type of moving averages. Its arithmetic weighting treats all input values (from the whole period) as equal. This indicator is commonly used to determine the trend's direction and strength.

$$SMA_t^N(Y) = \frac{\sum_{i=t-N+1}^{t} Y_i}{N}.$$

We included a few instances of Simple Moving Average in out set, using different period lengths.

Stochastic Oscillator (STS) is one of the most popular oscillators used to determine overbought and oversold areas. There are a few variants of this indicator. We implemented both raw (fast) and smoothed (slow) versions of this tool.

Triple Exponential Moving Average (TEMA) is another indicator based on the idea of moving averages. It utilizes three different averages to construct a single composite with a reduced time lag.

$$TEMA_t = 3 * EMA_t - 3 * EMA(EMA_t) + EMA(EMA(EMA_t)).$$

We implemented a few versions of this indicator along with a set of rules (similar to other sets related to moving averages) based on them.

True Range (TR) is a widely used volatility measure. The idea is based on bar ranges, but also takes price gaps into account. Average True Range (ATR) is commonly used to determine price volatility to be expected. We constructed a few versions of this indicator using different period lengths for averaging. We prepared a set of rules based on how volatile the market was in the recent days.

Vertical Horizontal Filter (VHF) is an indicator used to measure the strength of a trend. It is commonly used to determine whether prices are currently following a specific direction or remaining in the horizontal trend.

$$VHF_t = \frac{HCP_t(k) - LCP_t(k)}{\sum_{i=1}^{k} |CLOSE_i - CLOSE_{i-1}|},$$

where $HCP_t(k) = \max\{CLOSE_i : t - k \leq i \leq t\}, LCP_t(k) = \min\{CLOSE_i : t - k \leq i \leq t\}, 1 \leq k < t \leq T$. We constructed a few instances of this indicator (using different period values) and prepared a set of rules based on them. Our conditions utilize standard level values, commonly used to separate vertical and horizontal trends.

Weighted Moving Average (WMA) is another approach to constructing moving averages. It utilizes the linear weighting of input values in order to reduce the time lag associated with this kind of tools.

$$WMA_t^N(Y) = \frac{NY_t + (N-1)Y_{t-1} + \cdots + 2Y_{(t-N+2)} + Y_{(t-N+1)}}{N + (N-1) + \cdots + 2 + 1}.$$

We implemented a few versions of this indicator, using different period lengths.

The second classification (referred to as $Cl(B)$ classification) was based on the amount of price observations used by the computation procedure. We distinguished the following intervals (written for the shortness as R instruction)

Creating the Factor of Cl(B) Classification

```
ClB=cut(nlength, c(0,4,9,14,19,24,29,39,44,49,59,64,79,
                   84,89,99,104,109,159,199,Inf))
```

where *nlength* is the number of price observations taken (period). Every class in this classification will be denoted as the first element belonging to the range (1 for $(0, 4]$, 5 for $(4, 9]$ etc., cf. Table 7).

We will assume that at a certain moment t, $X_t v Y_t = 1$ when X_t crosses above $(v = ca)$, crosses bellow $(v = cb)$, is above $(v = a)$ or is below $(v = b)$ Y_t. Otherwise, $X_t v Y_t = 0$. For example, if $X_t = (t-5)^2 - 4, Y_t = 0, t = 1, 2, \ldots 10$, then

$$X_t a Y_t = \begin{cases} 1, & \text{if } t = 1, 2, 8, 9, 10, \\ 0, & \text{otherwise,} \end{cases} \quad X_t cb Y_t = \begin{cases} 1, & \text{if } t = 4, \\ 0, & \text{otherwise,} \end{cases}$$

$$X_t b Y_t = \begin{cases} 1, & \text{if } t = 4, 5, 6, \\ 0, & \text{otherwise,} \end{cases} \quad X_t ca Y_t = \begin{cases} 1, & \text{if } t = 8, \\ 0, & \text{otherwise.} \end{cases}$$

Furthermore, C_x denotes $CLOSE_{t-x}$ price, $C = C_0$, BBx^Z denotes upper $(x = u)$ or lower $(x = l)$ Bollinger Band based on SMA $(Z = S)$ or EMA $(Z = E)$. For Bollinger Bands, $X_{p,m}$ denotes a band with period p and multiplier m. For RSI and STS, $X_{p,s}$ denotes the respective indicator with period p and smoothing s. For DEM and TEM, $X_{p,m}$ denotes the respective indicator period p and momentum m. For $X = DEM$ or $X = TEM$, $X_{p,m} = MTM_m(X_p)$. For $X = RSI$ or $X = STS$, we will assume that $X_{p,1} = X_p$. For $X = EMA$, $X = HI$, $X = LI$ or $X = VHF$, we have only one parameter (period p). Thus, for example $CcbBBl_{80,3}^S = 1$ when CLOSE price crosses below the lower Bollinger Band based on the SMA with period 80 and multiplier 3. The indicator symbol informs us about the category of the series in $Cl(A)$ classification whereas the period value informs us about the category in $Cl(B)$ classification.

1.2   Details of Research Procedures

On the basis of described indicators we computed the Person's correlation matrix $\{\rho_{i,j}, 1 \leq i,j \leq 2657\}$. On the arbitrary chosen level 0.05, the series i and j with $-0.05 \leq \rho_{i,j} \leq 0.05$ were treated as uncorrelated, whereas the remaining pairs were treated as correlated. The function used to measure correlation-based distance was defined as $d_{i,j} = 1 - |\rho_{i,j}|, 1 \leq i,j \leq 2657$. Furthermore, for each series we divided the interval between the minimum and maximum value into 10 subintervals. If the amount of values in such a "window" was smaller than 5, it was joined with one of its neighbors. Using the R command we built the matrix of probability values of χ^2 test for each pair $1 \leq i,j \leq 2657$.

Computation of χ^2 Test p Value

```
dchi[i,j]<-chisq.test(table(data[,i],data[,j]),
          simulate.p.value=TRUE,B=2000)$p.value
```

The matrix $dchi_{i,j}$ was treated as the measure of pairwise independence between series. When $dchi_{i,j} > 0.05$ (significance level commonly used in statistics) we assumed that series i and j were independent. In other cases we treated them as dependent.

1.3   The Main Aim

In Sect. 3, we compared these two classifications with the distance functions $d_{i,j}$ and $dchi_{i,j}$. It is interesting in which group we get the most dependent indicators. Theoretically, if $\{X_i, i \geq 1\}$ are independent random variables, and f and g are continuous monotone functions, then $f(X_1, X_2, X_3, \ldots X_k)$ and $g(X_1, X_2, \ldots X_k)$ are dependent, but $f(X_1, X_2, \ldots X_k)$ and $f(X_1, X_2, \ldots X_k, X_{k+1}, \ldots X_{k+m})$ are independent. This suggests that distance $d(i,j)$ within every *Cl(B)* groups should be "higher" than that for *Cl(A)*. However, FOREX EUR/USD time series is not "white noise" (observations are not independent). Furthermore, if functions f and g are not monotone, from dependent random variables we may obtain independent result (cf. the Box-Muller method for constructing independent gaussian random variables from uniformly distributed independent random variables).

The general aim of this paper is to present an approach to market data analysis along with obtained results. This study can help researchers build models of FOREX market behavior. Section 2 of this paper presents the most interesting sets of pairwise uncorrelated indicators and pairwise independent indicators. The remaining sets can be found at matrix.umcs.lublin.pl/~akrajka/BDAS19. Tables 6 and 7 together with the described results for *Cl(B)* can also be helpful.

All calculations were implemented in R programming language and performed on RStudio platform. In the next section we present the distribution of correlations and dependency of market indicators and example subsets of independent and uncorrelated indicators obtained by us. In Sect. 3 we compared the classifications *Cl(A)* and *Cl(B)* with correlations and dependency, giving the hints for building correct FOREX models. The last section contains the main conclusions of our research.

2   The Sets of Uncorrelated and Independent Indicators

The distribution of correlation $\{\rho_{i,j}, 1 \leq i,j \leq 2657\}$ and probability values (quantiles) of χ^2 test can be seen in Figs. 1 and 2. The critical areas of uncorrelated and independent pairs are marked with blue color. The distribution of correlation values may seem gaussian. However, this is not true. Despite different "pessimistic" opinions, these results show a big number of uncorrelated pairs (51%) and a big number of pairwise independent pairs (40%) of series.

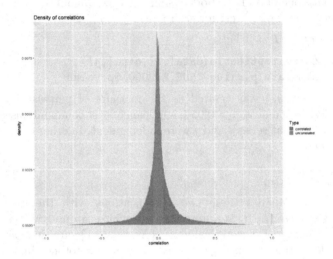

Fig. 1. Density function of indicators correlation. Blue color denotes correlation range [–0.05,0.05]. Such cases are treated here as uncorrelated. (Color figure online)

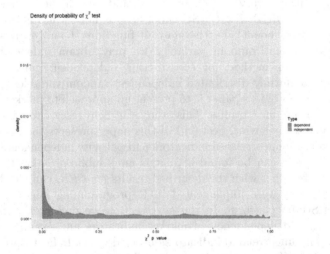

Fig. 2. Density function of p value from χ^2 tests for pairs of indicators. Blue color denotes p value greater then 0.05. Such cases are treated as pairwise independent. (Color figure online)

In order to build a FOREX market model, we need the set on uncorrelated or independent series. Therefore, we defined the following arrays:

$$D_{i,j} = \begin{cases} 1, & \text{if } |\rho_{i,j}| < 0.05 \\ 0, & \text{otherwise} \end{cases} \qquad Dchi_{i,j} = \begin{cases} 1, & \text{if } dchi_{i,j} > 0.05 \\ 0, & \text{otherwise} \end{cases}$$

The problem of finding the maximal subset of uncorrelated (pairwise independent) random variables may be formulated in the terms of graph theory as the problem of finding maximal clique for graphs described by matrix D and $Dchi$. The Bron-Kerbosch algorithm, first described in [2], can be used to solve this problem. There are two implementations of this algorithm in R language (`max_clique` in `igraph` and implementation in Bioconductor package `RBG`). Unfortunately, both turned out to take excessively long time and required too much memory. Therefore, using the theory developed in [4] we implemented the incremental version of Bron-Kerbosch algorithm in R language.

Table 2. Independent clique

$CcaBBl^{E}_{10,2}$	$CcaBBl^{E}_{40,3}$	$CcaBBu^{E}_{40,3}$	$CcaBBu^{E}_{100,3}$
$CcbBBl^{E}_{5,2}$	$CcbBBl^{E}_{10,3}$	$CcbBBl^{E}_{100,1}$	$CcbBBu^{E}_{5,3}$
$CcbBBu^{E}_{20,3}$	$CcbBBu^{E}_{80,3}$	$CaBBu^{E}_{5,2}$	$CcbBBl^{S}_{80,3}$
$BBl^{E}_{100,2}$	$DEM_{20}caDEM_{80}$	$DEM_{80}caDEM_{100}$	$DEM_{20}cbDEM_{80}$
$DEM_{40}cbDEM_{80}$	$DEM_{80}cbDEM_{100}$	$DEM_{20,5}cb0$	$EMA_{80}caEMA_{100}$
$MIN(HI_5, LI_5)b3$	$MIN(HI_{10}, LI_{10})a7$	C_5cbC_10	$RSI_{40}ca20$
$RSI_{20,5}ca80$	$STS_{10}caSTS_{20}$	$STS_{20}caSTS_{40}$	$STS_{20,5}ca40$
$STS_{20,5}ca20$	$STS_{100,5}ca90$	$STS_{100,5}cb80$	$TEM_{40}caTEM_{100}$
$TEM_{40,10}ca0$			

Table 3. Uncorrelated clique

$CcaBBl^{E}_{10,2}$	$CcaBBl^{E}_{40,3}$	$CcaBBu^{E}_{5,2}$	$CcaBBu^{E}_{100,2}$
$CcbBBl^{E}_{5,2}$	$CcbBBl^{E}_{10,3}$	$CcbBBl^{E}_{100,1}$	$CcbBBu^{E}_{5,3}$
$CcbBBu^{E}_{20,3}$	$CcbBBu^{E}_{80,3}$	$CaBBu^{E}_{40,3}$	$CcaBBu^{S}_{100,3}$
$CcbBBu^{S}_{10,1}$	$CcbBBl^{S}_{80,3}$	$CcbBBl^{S}_{100,2}$	$DEM_{20}caDEM_{80}$
$DEM_{80}caDEM_{100}$	$DEM_{20}cbDEM_{80}$	$DEM_{40}cbDEM_{80}$	$DEM_{80}cbDEM_{100}$
$DEM_{10,10}ca0$	$DEM_{20,5}cb0$	$EMA_{10}cbEMA_{40}$	$EMA_{80}caEMA_{100}$
$HI_{100}a95$	$MIN(HI_5, LI_5)b3$	$MIN(HI_{10}, LI_{10})a5$	$MAX(HI_{10}, LI_{10})b2$
C_5cbC_{10}	$RSI_{40}ca20$	$STS_{10}caSTS_{20}$	$STS_{20}caSTS_{40}$
$STS_{10}cbSTS_{20}$	$STS_{20,5}ca20$	$STS_{40,5}ca40$	$STS_{100,5}cb80$
$TEM_{20}caTEM_{80}$	$TEM_{40}caTEM_{100}$	VHF_{80}	

We used this algorithm to find the cliques in D and $Dchi$. The most interesting sets are presented in Tables 2 and 3. We could recommend these sets as input data for the process of preparing FOREX market behavior models. It is worth noting that in each clique we have only one non-binary time series ($BBl_{100,2}^{E}$ and VHF_{80}).

3   Classifications

This section contains the comparison of $Cl(A)$ and $Cl(B)$ classifications with d and $dchi$ distances. This was done in two ways. The first approach was based on comparing classifications built on d and $dchi$ with "true" classifications $Cl(A)$ and $Cl(B)$. We also calculated mean d and $dchi$ distances within and between classes $Cl(A)$ and $Cl(B)$ and compared them.

Let us describe the first method. We used the following methods of classifications: k-Nearest Neighbors Method (NNk) and Mean Nearest Center method (MCN), because these methods allow us to find the "geometrical" structure of $Cl(A)$ and $Cl(B)$ classifications for d and $dchi$ distances. They were applied to classifications $Cl(A)$ and $Cl(B)$, taking every element as unclassified and executing NNk and MCN method for this element. For NNk we considered two cases: $k = 16$ and $k = 64$.

Table 4. Percentage of correctly classified series (nearest 16 neighborhood, nearest 64 neighborhood, minimum class distance) in the classifications Cl(A) and Cl(B)

Distance	$Cl(A)$			$Cl(B)$		
	$NC16$	$NC64$	MCN	$NC16$	$NC64$	MCN
d	62.5	46.9	98.9	56.1	43.8	98.6
$dchi$	13.1	7.3	4.4	10.2	13.1	16.1

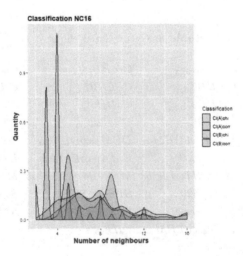

Fig. 3. The number of neighbors that influenced the affiliation to class in NC16 classification.

The number of neighbors on the basis of which the indicator was classified in NC16 classifications is shown in Fig. 3 (with smoothing for visual purpose). The percentage of true classified elements is shown in Table 4. Furthermore, in order to compare classifications $NN16, NN64$ and MCN with $Cl(A)$ and $Cl(B)$, we used the measures for multi-class classification described in [9] Table 3 p.430. We will use the following abbreviations: $Acc, Err, Pr, Rec, Fsc(\beta), Pr_M, Rec_M,$ $Fsc_M(\beta)$ for the *Average Accuracy, Error rate, Precision, Recall, Fscore(β), Precision$_M$, Recall$_M$, Fscore$_M$(β)*, respectively (index M for macro averaging, all details in [9]).

Table 5. The characterizations of procedure classifications based on correlation distance (nearest 16 neighborhood, nearest 64 neighborhood, minimum class distance) in the classifications Cl(A) and Cl(B).

Coefficient	$Cl(A)$			$Cl(B)$		
	$NC16$	$NC64$	MCN	$NC16$	$NC64$	MCN
Acc	0.9583	0.9410	0.9987	0.95611	0.94377	0.9987
Err	0.0417	0.0590	0.0013	0.0438	0.0562	0.0014
Pr	0.6251	0.4693	0.9887	0.5612	0.4377	0.9865
Rec	0.6251	0.4693	0.9887	0.5612	0.4377	0.9865
$Fsc(0.5)$	0.6251	0.4693	0.9887	0.5612	0.4377	0.9865
Pr_M	0.6118	0.3107	0.9792	0.4984	0.2588	0.9738
Rec_M	0.4994	0.2859	0.9894	0.3842	0.1952	0.9950
$Fsc_M(0.5)$	0.5854	0.3054	0.9812	0.4704	0.2430	0.9780

In order to construct the Table 5, for each method of classifications $(NC16, NC64, MCN)$ and each classification of indicators $(Cl(A),\ Cl(B))$, as well as each class of this classification (18 and 20 in $Cl(A)$ and $Cl(B)$ cases, respectively) we obtained two sets of indicators: A_i - a set of indicators belonging to the i-th class of classification, and M_i - a set of indicators classified by method that belongs to the i-th class of classification. We calculated $tp_i = |A_i \cap M_i|, tn_i = |\overline{A_i} \cap \overline{M_i}|, fp_i = |\overline{A_i} \cap M_i|$ and $fn_i = |A_i \cap \overline{M_i}|$ (named true positive, true negative, false positive and false negative, respectively), where $|A|$ and $\overline{A}$ denotes the cardinality and the compliment of the set AS. Further calculations were executed according to the formulas and procedures described in [9] (Table 3).

The results for dependence matrix *dchi* are very similar although the values *Acc* are smaller. We see that the big accuracy indicates the large effectiveness of classifiers but small precision and recall indicates the large differences between Cl(A) and Cl(B) classifications.

As the next step, for two classes of Cl(A) or Cl(B) we defined the measure of consistency as

$$\eta(c_i, c_j) = \frac{\sum_{k \in c_i, l \in c_j} d_{k,l}}{|c_i||c_j|}, 1 \leq i, j \leq l. \tag{1}$$

Because the d is measure of uncorrelatedness, the greater values indicate the greater possibility that two chosen elements from two classes will be uncorrelated. Full results are presented in Tables 6 and 7.

Table 6. The average $1 - |\rho_{i,j}|$ function within and between classes of Cl(A) classification

	BB	CCI	DEM	DI	DMI	EMA	HLI	MTM	PRI	PCH	RAN	RSI	SMA	STS	TEM	TR	VHF	WMA
BB	0.91	0.77	0.91	0.91	0.90	0.89	0.92	0.89	0.95	0.88	0.95	0.90	0.94	0.89	0.91	0.95	0.94	0.94
CCI		0.14	0.83	0.79	0.93	0.78	0.75	0.63	0.85	0.70	0.98	0.78	0.93	0.65	0.86	0.98	0.97	0.93
DEM			0.87	0.88	0.92	0.88	0.93	0.90	0.93	0.88	0.95	0.87	0.95	0.88	0.87	0.95	0.95	0.95
DI				0.72	0.95	0.89	0.92	0.88	0.9	0.85	0.97	0.87	0.98	0.88	0.88	0.97	0.97	0.98
DMI					0.51	0.88	0.93	0.95	0.97	0.84	0.96	0.92	0.97	0.91	0.94	0.96	0.84	0.97
EMA						0.86	0.91	0.88	0.93	0.86	0.96	0.88	0.93	0.86	0.9	0.96	0.93	0.93
HLI							0.87	0.88	0.94	0.88	0.96	0.92	0.94	0.88	0.93	0.96	0.94	0.94
MTM								0.79	0.91	0.86	0.97	0.88	0.94	0.83	0.92	0.97	0.96	0.93
PRI									0.91	0.93	0.97	0.93	0.96	0.92	0.93	0.97	0.97	0.95
PCH										0.67	0.91	0.87	0.98	0.85	0.89	0.91	0.91	0.98
RAN											0.66	0.96	0.93	0.96	0.95	0.66	0.94	0.94
RSI												0.85	0.96	0.87	0.88	0.96	0.94	0.96
SMA													0.01	0.96	0.96	0.93	0.96	0.01
STS														0.83	0.90	0.96	0.93	0.95
TEM															0.87	0.95	0.96	0.96
TR																0.66	0.94	0.94
VHF																	0.77	0.97
WMA																		0.01

We can see the large correlation within BB, DEM, EMA, PRI, RSI and TEM groups of Cl(A) classification. In order to build a market model, it could be recommended to use the indicators related to large values in Table 6. For example, we may take one BB (BOLLINGER) based series, the second from RAN (RANGE) group (mean distance 0.95) and the third from DEM (DEMA) group (0.91 and 0.95).

Table 7. The average $1 - |\rho_{i,j}|$ function within and between classes of Cl(B) classification

	1	5	10	15	20	25	30	40	45	50	60	65	80	85	90	100	105	110	160	200
1	0.65	0.88	0.93	0.95	0.96	0.93	0.98	0.91	0.99	0.94	0.96	0.92	0.94	0.85	0.96	0.97	0.96	0.93	0.95	0.97
5		0.86	0.92	0.91	0.92	0.87	0.95	0.88	0.95	0.88	0.89	0.90	0.89	0.85	0.9	0.93	0.85	0.91	0.90	0.91
10			0.89	0.89	0.89	0.92	0.93	0.91	0.92	0.91	0.91	0.91	0.89	0.91	0.91	0.86	0.76	0.90	0.89	0.90
15				0.85	0.87	0.89	0.93	0.90	0.92	0.88	0.89	0.90	0.86	0.91	0.88	0.84	0.69	0.89	0.85	0.87
20					0.78	0.89	0.96	0.9	0.96	0.88	0.84	0.89	0.86	0.93	0.83	0.85	0.66	0.90	0.90	0.81
25						0.84	0.93	0.88	0.93	0.86	0.85	0.90	0.87	0.88	0.87	0.91	0.80	0.91	0.89	0.88
30							0.66	0.93	0.68	0.91	0.96	0.92	0.91	0.97	0.96	0.92	0.96	0.91	0.92	0.96
40								0.88	0.94	0.88	0.88	0.89	0.88	0.88	0.89	0.91	0.82	0.91	0.90	0.90
45									0.66	0.92	0.96	0.92	0.91	0.97	0.96	0.91	0.95	0.91	0.91	0.96
50										0.84	0.85	0.89	0.85	0.89	0.86	0.89	0.75	0.91	0.88	0.87
60											0.74	0.89	0.86	0.91	0.79	0.89	0.74	0.91	0.88	0.83
65												0.89	0.88	0.89	0.89	0.89	0.78	0.9	0.89	0.89
80													0.83	0.89	0.85	0.86	0.69	0.89	0.85	0.85
85														0.82	0.92	0.94	0.88	0.91	0.91	0.93
90															0.75	0.88	0.69	0.90	0.87	0.81
100																0.62	0.63	0.87	0.85	0.85
105																	0.04	0.78	0.68	0.66
110																		0.89	0.89	0.90
160																			0.84	0.86
200																				0.78

4   Conclusions

FOREX analysts, as well as scientific researches interested in building market models could consider our following remarks:

(i) For the purpose of building a prediction model, we could recommend one of the cliques presented in the paper (Tables 2 and 3) or available at matrix.umcs.lublin.pl/~akrajka/BDAS19.

(ii) As we can see in Table 4, there is a big difference between Pearsons's correlation and χ^2-based measurement of independence. Uncorrelated series are often not independent, but the applied measure of independence is not perfect. The different classes in both classifications Cl(A) and Cl(B) were distinguished more precisely when the correlation coefficient was used.

(iii) The set of inputs should consist of indicators and rules built on a highly varied number of observations. The values in Tables 6 and 7 are smaller on the diagonal (means 0.66 and 0.76) then that out diagonal (means 0.91 and 0.88) but difference is greater in the case Cl(A) classification. The series of FOREX observations behave here as the series of independent observations.

(iv) In order to build a set of inputs, we could recommend to look up the values in Tables 6 and 7 and use indicators and rules from Cl(A) and Cl(B) class with the highest values.

(v) There are some contradictions between the results in Table 4 (the big number of positively classified elements in the case of correlation) and the results presented in Tables 6 and 7, where values between different classes are greater. This may suggest that there are small, maximal 4–10 (cf. Fig. 3) clusters of series belonging to the same class, strongly intermixed with similar clusters from other classes.

References

1. Box, G.E.P., Jenkins, G.M.: Time Series Analysis: Forecasting and Control. Holden-Day, San Francisco (1976). Düsseldorf-Johannesburg, revised edn., Holden-Day Series in Time Series Analysis https://books.google.pl/books?id=1W VHAAAAMAAJ
2. Bron, C., Kerbosch, J.: Algorithm 457: finding all cliques of an undirected graph. Commun. ACM **16**(9), 575–577 (1973). https://doi.org/10.1145/362342.362367
3. Hagan, M.T., Demuth, H.B., Bealey, M.H., Jesús, O.d.: Neural Network Design. http://hagan.okstate.edu/NNDesign.pdf
4. Kovács, L., Szabó, G.: Conceptualization with incremental Bron-Kerbosch algorithm in big data architecture. Acta Polytech. Hung. **13**(2), 139–158 (2016). https://doi.org/10.12700/APH.13.2.2016.2.8
5. Murphy, J.: Technical analysis of the financial markets: a comprehensive guide to trading methods and applications. New York Institute of Finance Series, New York Institute of Finance (1999). https://books.google.pl/books?id=5zhXEqdr_IcC
6. Murphy, K.P.: Machine Learning. A Probabilistic Perspective. Massachusetts Institute of Technology, Cambridge (2012)
7. Schwager, J.D.: Schwager on Futures: Technical Analysis. Wiley, Hoboken (1996)

8. Singh, R., Srivastava, S.: Stock prediction using deep learning. Multimedia Tools Appl. **76**(18), 18569–18584 (2017). https://doi.org/10.1007/s11042-016-4159-7
9. Sokolova, M., Lapalme, G.: A systematic analysis of performance measures for classification tasks. Inf. Process. Manage. **45**(4), 427–437 (2009). https://doi.org/10.1016/j.ipm.2009.03.002

The Comparison of Processing Efficiency of Spatial Data for PostGIS and MongoDB Databases

Dominik Bartoszewski[1], Adam Piorkowski[2], and Michal Lupa[1(✉)]

[1] Department of Geoinformatics and Applied Computer Science, AGH University of Science and Technology, al. Mickiewicza 30, 30-059 Cracow, Poland
mlupa@agh.edu.pl
[2] Department of Biocybernetics and Biomedical Engineering, AGH University of Science and Technology, A. Mickiewicza 30 Av., 30-059 Cracow, Poland

Abstract. This paper presents the issue of geographic data storage in NoSQL databases. The authors present the performance investigation of the non-relational database MongoDB with its built-in spatial functions in relation to the PostgreSQL database with a PostGIS spatial extension. As part of the tests, the authors were designed queries simulating common problems in the processing of point data. In addition, the main advantages and disadvantages of NoSQL databases are presented in the context of the ability to manipulate spatial data.

Keywords: Spatial databases · NoSQL · GIS · PostGIS · MongoDB · Query performance

1 Introduction

According to IBM estimates, 90% of the world's data has been created in the last two years. Other forecasts say that in 2025 there will be 175 ZB of data in the world, which means an increase from 33 ZB in 2018. This data is big data. This increase also applies to spatial data, which have particularly gained importance due to mobile devices equipped with geolocations [18,28,32], constellations of various geostationary and non-geostationary satellites [20,23,25,27] Volunteer Geographic Information [9], or finally Global Positioning System [14]. It should also be added that spatial data have a decidedly different character than alphanumeric information, hence the increasing amount of this type of data forces the use of a dedicated approach to handle it [7,11,12,17,19,24,26,31].

The processing of spatial data has been discussed in the literature since the beginnings of GIS [22]. Analyzing the solutions available in the literature, it is worth paying attention to two issues. First, the strategies for optimizing geospatial queries are very different from the classic optimization methods [8,10]. The second issue is that all classical methods of geospatial queries are tested on

© Springer Nature Switzerland AG 2019
S. Kozielski et al. (Eds.): BDAS 2019, CCIS 1018, pp. 291–302, 2019.
https://doi.org/10.1007/978-3-030-19093-4_22

data whose size is significantly different from the size considered in the context of Big Data.

The problem of large data amount is inseparably connected with the NoSQL databases [13,16,30]. As recent years have shown, NoSQL is a great alternative for RDBMS in the context of web applications based on large data sets [5, 15]. However, it should be borne in mind that the price for this is the lack of fulfillment of assumptions ACID. In the field of NoSQL databases and spatial data, it is worth quoting the authors' research [33], where MongoDB capabilities were tested for processing data from shapefiles. The article [21] presents the method of indexing spatial data in document based NoSQL.

Analyzing the above issues, the question arises: what strategy do we have to adopt when there is a need to process a dozen TB of GPS logs saved in the form of PointGeometry? Is the replacement of RDBMS by NoSQL crucial in this case? The natural answer to this question seems to be transferring the entire database to the NoSQL. Nevertheless, as mentioned earlier, spatial data and the way they are processed require a completely different approach. Moreover, as shown in [29], even well-known RDBMS systems have problems with the implementation of geospatial functions. Therefore, in addition to the profits related to queries speed, there also should be examined the available geospatial functions offered by NoSQL systems.

In this work we are clearing the above mentioned issues. First of all, the performance of RDBMS and NoSQL for classic spatial queries, which are based on point and polygon data, were compared. What is more, the functionalities offered by the most popular free RDBMS and NoSQL systems were also verified.

2   Test Environment

According to the DB-Engines ranking, the most popular open source and non-relational data base for handling spatial information is the MongoDB document database [1]. It is written in C++ and is licensed under the GNU AGPL open license [6]. It uses objects in the GeoJSON format to store spatial data. MongoDB supports also geospatial indexes as (2D indexes) and spherical indexes (2D Sphere). The RDBMS that has been selected for research purposes is PostgreSQL with the PostGIS spatial extension. The PostgreSQL database is a popular object-relational database management system (ORDBMS).

The PostGIS extension provides more than a thousand geospatial functions and contains all of the 2D and 3D spatial data types. Compared to PostGIS, MongoDB has only three geospatial functions: *geoWithin*, *geoIntersects* and *nearSphere*. The function *geoWithin* corresponds to the function *S_Within* in PostGIS, where the *geoIntersects* is related to the *ST_Intersection*. The *nearSphere* function, combined with the *maxDistance* parameter, returns all of the geometries at a certain distance sorted by distance. PostGIS is able to perform an analogous operation using the *ST_DWithin* function (Table 1).

PostGIS has a huge variety of geospatial functions. A full list and description of these functions can be found in the PostGIS documentation [3].

Table 1. Geospatial functions in the PostGIS and MongoDB databases

PostGIS	MongoDB
ST_Within	$\$geoWithin$
$ST\_Intersection$	$\$geoIntersects$
$ST\_DWithin$	$\$nearSphere + \$maxDistance$

3   Experiments

The document database MongoDB v3.6.3 and thePostgreSQL version 10.1 were selected for the experiments. The queries were counted on a computer with the following specification:

- Intel Core i7 2600 3,4 GHz,
- 4 GB RAM,
- 1000 GB HDD,
- Windows 7 Professional.

3.1   The Experiments Methodology

For both database systems, the authors were prepared scripts that performed the appropriate commands after running. The purpose of the tests was to measure the time of performing the queries. During the tests, all the processes requiring high computing power and background systems tasks were excluded. Test scripts were two .bat files - DOS/Windows shell scripts executed by the command interpreter (cmd). The first of them was used to test the PostgreSQL performance, while the second one was used to test the performance of the MongoDB database. The performance tests of both databases were done in the following way:

- each query has been repeated 10 times,
- the average and standard deviation were calculated for each query,
- the result of each script has been saved to a text file.

3.2   Query Efficiency Tests

For the tests purposes were used the basic elements provided with a database system. In the software downloaded with the system, you can find the pgbench application, which provides the following possibilities:

- repeating the query a specified number of times,
- calculation of the average and standard deviation from the query times,
- counting the number of queries performed per unit of time,
- calculation of the query execution time,
- using the specified number of threads to process the query.

3.3   The MongoDB Experiments

In the MongoDB database, the explain() method is used to check the performance of the queries. It provides a lot of information about the query. The explain() method can be called with or without a parameter. In the case of the second option, we have three parameters available the "queryPlanner", the "executionStats" and the "allPlansExecution". They determine the level of detail of the displayed information. The default mode is the "queryPlanner" [2].

3.4   The Comparison of the Query Times

Test 1. The first test was to check whether the points are within a certain polygon feature. The coordinates of the points were determined randomly, according to the uniform distribution. The tests included collections that contained 1000, 5000, 10000, 50000, 100000, 500000 or 1,000,000 points. The code for both queries is presented in the Table 2. The visualization of the query is shown in Fig. 1. Query times were collected in Table 3.

Table 2. Query #1 - selecting the points contained in the previously selected polygon

MongoDB	`stats0t = db.otpoints.find({ geometry:{$geoWithin: {$geometry : alaska.geometry }}}).explain (executionStats'');`
PostGIS	`SELECT * FROM otpoints,alaskasimple WHERE ST_Within (otpoints.geom, alaskasimple.geom);`

This case gives a clear advantage to the NoSQL database. In MongoDB, the same queries run on average 3× faster than in PostGIS. The standard deviation is very low.

Test 2. The second test was to the points located within a given distance from the selected coordinates. The collections included 50,000, 100,000, 500,000, 1,000,000 points. The authors were determined four test scenarios:

- points within a maximum distance of 100 km from centroid,
- points within a maximum distance of 200 km from centroid,
- points within a maximum distance of 500 km from centroid,
- points within a maximum distance of 1000 km from centroid.

The visualization of the query is shown in Fig. 2, the code of queries in the Table 4. Query times are collected in the Table 5.

The test performed faster in MongoDB, taking into account the radius of 100, 200 and 500 km. In the case of the 1000 km circle radius, the PostGIS proved to be about 3 times faster. We observed very clearly that the query time in the MongoDB database grows much faster with more and more points to count. This situation presents completely different conclusions in relation to the theoretical use of the NoSQL databases, which are tailored to the processing of large amounts of data. The standard deviation is very low for both database systems.

Table 3. The comparison of the test 1 queries

	MongoDB		PostGIS	
Point number	Time [ms]	Std. dev.	Time [ms]	Std. dev.
TEST 1				
1 000	**51.41**	0.51	112.21	0.51
5 000	**196.77**	1.33	496.53	2.58
10 000	**396.23**	2.29	987.61	3.75
50 000	**1855.42**	8.14	5282.78	29.58
100 000	**3793.71**	17.60	11895.12	55.90
500 000	**18520.42**	233.80	61195.34	257.01
1 000 000	**38128.00**	192.35	131895.52	725.42
TEST 4				
5 000	**52.27**	0.63	304.66	5.35
10 000	**103.71**	0.94	612.33	32.44
50 000	**515.84**	1.87	3017.80	24.14
100 000	**1037.41**	3.50	6022.55	72.56
500 000	**5166.74**	14.62	30525.22	770.25

Fig. 1. Visualization of the used polygon and randomly drawn points using the QGIS
[4]. The picture shows a case with a thousand points. The query will return only the
points inside the blue polygon. (Color figure online)

Table 4. Query #2 - selecting points within a certain distance from point

MongoDB	`var statsOt = db.otpoints.find({ geometry : { $nearSphere:`
	`$geometry: centroid.geometry, $maxDistance: 100000} } }).`
	`    explain`
	`(executionStats'');`

PostGIS	`SELECT * FROM otpoints WHERE ST_DWithin`
	`(otpoints.geom, ST_GeographyFromText`
	`('SRID=4326;␣POINT(-153.138␣64.731)'), 100000);`

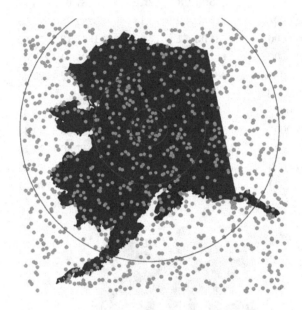

Fig. 2. Visualization of the query using the QGIS [29]. The first circle represents 100 km, the second 200 km, the third 500 km, and the last 1000 km from the central point. Visible case concerns a thousand points.

Test 3. The third query is based on a very similar scheme to the previous ones, but now random polygons are examined instead of random points - each of them is a square about 50 km. The sets of 5000, 10000, 50000, 100000 and 500000 polygons were taken into account. The authors were determined four test scenarios:

- polygons within a maximum distance of 100 km from centroid,
- polygons within a maximum distance of 200 km from centroid,
- polygons within a maximum distance of 500 km from centroid,
- polygons within a maximum distance of 1000 km from centroid.

The visualization of the query is shown in Fig. 3, the code of queries in the Table 6. Query times are collected in the Table 5.

An analogous situation as in the previous test with points. When the number of polygons is smaller, MongoDB has the advantage. However, when there is a larger radius of a circle, MongoDB begins to clearly slow down.

Table 5. Test 3: query execution times comparison

Circle radius/ Num. of points	MongoDB Query time [ms]	Std. dev.	PostGIS Query time [ms]	Std. dev.
TEST 2				
100 km				
50 000	**5.11**	0	12.17	0.03
100 000	**9.94**	0	53.97	0.04
500 000	**47.12**	0.31	104.75	0.21
1 000 000	**93.03**	0.52	500.43	0.67
200 km				
50 000	**4.20**	0.32	13.24	1.36
100 000	**18.77**	0.42	62.30	1.33
500 000	**37.54**	0.48	124.28	1.42
1 000 000	**194.56**	1.05	643.14	1.44
500 km				
50 000	**22.21**	0	29.38	1.34
100 000	**107.44**	0.37	142.81	1.43
500 000	**227.17**	0.43	285.86	1.37
1 000 000	**1208.36**	1.50	1424.83	2.69
1000 km				
50 000	320.21	0.42	**56.37**	0.18
100 000	677.45	0.51	**275.75**	0.32
500 000	3856.75	4.05	**554.95**	1.51
1 000 000	8489.68	11.95	**2711.01**	2.27
TEST 3				
100 km				
5000	**5.13**	0.14	29.48	0.62
10000	**8.81**	0.21	49.46	5.14
50000	**44.17**	0.91	250.95	11.31
100000	**86.12**	2.11	495.29	18.66
500000	**708.45**	18.24	2535.06	108.24
200 km				
5000	**15.37**	0.67	46.37	0.94
10000	**29.51**	1.26	85.30	1.56
50000	**163.15**	4.40	362.23	7.34
100000	**354.37**	7.63	737.35	14.50
500000	**2531.62**	45.34	3629.29	72.44
500 km				
5000	**93.88**	1.81	104.03	2.50
10000	**184.11**	7.13	209.69	4.18
50000	**979.57**	23.14	1029.64	20.36
100000	2136.14	64.69	**2094.75**	51.15
500000	12717.45	523.85	**10329.26**	210.67
1000 km				
5000	248.67	3.54	**104.03**	3.22
10000	516.77	9.21	**209.69**	8.66
50000	2862.18	40.52	**1029.64**	108.51
100000	6041.14	112.47	**2094.75**	210.32
500000	36434.61	727.81	**10329.26**	410.22

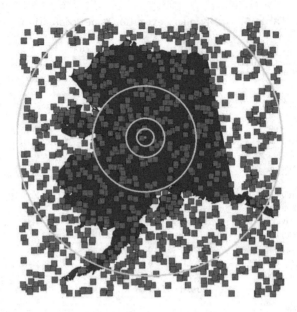

Fig. 3. Visualization of the query using the QGIS [29]. The first circle represents 100 km, the second 200 km, the third 500 km, and the last 1000 km from the central point. Visible concerns a thousand polygons.

Test 4. The fourth and last test case is a compound query. First, in the inner query, the polygon closest to the centroid is selected, then it is investigated whether the returned polygon does not intersect with other polygons. The visualization of the query is shown in Fig. 4, the code of queries in the Table 7. Query times are collected in the Table 3.

The tests gave results similar to those in test 1 - a clear advantage in favor of MongoDB. The NoSQL queries take about 6 times shorter. The standard deviation is too low to be visible on the graph (except for one result in PostGIS for the 500,000 points).

Table 6. Query #3 - selecting polygons within a set distance from point

MongoDB	```var statsOt = db.otpolygons.find({ geometry : { $nearSphere: { $geometry: centroid.geometry, $maxDistance: 100000} } }).explain (executionStats'');```
PostGIS	```SELECT * FROM otpolygons WHERE ST_DWithin (otpolygons.geom, ST_GeographyFromText ('SRID=4326;␣POINT(-153.138␣64.731)'), 100000);```

Table 7. Query #4 - Compound query: selecting polygon closest to the certain point and next checking if the polygon is intersecting with other polygons.

| MongoDB | ```
var centroid = db.centroid.findOne();
var statsOt = db.otpolygons.find({ geometry : {
    $geoIntersects :
{ $geometry : db.otpolygons.findOne( { geometry : {
$geoNear : { $geometry : db.centroid.findOne()
.geometry, $maxDistance: 2500 } } } ).geometry } } )
    .explain(executionStats'');
``` |
| PostGIS | ```
SELECT ST_Intersection
(otpolygons.geom, (SELECT otpolygons.geom FROM otpolygons
    WHERE ST_DWithin
(otpolygons.geom, ST_GeographyFromText
('SRID=4326; POINT(-153.138 64.731)'), 2500)
LIMIT 1)) FROM otpolygons;
``` |

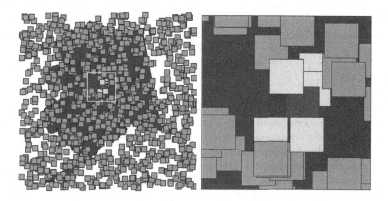

Fig. 4. Visualization of the query using the QGIS [29]. The green polygon represents the result of the internal query (search polygon closest to centroid - here marked with a green point), yellow polygons represent the result of an external query (searching for the intersecting polygons). (Color figure online)

4   Conclusions

NoSQL databases are a relatively new technology in the context of spatial data processing. There are only a few such systems available that provide this kind of data. This paper shows that the mechanisms for handling spatial data, compared to relational systems, are much more limited. Support for geographical features includes only the most basic functionalities. Performance tests show that queries concerning within and intersection operations take less time in MongoDB, as opposed to operations to select objects in the neighbourhood of another object, where PostGIS has a big advantage.

This shows that many criteria should be used to work with geographic data. One of them is to determine the operations that will have to be done using the database. In the case of many different spatial analyzes, the most obvious choice will be to use RDBMS, where there are several hundred geospatial functions.

Much smaller possibilities in the area of geospatial functions are provided by the non-relational databases, where in the case of MongoDB (which seems to best support the spatial data), there are only a few of these functions. Relational databases also have the advantage of many years of presence on the market, which led to their enormous popularity and the presence of many qualified professionals familiar with this subject.

Non-relational databases can be an alternative when working in dispersed environments that process a huge amount of data simultaneously. What is more, it should be noted that non-relational systems are constantly evolving, which in the future will probably result in an increase in the number of available geospatial functions. This may cause non-relational bases to take over part of the spatial data market.

Acknowledgements. This work was financed by the AGH - University of Science and Technology, Faculty of Geology, Geophysics and Environmental Protection as a part of a statutory project.

References

1. Db-engines ranking. https://db-engines.com/en/ranking/
2. MongoDB Docs - geospatial query operators. https://docs.mongodb.com/manual/reference/operator/query-geospatial/
3. PostGIS 2.5.2 dev manual. https://postgis.net/docs/
4. QGIS documentation. https://qgis.org/en/docs/
5. Burzańska, M., Wiśniewski, P.: How poor Is the "poor man's search engine"? In: Kozielski, S., Mrozek, D., Kasprowski, P., Małysiak-Mrozek, B., Kostrzewa, D. (eds.) BDAS 2018. CCIS, vol. 928, pp. 294–305. Springer, Cham (2018). https://doi.org/10.1007/978-3-319-99987-6_23
6. Agarwal, S., Rajan, K.: Performance analysis of MongoDB versus postGIS/postgreSQL databases for line intersection and point containment spatial queries. Spat. Inf. Res. **24**(6), 671–677 (2016)
7. Akulakrishna, P.K., Lakshmi, J., Nandy, S.: Efficient storage of big-data for real-time GPS applications. In: 2014 IEEE Fourth International Conference on Big Data and Cloud Computing (BdCloud), pp. 1–8. IEEE (2014)
8. Bajerski, P., Kozielski, S.: Computational model for efficient processing of geofield queries. In: Cyran, K.A., Kozielski, S., Peters, J.F., Stańczyk, U., Wakulicz-Deja, A. (eds.) Man-Machine Interactions. AISC, vol. 59, pp. 573–583. Springer, Heidelberg (2009). https://doi.org/10.1007/978-3-642-00563-3_60
9. Chmielewski, S., Samulowska, M., Lupa, M., Lee, D.J., Zagajewski, B.: Citizen science and WebGIS for outdoor advertisement visual pollution assessment. Comput. Environ. Urban Syst. **67**, 97–109 (2018)
10. Chromiak, M., Stencel, K.: A data model for heterogeneous data integration architecture. In: Kozielski, S., Mrozek, D., Kasprowski, P., Małysiak-Mrozek, B., Kostrzewa, D. (eds.) BDAS 2014. CCIS, vol. 424, pp. 547–556. Springer, Cham (2014). https://doi.org/10.1007/978-3-319-06932-6_53
11. Chuchro, M., Franczyk, A., Dwornik, M., Lesniak, A.: A big data processing strategy for hybrid interpretation of flood embankment multisensor data. Geol. Geophys. Environ. **42**(3), 269–277 (2016)

12. Czerepicki, A.: Perspektywy zastosowania baz danych nosql w inteligentnych systemach transportowych. Prace Naukowe Politechniki Warszawskiej. Transport **92**, 29–38 (2013)

13. Fraczek, K., Plechawska-Wojcik, M.: Comparative analysis of relational and nonrelational databases in the context of performance in web applications. In: Kozielski, S., Mrozek, D., Kasprowski, P., Małysiak-Mrozek, B., Kostrzewa, D. (eds.) BDAS 2017. CCIS, vol. 716, pp. 153–164. Springer, Cham (2017). https://doi.org/10.1007/978-3-319-58274-0_13

14. Goodchild, M.F.: Citizens as sensors: the world of volunteered geography. GeoJournal **69**(4), 211–221 (2007)

15. Harezlak, K., Skowron, R.: Performance aspects of migrating a web application from a relational to a NoSQL database. In: Kozielski, S., Mrozek, D., Kasprowski, P., Małysiak-Mrozek, B., Kostrzewa, D. (eds.) BDAS 2015. CCIS, vol. 521, pp. 107–115. Springer, Cham (2015). https://doi.org/10.1007/978-3-319-18422-7_9

16. Hricov, R., Šenk, A., Kroha, P., Valenta, M.: Evaluation of XPath queries over XML documents using sparkSQL framework. In: Kozielski, S., Mrozek, D., Kasprowski, P., Małysiak-Mrozek, B., Kostrzewa, D. (eds.) BDAS 2017. CCIS, vol. 716, pp. 28–41. Springer, Cham (2017). https://doi.org/10.1007/978-3-319-58274-0_3

17. Inglot, A., Koziol, K.: The importance of contextual topology in the process of harmonization of the spatial databases on example BDOT500. In: 2016 Baltic Geodetic Congress (BGC Geomatics), pp. 251–256 (2016)

18. Kopec, A., Bala, J., Pieta, A.: WebGL based visualisation and analysis of stratigraphic data for the purposes of the mining industry. Procedia Comput. Sci. **51**, 2869–2877 (2015)

19. Kozioł, K., Lupa, M., Krawczyk, A.: The extended structure of multi-resolution database. In: Kozielski, S., Mrozek, D., Kasprowski, P., Małysiak-Mrozek, B., Kostrzewa, D. (eds.) BDAS 2014. CCIS, vol. 424, pp. 435–443. Springer, Cham (2014). https://doi.org/10.1007/978-3-319-06932-6_42

20. Krawczyk, A.: A concept for the modernization of underground mining master maps based on the enrichment of data definitions and spatial database technology. In: E3S Web of Conferences, vol. 26, p. 00010. EDP Sciences (2018)

21. Li, Y., Kim, G., Wen, L., Bae, H.: MHB-tree: a distributed spatial index method for document based nosql database system. In: Han, Y.H., Park, D.S., Jia, W., Yeo, S.S. (eds.) Ubiquitous Information Technologies and Applications. LNCS, vol. 214, pp. 489–497. Springer, Dordrecht (2013). https://doi.org/10.1007/978-94-007-5857-5_53

22. Longley, P.A., Goodchild, M.F., Maguire, D.J., Rhind, D.W.: Geographic Information Systems and Science. Wiley, Hoboken (2005)

23. Salazar Loor, J., Fdez-Arroyabe, P.: Aerial and satellite imagery and big data: blending old technologies with new trends. In: Dey, N., Bhatt, C., Ashour, A.S. (eds.) Big Data for Remote Sensing: Visualization, Analysis and Interpretation, pp. 39–59. Springer, Cham (2019). https://doi.org/10.1007/978-3-319-89923-7_2

24. Lupa, M., Kozioł, K., Leśniak, A.: An attempt to automate the simplification of building objects in multiresolution databases. In: Kozielski, S., Mrozek, D., Kasprowski, P., Małysiak-Mrozek, B., Kostrzewa, D. (eds.) BDAS 2015. CCIS, vol. 521, pp. 448–459. Springer, Cham (2015). https://doi.org/10.1007/978-3-319-18422-7_40

25. Ma, Y., et al.: Remote sensing big data computing: challenges and opportunities. Future Gener. Comput. Syst. **51**, 47–60 (2015)

26. Martins, P., Cecílio, J., Abbasi, M., Furtado, P.: GISB: a benchmark for geographic map information extraction. In: Kozielski, S., Mrozek, D., Kasprowski, P., Małysiak-Mrozek, B., Kostrzewa, D. (eds.) BDAS 2015-2016. CCIS, vol. 613, pp. 600–609. Springer, Cham (2016). https://doi.org/10.1007/978-3-319-34099-9_46

27. Mirek, K., Mirek, J.: Non-parametric approximation used to analysis of psinsar[tm] data of upper silesian coal basin, poland. Acta Geodynamica et Geomaterialia 6(4), 405–410 (2009)

28. Pavlicek, A., Doucek, P., Novák, R., Strizova, V.: Big data analytics – geolocation from the perspective of mobile network operator. In: Tjoa, A.M., Zheng, L.-R., Zou, Z., Raffai, M., Xu, L.D., Novak, N.M. (eds.) CONFENIS 2017. LNBIP, vol. 310, pp. 119–131. Springer, Cham (2018). https://doi.org/10.1007/978-3-319-94845-4_11

29. Piorkowski, A.: MySQL spatial and PostGIS-implementations of spatial data standards. EJPAU 14(1), 03 (2011)

30. Płuciennik, E., Zgorzałek, K.: The multi-model databases – a review. In: Kozielski, S., Mrozek, D., Kasprowski, P., Małysiak-Mrozek, B., Kostrzewa, D. (eds.) BDAS 2017. CCIS, vol. 716, pp. 141–152. Springer, Cham (2017). https://doi.org/10.1007/978-3-319-58274-0_12

31. Wyszomirski, M.: Przeglad mozliwosci zastosowania wybranych baz danych nosql do zarzadzania danymi przestrzennymi. Roczniki Geomatyki-Annals of Geomatics 16(1 (80)), 55–69 (2018)

32. Xu, G., Gao, S., Daneshmand, M., Wang, C., Liu, Y.: A survey for mobility big data analytics for geolocation prediction. IEEE Wirel. Commun. 24(1), 111–119 (2017)

33. Zhang, X., Song, W., Liu, L.: An implementation approach to store GIS spatial data on NoSQL database. In: 2014 22nd International Conference on Geoinformatics (GeoInformatics), pp. 1–5. IEEE (2014)

Fuzzy Modelling of the Methane Hazard Rate

Dariusz Felka[1(✉)], Marcin Małachowski[1], Lukasz Wróbel[1,2],
and Jarosław Brodny[2]

[1] Institute of Innovative Technologies EMAG, Katowice, Poland
{dariusz.felka,marcin.malachowski}@ibemag.pl
[2] Silesian University of Technology, Gliwice, Poland
{lukasz.wrobel,jaroslaw.brodny}@polsl.pl

Abstract. Methane hazard assessment is an important aspect of coal mining, influencing the safety of miners and work efficiency. Calculation of the methane hazard rate requires specialized equipment, which might be not fully available at the monitored place and consequently might require manual measurements by miners. The lack of measurements can be also caused by a device failure, thus hampering continuous evaluation of the methane rate. In this paper we address this problem by constructing a fuzzy system being able to calculate the methane hazard rate in a continuous manner and deal with data incompleteness. We examine the effectiveness of different fuzzy clustering algorithms in our system and compare the proposed system to other state-of-the-art methods. The extensive experiments show that the proposed method is characterized by superior accuracy when compared to other methods.

Keywords: Fuzzy modelling · Data clustering · Methane hazard

1 Introduction

The concept of artificial intelligence refers to the characteristics of human intelligence that has the ability to think and learn, enabling the use of acquired knowledge to solve problems. The created algorithms imitate more or less these features. Thanks to artificial intelligence it is possible to analyze and to model complex phenomena, which are difficult or even impossible to create with the use of classical algorithms. On the basis of input/output data, an intelligent model can be developed that would be able to create rules of inference, generalize knowledge and classify data and at the same time, would be resistant to the lack of precision and noisy data [4].

The theory of fuzzy sets was developed as an alternative to the classical concepts of the set theory and logic. It describes phenomena whose concepts are defined in an imprecise, unclear way. Every object can be described by the degree of membership between total membership and non-membership. In fuzzy logic, there is no sharp border between elements belonging or not belonging to a

© Springer Nature Switzerland AG 2019
S. Kozielski et al. (Eds.): BDAS 2019, CCIS 1018, pp. 303–315, 2019.
https://doi.org/10.1007/978-3-030-19093-4_23

given set. This approach is significantly different from classical logic, where the set of logical values is a two-element set.

The main issue of developing a model of the studied phenomenon is its generation from numerical data. The successful functioning of knowledge extraction algorithms, such as clustering of measurement data, makes it possible to determine the parameters of the model together with the organization of its structure.

The methane hazard is crucial for the safety in Polish hard coal mines. It consists in the release of methane from coal deposits and rocks [22]. The reason for the release of methane is mining excavation, disturbing the equilibrium in the rock mass. Among the basic criteria of the methane hazard assessment, one can distinguish the forecasts of ventilation, absolute and criterial methane-bearing capacity. Due to the fact that measurements necessary for risk assessment are collected manually by the mine personnel, the detailed measuring data processing is very delayed. The current assessment of the methane hazard, based on the analysis of the course of the absolute methane-bearing capacity related to the criterial methane-bearing capacity, will allow better control of ventilation parameters [5]. The proposed solution to this problem is based on the application of fuzzy logic to assess the methane hazard on the basis of measuring data recorded by the sensors of the monitoring system. The status of the methane hazard is also analyzed by numerical methods based on Computational Fluid Dynamics (CFD) [1,2,20,21]. They may offer an alternative to the methods presented in the article.

The paper is organized as follows. Section 2 describes the basic concepts related to fuzzy modelling and clustering methods. Section 3 outlines details on calculating methane hazard in coal mines and presents our proposal for solving this problem by means of the fuzzy system. Section 3 also includes an experimental evaluation of our method on real-life datasets and comparison with other approaches. Section 4 gives conclusions of the work.

2   Fuzzy Modelling

2.1   Basic Concepts

Fuzzy reasoning is used where it is difficult to describe the investigated phenomenon using conventional relations. The fuzzy reasoning process can be divided into three stages [14]. The first stage is fuzzification that consists in computing a membership degree of input values. The second stage is inference that is responsible for computing a resulting membership function on the basis of input values. The calculations are made with the use of an inference mechanism and an implemented base of rules. The third stage is sharpening that consists in transforming the resulting membership function into a numeric value. Figure 1 presents the stages of fuzzy reasoning for a model with n inputs and 1 output.

The most frequently used fuzzy models are Mamdani and Takagi–Sugeno (Sugeno) [13]. Mamdani model is based on a fuzzy rules, each defined as:

$$R = \{\text{IF } x_n = A_n, \text{ THEN } y = B\} \tag{1}$$

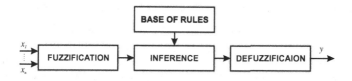

Fig. 1. Stages of fuzzy reasoning for a model with n inputs and 1 output

where: x_n – input linguistic variable, y – output linguistic variable, A_n, B – linguistic values.

Takagi–Sugeno (Sugeno) model is a connection of a model based on the linguistic description and polynomial functions. The base rule of knowledge is described as:

$$R = \{IF \ x_n = A_n, \ THEN \ y = f(x_n)\} \tag{2}$$

where: x_n – input linguistic variable, A_n – premise value, $y = f(x_n)$ – function in conclusion of the rule.

The main difference between fuzzy models are the types of conclusions in fuzzy conditional rules. The output of the fuzzy model may have a linguistic or numeric value.

The organization of the model structure is a crucial problem in the process of creating an object representing a selected issue. In fuzzy modelling, two approaches to knowledge acquisition are most frequently used. The first one is to obtain knowledge from an expert. A structure and parameters are determined arbitrarily on the basis of the expert's knowledge. The problem with this approach may be the difficulty of precise parameterization in the case of complex mapping. Therefore, a frequent solution is to make a preliminary model, which is then fine-tuned on the basis of the measuring data [13].

The second method is automatic getting rules based on numerical data. Clustering algorithms (grouping the input/output area) are most frequently used here. The results of such an algorithm are the parameters of the system together with the organization of its structure. This approach also accepts the use of knowledge from an expert. In this case, a part of the knowledge will come from the expert and other from the automatic selection of rules.

To create the fuzzy model it is necessary to aim at maximal simplification of a structure and a base of rules. A complicated and extended base may not give satisfactory results at all. The training of this model may be difficult or simply impossible.

2.2 Clustering of Measuring Data

Data clustering is one of the unsupervised methods of data analysis and consists in splitting a set of elements into subsets in such a way that the individual elements of the clusters would be similar and the clusters would be different from each other as much as possible [7,13]. A similarity of the elements can be described in different ways depending on the clustering method. Distances among the elements are the most commonly used measures [15]. The data can form clusters of different sizes, shapes or densities. The representation of a cluster is its central point. The fuzzy clustering creates an association in an uncertain way. One measuring element may belong, simultaneously and to a certain extent, to several clusters.

Clustering methods can be used in the absence of a database of rules to develop a preliminary model based on a set of numerical data. The final stage of parameters selection of a fuzzy model can be its transformation into an equivalent multi-layer neural network. Such a network can be additionally subjected to the process of learning on the basis of a learning dataset.

Fuzzy C-means Clustering. One of the most popular methods of clustering is the fuzzy c-means method. This method consists in creating groups to which objects with appropriate degrees of membership are associated. In each iteration, the algorithm compares all elements with each other, determines the centres of clusters and the degree of membership of all elements of the dataset to individual groups. The limitation of the c-means method consists in the need of arbitrary establishment of the number of clusters to be determined as a result of the algorithm. The compromise solution, in this case, is to create an algorithm for different parameter values and then select the best option.

Subtractive Clustering Algorithm. Subtractive clustering is another widely used method of data analysis. This method is based on a measure of data density. The method assumes that the potential candidates for the centre of the cluster are the actual training set points. The algorithm operates by finding a high-density data area. The element with the highest number of "neighbours" is selected as the centre of the cluster. The points in the closest vicinity of the centre are "subtracted" from the training set. The algorithm then searches for the next points with the highest density measurement value. The above steps are repeated until all data have been checked. The disadvantage of this solution is that it is effective with a relatively small amount of training data. In spaces with a larger number of dimensions, it is possible to increase the number of possible solutions [7]. The advantage of subtractive clustering is that it is not necessary to define the number of clusters, which is determined during the operation of the algorithm.

3   Methods and Experiments

3.1   Diagnosis of the Methane Hazard Index in the Area of Mining Operations

An integral part of the conducted mining operations is the occurrence of natural hazards in the production area. One of the most important is the methane hazard that occurs mainly in hard coal seams. It is induced by a transparent and odourless gas – methane. It is a flammable gas that forms an explosive mixture with the air. In the rock mass it is in the free state or associated with coal [22]. It may occur in cracks and other hollow spaces of the rock mass. It is emitted into excavations from exposed seams and spaces as well as from goafs.

The analysis of methane hazard in the working face area is carried out on the basis of averaged values of methane concentration and volume airflow rates over longer periods of time. Such a solution does not take into account the actual methane hazard that varies over time. Increased control of working conditions in the longwall area can be ensured by the current calculation of absolute and criterial methane. Better control of working conditions in the longwall area can be provided by the current calculation of absolute and criterial methane-bearing capacity.

The problem of determining the methane hazard rate can be considered as follows. The [11] describes the rules of mining operations in the longwalls under methane hazard conditions. The value describing an excavation in the area of a longwall is its methane-bearing capacity. It is calculated on the basis of measured values of methane concentration and air flow velocity (volume airflow rates) after averaging. In order to characterize the methane-bearing capacity content in the working face area, the following terms are defined: ventilation, absolute and criterial methane-bearing capacity. In addition, the conditions for action to be taken in the event of a potentially hazardous situation are specified.

A value of the criterial methane-bearing capacity is determined in accordance with the guidelines, depending on the longwall ventilation system. For longwalls with fresh air flow, the criterial methane-bearing capacity is determined from the following relation:

$$V_{\mathrm{KR}} = \frac{C_m \cdot V_s \cdot k}{100 \cdot n} + \frac{V_p \left(\frac{C_m}{n} - C_p \right)}{100 - \frac{C_m}{n}} - V_D \quad \left[\frac{\mathrm{m}^3}{\mathrm{min}} \right] \tag{3}$$

In the case of the methane drainage process conducted in the longwall area, the calculated value of methane-bearing capacity V_{KR} should be substituted to the following dependency:

$$V_{\mathrm{KR\text{-}O}} = \frac{100 \cdot V_{KR}}{100 - E} \quad \left[\frac{\mathrm{m}^3}{\mathrm{min}} \right] \tag{4}$$

where the value of methane drainage E is determined from the following formula:

$$E = \frac{100 \cdot V_O}{V_O + V_W} \quad [\%] \tag{5}$$

On this basis, the methane hazard rate has been defined:

$$W = \frac{V_B}{V_{\mathrm{KR}}} \tag{6}$$

The symbols in the formulas (3)–(6) stand for the following:

- C_m – admissible methane content [%],
- C_p – methane concentration in refreshing air current [%],
- V_P – volume stream of refreshing air current [m³/min],
- V_s – air volume stream in the longwall [m³/min],
- V_D – methane volume stream coming to the longwall [m³/min],
- V_O – volume of methane drained away by the drainage process [m³/min],
- V_W – volume of methane emitted to the excavations [m³/min],
- V_B – absolute-methane-bearing capacity – the total amount of methane emitted to the excavation (ventilation-methane-bearing capacity V_W) and the methane drainage value V_O [m³/min],
- k – irregularity coefficient of air distribution in the longwall,
- n – irregularity coefficient of methane emission.

The aim of determining the rate is to diagnose the methane hazard on the basis of current estimates of absolute methane-bearing capacity in relation to criterial methane-bearing capacity. Input values are the measurements of sensors in the monitoring system, and the output value is the methane hazard rate, defined as the ratio of the absolute and criterial methane-bearing capacities estimated on the basis of measurements. A value of '0' indicates the highest level of safety and a value of '1' indicates a real methane hazard, which requires appropriate actions to be taken to reduce the amount of methane in the ventilation air.

3.2 Experiments in the Longwall Area

The experiments were carried out on data coming from the working face area [8]. This is a longwall with high methane-bearing capacity with active methane drainage operations. The longwall has been designed with "Y" type ventilation. The data presented in the longwall layout show that the air current in the longwall area N-2 was very strong. The intake air was supplied through the incline N-1 and next through the main gate N-2. A part of the air from the incline N-1 flowed through the main gate N-3 and then was connected with the return air from the longwall N-2. The return air was exhausted from the longwall through the main gate N-3 and the raise N-3.

To calculate the value of volume airflow rates and methane concentrations in individual roadways and the volume flow rate of methane in the methane drainage pipeline, the measurements recorded and archived in the database of the mining monitoring system were used. The measurement process employed the following devices:

- AS038 – air flow velocity (roadway N-2) [m/s];
- AS099 – air flow velocity (roadway N-3) [m/s];
- AS072 – air flow velocity (raise N-3) [m/s];
- MM137 – methane concentration (roadway N-2) [%];
- MM116 – methane concentration (roadway N-3) [%];
- MM104 – methane concentration (raise N-3) [%];
- WM46 – volume flow rate of methane in the methane drainage pipeline [m³/min].

3.3 Fuzzy Model Diagnosing the Level of Methane Hazard

According to the assumptions presented previously, five training and test sets were prepared on the basis of measurements archived in the database of the monitoring system of the mine.

The fuzzy reasoning technique presented in the publication was used to develop the model. The advantage of this solution is the possibility to make such a model based on measurement data and expert knowledge, data approximation, and the possibility of reasoning on the basis of inaccurate, incorrect or incomplete data. Fuzzy reasoning has already been used in the subject of natural hazards occurring in mining, e.g. in [3,6,16]. It is also worth referring to some other machine learning approaches to prediction of methane concentration, such as regression rules [10,17], meta-learning [9] as well as examples of complex decision systems supporting methane forecasting [19].

There are many options to use fuzzy modelling, e.g. in combination with artificial neural networks, as described, inter alia, in [5,18], while in the following sections the model based only on fuzzy reasoning is described.

The aim of the model is to diagnose the methane hazard on the basis of current estimates of absolute methane-bearing capacity in relation to criterial methane-bearing capacity. Input values are the measurements of sensors in the monitoring system, and the output value is the methane hazard rate.

The process of developing the whole system can be divided into two stages. After loading the set of measurement data, they were clustered with the use of two methods: subtractive and c-means methods. Grid partition structures were also generated. As a result, the parameters of the system describing its structure and mode of operation were obtained. The algorithm showed the location of cluster centres and optionally the number of clusters. This parameter is very important because the number of clusters is equal to the number of rules of the knowledge base, and the coordinates of cluster centres correspond to the location of the centres of the membership function in the fuzzy model.

The next part of this section features the design and operation of a fuzzy model based on the analysis of measurement data with the use of a clustering algorithm. As a result of subtractive clustering of one of the datasets, 8 clusters were obtained. The method determined the number of groups during the operation of the algorithm. The coordinates of the means are projected on the spatial axes of the individual input data in order to determine the parameters of the membership function. These coordinates correspond to the maximum values of the function. For further description, the Gauss function was selected because of its shape and description using two parameters. The number of clusters also corresponds to the number of rules of the base rule of the model. Figure 2 shows selected membership functions for model inputs on the basis of projected cluster means.

On the basis of the results obtained in the process of clustering, a structure of a fuzzy model was developed. The type of the fuzzy model (Sugeno), the number of inputs (7) and outputs (1) and the mechanisms of fuzzy reasoning, such as the implication (AND) or rule aggregation (OR) method were determined. Then a

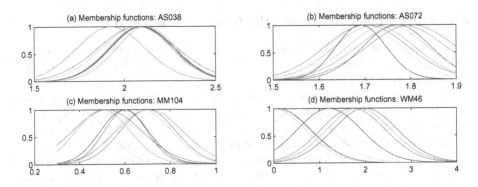

Fig. 2. Dataset *201401*: selected membership functions of the model inputs

list of fuzzy rules in the base of rules was made. Every cluster corresponded to one fuzzy rule. The membership functions in the input space are assigned to the corresponding rules and the function parameters are assigned in the output space.

Figure 3 shows the structure of the developed METHANE model with its most important parameters, while Fig. 4 shows an example of reasoning. The way of inference for the created network is as follows. In the first place the inputs of the model are fuzzified by the calculation of membership functions. The values are connected by means of a product, this way making levels of activation of every rule. The functions in conclusions of fuzzy rules are determined also on the basis of measurements. The final output value of the model is a weighted average of all elements of a rule base of the model. The weights are the levels of activation and the elements are the functions in rule conclusions.

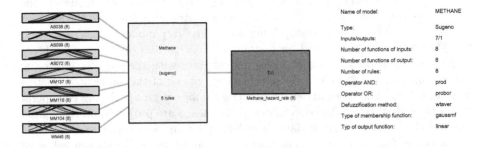

Fig. 3. Fuzzy model based on the dataset *201401*: structure and parameters

Depending on the value of the methane hazard rate, miners must decide to take measures to reduce the methane content in the return air current. According to [12], rates below 0.2 indicate no methane hazard and values above 0.8 indicate a high hazard. It is possible that a rate of more than 1.0 may occur. This indicates a dangerous longwall requiring an immediate action to reduce the amount of methane in the excavation area.

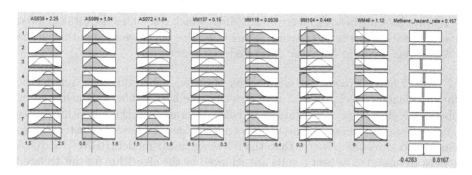

Fig. 4. Example of reasoning for a fuzzy model developed on the basis of the dataset *201401*

The time of the decision-making is also important. The current estimation of the value of the methane hazard rate allows an almost immediate response to an emerging hazard. This would not be possible with manual measurements made by the mining personnel.

Figure 5 shows a graph of the methane hazard rate for the training and test datasets. Good clustering results were reflected in the effective determination of the methane hazard rate for test data.

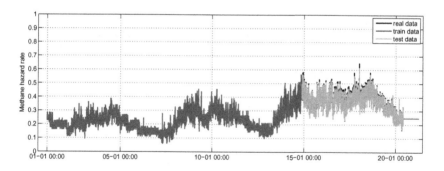

Fig. 5. Course of the methane hazard rate for the dataset *201401*

3.4   Results

In the first stage of the tests, an analysis was carried out to check the effectiveness of fuzzy models generated on the basis of measuring data obtained in the hard coal production process. The results were compared with other methods. The verification of the system performance was carried out on the basis of:

- mean absolute percentage error (MAPE);
- root mean squared error (RMSE);

– relative root mean squared error (RRMSE) which calculates the total squared error relative to the error which is made by the reference model, that is the linear regression in our case, expressed as a percentage.

Tables 1 and 2 present the comparison of tested algorithms according to the aforementioned criteria. Table 1 indicates that the best results were achieved by subtractive clustering. The number of rules generated by the algorithm varies between 5 and 8. Average values of errors are the lowest. C-means clustering also achieved good results – better than the reference linear model (RRMSE < 100% on most datasets). The best results are achieved by the algorithm for the declared number of clusters from 10 to 15. The advantage of c-means method is that the running time of the algorithm is much better than in the case of the subtractive method. The other methods – regression tree, random forest, and boosting tree – achieved worse results than the reference linear model (RRMSE > 100% on all datasets). The worst results were obtained for the application of grid partition. A large number of generated rules do not correspond to the effectiveness of model learning.

Table 1. Comparison of fuzzy models with different clustering methods (grid partition, fuzzy c-means and subtractive clustering) according to the number of rules and regression errors (MAPE, RMSE, and RRMSE)

Dataset	Grid				Fuzzy c-means				Subtractive			
	rules	mape	rmse	rrmse	rules	mape	rmse	rrmse	rules	mape	rmse	rrmse
Dataset 1	128	35.3	.1388	499.3	10	1.9	.0115	41.4	5	1.1	.0072	25.9
Dataset 2	128	215.0	.3134	877.9	10	5.0	.0160	44.8	5	2.4	.0068	19.0
Dataset 3	128	232.6	.3680	2389.6	10	4.4	.0109	70.8	6	1.5	.0029	18.8
Dataset 4	2187	165.6	.2961	6886.0	10	1.1	.0033	76.7	6	0.5	.0012	27.9
Dataset 5	128	101.4	.3653	2111.6	15	4.3	.0182	105.2	8	2.2	.0104	60.1
Average	539.8	150.0	.2963	2552.9	11.0	3.3	.0120	67.8	6.0	1.5	.0057	30.3

In the second phase of the analysis, a simulation was carried out on the basis of incomplete measurement data, i.e. lack of one of the recording devices in the area of the mining production. Table 3 presents a comparison of the effectiveness of tested approaches.

The best results were observed for models based on fuzzy clustering of measuring data. The average values of the errors are lower than for the reference linear model. Similar conclusions can be drawn from the analysis of individual tests. Fuzzy models have shown significantly higher effectiveness for most test cases. Only a simulation of the lack of an anemometer AS072 and methane meter MM104 in the return air stream made a problem for the fuzzy models. The reason for that situation is the great importance of these measuring devices in the

Table 2. Comparison of linear regression, regression tree, random forest, and gradient boosted tree in methane hazard rate estimation problem according to regression errors (MAPE, RMSE, and RRMSE)

Dataset	Linear regression			Regression tree			Random forest			Boosting tree		
	mape	rmse	rrmse	mape	rmse	rrmse	mape	rmse	rrmse	mape	rmse	rrmse
Dataset 1	3.7	.0278	100.0	5.1	.0390	140.3	7.7	.0633	227.7	8.2	.0558	200.7
Dataset 2	12.4	.0357	100.0	11.9	.0502	140.6	11.5	.0383	107.3	19.3	.0593	166.1
Dataset 3	6.5	.0154	100.0	9.8	.0180	116.9	8.7	.0166	107.8	13.7	.0260	168.8
Dataset 4	1.8	.0043	100.0	5.6	.0127	295.3	5.7	.0122	283.7	7.6	.0185	430.2
Dataset 5	4.1	.0173	100.0	10.3	.0435	251.4	15.0	.0634	366.5	16.1	.0626	361.8
Average	5.7	.0201	100.0	8.5	.0327	188.9	9.7	.0388	218.6	13	.0444	265.5

process of estimating the level of the methane hazard. However, all tested models showed the difficulty in modelling the problem in case of the lack of measuring devices in the return air. The models that not based on fuzzy clustering achieved worse results than the reference linear model.

Table 3. Comparison of models learned on the dataset with missing sensors

Missing sensor	Fuzzy c-means		Subtractive clustering		Linear regression		Regression trees		Random forests		Boosting trees	
	rmse	rrmse	rmse	rrmse	rmse	rrmse	rmse	rrmse	rmse	rrmse	rmse	rrmse
AS038	.0375	90.6	.0406	98.1	.0414	100.0	.0433	104.6	.0634	153.1	.0544	131.4
AS099	.0188	55.0	.0177	51.8	.0342	100.0	.0407	119.0	.0703	205.6	.0540	157.9
AS072	.1324	126.5	.1204	115.0	.1047	100.0	.1230	117.5	.1118	106.8	.1181	112.8
MM137	.0166	57.6	.0125	43.4	.0288	100.0	.0400	138.9	.0625	217.0	.0558	193.8
MM116	.0135	48.7	.0103	37.2	.0277	100.0	.0401	144.8	.0666	240.4	.0555	200.4
MM104	.1065	100.0	.1147	107.7	.1065	100.0	.1364	128.1	.1250	117.4	.1449	136.1
WM46	.0104	38.0	.0053	19.3	.0274	100.0	.0403	147.1	.0448	163.5	.0554	202.2
Average	.0480	73.8	.0459	67.5	.0530	100.0	.0663	128.6	.0778	172.0	.0769	162.1

In diagnosing the methane hazard, it is important that, depending on its level, miners have to make a decision to initiate actions specified in the regulations, therefore the value of the error is important for diagnosing the hazard. A more accurate diagnosis allows a quicker and more adequate response to the situation. The test results showed that the system can diagnose the hazard even when one sensor is disabled. The model can therefore be used in support activities, such as in the case of failure of a measuring sensor or during calibration of methane meters.

4 Conclusions

The paper presents the issue of modelling phenomena with the use of fuzzy reasoning. An analysis of the effectiveness of clustering algorithms and their comparison with other techniques of acquiring knowledge from numerical data have been made. A fuzzy model is presented. Its task is to assess the current level of methane hazard in the region of hard coal extraction on the basis of measurements coming from devices of the monitoring system.

The process of developing this model is complex and depends on many factors, such as the choice of the method of dividing the input-output data space, the type of the fuzzy model and its parameters, as well as on the properties of the measuring dataset. The application of artificial intelligence methods provides great opportunities to develop models reproducing the phenomena under investigation. Their description by means of mathematical dependencies would be difficult. Another advantage of modelling is the ability to develop systems based on incomplete data, as shown in this paper.

The problem of using an artificial intelligence technique in modelling the level of the methane hazard presented in the article is of great practical importance. This hazard is one of the most dangerous in coal mining. The article shows that the current assessment of the methane hazard concluded on the basis of the analysis of the course of absolute methane-bearing capacity in relation to criterial methane-bearing capacity may allow better control of parameters, including the level of the methane hazard in the area of mining production, especially in situations of dynamic changes and exceeding the limit values. The analysis presented in the paper may be the basis for the mine personnel to take action in the event of a dangerous situation. Therefore, these methods can usefully complement conventional methods of knowledge acquisition and data processing. The analysis also gives great opportunities for practical use and improvement in mining safety.

Acknowledgements. The work was carried out within the statutory research project of the Institute of Innovative Technologies EMAG.

References

1. Brodny, J., Tutak, M.: Determination of the zone endangered by methane explosion in goaf with caving of operating longwalls. In: International Multidisciplinary Scientific GeoConference SGEM: Surveying Geology & mining Ecology Management, vol. 2, pp. 299–306 (2016)
2. Brodny, J., Tutak, M.: Analysis of methane hazard conditions in mine headings. Tehnički vjesnik - Technical Gazette **25**(1), 271–276 (2018)
3. Brzychczy, E., Kęsek, M., Napieraj, A., Sukiennik, M.: The use of fuzzy systems in the designing of mining process in hard coal mines. Arch. Min. Sci. **59**(3), 741–760 (2014)
4. Felka, D.: Metody budowy inteligentnych modeli na bazie danych numerycznych. Konkurs Młodzi Innowacyjni. Innowacyjne rozwiązania w obszarze automatyki, robotyki i pomiarów, pp. 75–88 (2012)

5. Felka, D., Brodny, J.: Application of neural-fuzzy system in prediction of methane hazard. In: Burduk, A., Mazurkiewicz, D. (eds.) ISPEM 2017. AISC, vol. 637, pp. 151–160. Springer, Cham (2018). https://doi.org/10.1007/978-3-319-64465-3_15

6. Grychowski, T.: Multi sensor fire hazard monitoring in underground coal mine based on fuzzy inference system. J. Intell. Fuzzy Syst. **26**(1), 345–351 (2014)

7. Jang, J.S.R., Sun, C.T., Mizutani, E.: Neuro-Fuzzy and Soft Computing: A Computational Approach to Learning and Machine Intelligence, 1st edn. Pearson, Upper Saddle River (1997)

8. Jastrzębska Spółka Węglowa S.A.: Projekt techniczny eksploatacji ściany N-2 w pokładzie 404/2 w KWK Pniówek. Technical report, JSW S.A., Pawłowice (2013)

9. Kozielski, M.: A meta-learning approach to methane concentration value prediction. In: Kozielski, S., Mrozek, D., Kasprowski, P., Małysiak-Mrozek, B., Kostrzewa, D. (eds.) BDAS 2015–2016. CCIS, vol. 613, pp. 716–726. Springer, Cham (2016). https://doi.org/10.1007/978-3-319-34099-9_56

10. Kozielski, M., Skowron, A., Wróbel, Ł., Sikora, M.: Regression rule learning for methane forecasting in coal mines. In: Kozielski, S., Mrozek, D., Kasprowski, P., Małysiak-Mrozek, B., Kostrzewa, D. (eds.) BDAS 2015. CCIS, vol. 521, pp. 495–504. Springer, Cham (2015). https://doi.org/10.1007/978-3-319-18422-7_44

11. Krause, E., Łukowicz, K.: Regulations of longwalls management in the conditions of methane hazard. Instruction No 17. Central Mining Institute (2004)

12. Krzystanek, Z., Mróz, J., Trenczek, S.: Integrated system for monitoring and analysis of methane hazards in the longwall area. Min. Inf. Autom. Electr. Eng. **525**(1), 21–32 (2016)

13. Lęski, J.: Systemy neuronowo-rozmyte. Wydawnictwa Naukowo-Techniczne (2008)

14. Piegat, A.: Modelowanie i sterowanie rozmyte. Akademicka Oficyna Wydawnicza "Exit" (1999)

15. Rutkowski, L.: Metody i techniki sztucznej inteligencji: inteligencja obliczeniowa. Wydawnictwo Naukowe PWN (2005)

16. Sikora, M., Krzystanek, Z., Bojko, B., Śpiechowicz, K.: Application of a hybrid method of machine learning for description and on-line estimation of methane hazard in mine workings. J. Min. Sci. **47**(4), 493–505 (2011)

17. Sikora, M., Sikora, B.: Improving prediction models applied in systems monitoring natural hazards and machinery. Int. J. Appl. Math. Comput. Sci. **22**(2), 477–491 (2012)

18. Siminski, K.: Improvement of precision of neuro-fuzzy system by increase of activation of rules. In: Kozielski, S., Mrozek, D., Kasprowski, P., Małysiak-Mrozek, B., Kostrzewa, D. (eds.) BDAS 2015-2016. CCIS, vol. 613, pp. 157–167. Springer, Cham (2016). https://doi.org/10.1007/978-3-319-34099-9_11

19. Ślęzak, D., et al.: A framework for learning and embedding multi-sensor forecasting models into a decision support system: a case study of methane concentration in coal mines. Inf. Sci. **451–452**, 112–133 (2018)

20. Tutak, M., Brodny, J.: Analysis of influence of goaf sealing from tailgate on the methane concentration at the outlet from the longwall. In: IOP Conference Series: Earth and Environmental Science. vol. 95, p. 042025. IOP Publishing (2017)

21. Tutak, M., Brodny, J.: Analysis of the impact of auxiliary ventilation equipment on the distribution and concentration of methane in the tailgate. Energies **11**(11), 1–28 (2018)

22. Wacławik, J.: Wentylacja kopalń. Wydawnictwa AGH (2010)

Networks and Security

Application of Audio over Ethernet Transmission Protocol for Synchronization of Seismic Phenomena Measurement Data in order to Increase Phenomena Localization Accuracy and Enable Programmable Noise Cancellation

Krzysztof Oset[(⊠)], Dariusz Babecki, Sławomir Chmielarz, Barbara Flisiuk, and Wojciech Korski

Institute of Innovative Technologies EMAG, Leopolda 31, 40-189 Katowice, Poland
{krzysztof.oset,dariusz.babecki,slawomir.chmielarz,barbara.flisiuk,
wojciech.korski}@ibemag.pl

Abstract. Seismic phenomena, in particular underground tremors, as extremely dangerous, require activities related to their location and prediction. The registration of high-energy phenomena is not a technological challenge at present, while recognizing their precursors as low-energy phenomena, and in particular the associated precise location and isolation from the seismic (acoustic) background, is relatively problematic. The article presents the concept and work related to the elimination of desynchronization of measurements in the seismoacoustic band and ensuring the compliance of the measurement phase in order to increase the accuracy of phenomena location and the possibility to use standard programming tools to eliminate noise (seismoacoustic background).

Keywords: Audio over Ethernet · Seismic Data ·
Geophysical phenomena · Synchronic transmission ·
Programmable noise cancellation

1 Introduction

Seismic measurements are based on processing signals from sensors, such as seismometers, geophones and accelerometers. The sensors process relevant physical quantities of the phenomena (proper relocation, velocity and acceleration of a seismic wave) which take place in the medium where a sensor was installed. Obviously, the output quantity of sensor is voltage which is conventionally processed into a digital form by means of analogue-digital converters. In this form, after preliminary treatment, the measurement is transmitted to either local or central seismic databases.

© Springer Nature Switzerland AG 2019
S. Kozielski et al. (Eds.): BDAS 2019, CCIS 1018, pp. 319–340, 2019.
https://doi.org/10.1007/978-3-030-19093-4_24

Unfortunately, it is not possible to measure the parameters of a (natural) tremor directly in its epicentre as, for now, nobody can predict exactly where the tremor will happen. For this reason, the area of the tremor potential occurrence is monitored and, somewhat in advance, online measurement is assumed [10]. Thus, it turns out that one sensor is not enough. Moreover, the structure of the sensor has to take into account the necessity to measure relatively small quantities, due to possible long distances between the sensor and the tremor epicentre. In the case of earthquakes, these distances may be up to several hundred kilometres but, at the same time, their energy is relatively high and it is not necessary to determine the source with the accuracy of, say, several hundred metres [13]. A different situation occurs while monitoring tremors caused by the exploitation of underground deposits. Here the localization of the source may require the accuracy of several meters and in the case of monitoring micro seismic phenomena in a seismic-acoustic band (the so called cracks) – even centimetres. A separate issue is to isolate ultra-low energy phenomena from the seismic noise surrounding the measured area (e.g. working machines) and these phenomena play the key role in tremors prediction algorithms and risk assessment [12].

In general, the localization of seismic phenomena is based on measuring the time of the first break in all sensors which surround the measured area [8]. The algorithms which process these signals allow to determine the tremor epicentre based on exactly this time measurement. In reality, the absolute time of the measurement is not as important as the relative time shift. If the shift depends only on the distance from the source to the sensor (the wave velocity in the elastic medium is explicitly determined), then it is very easy to locate the source. However, firstly, it is not recommended to assume that the wave propagation medium is homogeneous. Secondly, in the case of analogue-digital processing the notion of "time" will be strictly related to the structure of the system for sampling the analogue signal. Due to the geological structure, it is rather not possible to overcome the first disadvantage. Obviously, being aware of it and increasing the number of sensors in the given area, it is possible to diminish the measurement ambiguity quite efficiently. The second, time-related variable generates extra problems. Such a situation happens when, while developing the device, we assume that it will be active only in the time of the event occurrence – the reasons are the need to save energy or to simplify the device structure and communication protocol. No matter what the structure is like, the target is to synchronize the operations of all sensors in the system.

The synchronization of A/D processing, in the case of digital transmission of signals, is practically based on the synchronization of the clocks of processing modules in the measurement site (coupled with the sensor). This solution is applied in most monitoring (measurement) systems. Still, it is important to note that such systems are vulnerable to desynchronization, for example due to the following [15]:

– ageing of quartz elements,
– temperature drift,
– unstable firmware of modules.

For this reason the modules have to be synchronized quite frequently. In the case of discreet measurements, where the measured values are sampled every one second or less frequently, the accuracy of synchronization does not have much impact, in fact. All the more so because such a frequency of measurements is rather related to very slowly changing phenomena. The problem arises with more frequent sampling of fast changing phenomena, for example with accurate mapping of processes in an acoustic band (up to a dozen or so kHz). More particularly as we need to locate a phenomenon to be measured by means of several sensors distributed in space. Of course, for slow changing processes the processing desynchronization phenomenon is not as important as for fast changing ones – in the case of frequency analysis or phase analysis. If we want to measure the velocity of the wave propagation in a medium, the measurement accuracy will depend, first of all, on synchronization, regardless of the frequency spectrum of the phenomenon. The solutions that have been applied so far require synchronization in each local measurement network. Even if we use the NTP protocol for time synchronization of computers, or GPS, this synchronization may be lost, e.g. due to the lack of access to a satellite signal or lack of proper IT maintenance of the system.

The paper features the possibility to eliminate, almost completely, the measurement error of relative time. This advantage considerably impacts the accuracy of the tremor localization. In addition, the described technology, used in an experimental system, allows to apply standard, programming methods of noise reduction.

2   Related Works

The works related to the collection of seismic and seismic-acoustic data have been carried out in the EMAG Institute since 1980s. EMAG's most recognized and most frequently applied systems, in Poland and abroad, are ARAMIS [5] for collecting seismic data and ARES [11] for recording seismic-acoustic data. The systems have been developed and updated till now.

ARES is equipped with analogue transmission of data which, unfortunately, results in very high sensitivity to electromagnetic and electrostatic disturbances of the transmission channel. Moreover, the system does not have any solutions to reduce the seismic background (noise). On the other hand, as an analogue system, ARES is free from the measurement desynchronization error – provided that we do not compare data from several different systems. ARAMIS allows digital data transmission. However, such a structure makes the measurement synchronization process quite difficult, first of all due to a specific transmission protocol. Here the synchronization depends, to a large extent, on proper functioning of IT infrastructure in the data collection centre. During the last few years the EMAG Institute launched some works to improve the synchronization process in this very case. These works were followed by a patent [9].

Due to considerable market demand for security systems (including those for tremor hazards assessment), some projects were launched to solve the synchronization problem, i.e. to fulfill high quality requirements of such systems and to apply a solution that would shorten the time-to-market period for the developed system. The last few years saw state-of-the-art solutions based on a data transmission protocol which uses the Ethernet and the IP layer. The examples are Quanterra Q330 [2] made by Kinematrix, USA, or a Swiss system GeoSIG GMSplus [1]. Still, in spite of a very advanced structure using the NTP and GPS synchronization which ensures an adequate time-stamp, these solutions not grant the phase compatibility of the measured signals, which is indispensable to first-break and phase analysis of a seismic wave and the application of software noise cancellation procedures.

Searching for a suitable solution eventually resulted in a project to implement a typical Audio over Ethernet technology, used in the audio industry, in a relatively untypical application, i.e. geophysics.

In conventional audio systems, synchronization with accuracy to one sample is not as important as the delay of the signal reaching e.g. the recorder from many channels, especially in connection with the source of the signal directly connected to the recorder input. In practice, a delay of 20 ms is unacceptable for most users. For this reason, for example the Ethersound system resigns from establishing a connection in such conditions. In seismics, the absolute time of receiving the measurement result is not important (20 ms and more is not a problem), but the inter-channel desynchronization creates serious problems in the relative measurement. If we reduce the inter-channel shift to near zero (compliance with the A/D processing sample), the first-break shift of the measured seismic wave corresponds only to the delay associated with the wave travel distance or change of the material stress (i.e. the wave velocity change) [4]. The reduction of this shift is also essential for possible further processing of data related to the reduction of ambient noise.

3   Audio over Ethernet

3.1   Audio over Ethernet Solutions

In contrast to the aforementioned data transmission solutions which use the Ethernet IP technology [3], Audio over Ethernet has one important feature to be used in the tested configuration. Synchronization, so important in the case of the audio technology, must be properly continuous. This is guaranteed by the Master Clock signal which ensures the synchronization of local Word Clock generators, responsible for sampling the analogue signal. The Word Clock signal is part of the transmission protocol and this way ensures synchronization of all A/D converters of the measuring network with accuracy to a single sample. In this case, it is entirely possible to omit the construction of specialized synchronization circuits, coupled with NTP protocols or using GPS signals. The use of NTP/GPS boils down to determining the timestamp for the entire measurement network (and of course to generate the Master Clock signal (Fig. 1).

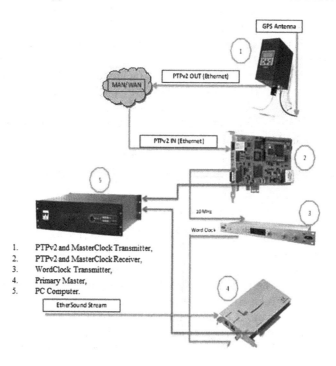

Fig. 1. Principle of Master Clock signal distribution

The operation principle of Audio over Ethernet (AoE) is similar to Voice over IP (VoIP), however, it provides high-fidelity sound with low latency. Due to the need to preserve high quality of the sound, no signal compression is used. AoE needs higher transmission speed (typically 1 Mbit per channel) and much lower latency (typically less than 10 ms) than VoIP. Audio over Ethernet requires a high performance network. This is due to the provision of a dedicated LAN or virtual LAN (VLAN). There are many AoE protocols (usually different from and incompatible with each other). For example, using the Cat5 twisted pair and the 100BASE-TX standard (100 Mbits/s), a single link can transmit 32 to 64 channels with 48 kHz sampling. Some can handle other frequencies such as 44.1 kHz (CD), 88.2 and 96 kHz (2x oversampling), and even 192 kHz (4x), as well as 32-bit samples, which, however, significantly reduces the channel capacity. AoE is not likely to be used for working in a wireless network (e.g. WiFi).

The protocols can be divided according to the Ethernet protocol layers in which they work.

*Layer*1: The protocols use Ethernet cabling and signal components but they do not use the Ethernet frame structure. They often have their own means of access control (MAC) instead of the original, which may cause compatibility problems with standard PC hardware.

*Layer*2: The protocols transfer audio data using standard Ethernet packets. Most of them can use standard hubs and switches but some require a dedicated LAN or a virtual VLAN.

*Layer*3: Actually, these protocols are included in the AoIP technology (Audio over IP) and they use standard IP packets (usually UDP/IP or RTP/UDP/IP). The use of these packages increases the interoperability of devices with multiple operating systems and, in many cases, the system scalability is available. They are not designed for use on the Internet.

To study the newly developed seismic data transmission system, the Ethersound system made by a French company Digigram was chosen.

3.2   Digigram's Ethersound Systems

Due to the confidentiality agreement (NDA) signed by the EMAG Institute with Digigram, the system presentation cannot contain more detailed information. The general operation principle of the transmission network is presented below.

For the first time, the EtherSound system was presented by Digigram in 2001. The company has developed a logical layer of the network and produces FPGA chips with the implemented EtherSound technology. This system provides two-way communication via Ethernet. It can also use VLAN (Virtual LAN) as part of an existing network. EtherSound comes in two versions: ES-100 and ES-Giga [6].

The ES-100 protocol uses a 100-Mbit Ethernet network [7] to transfer data and takes up the whole available bandwidth (part of the band is used for control). There are 64 bi-directional channels available at 48 kHz sampling. However, depending on the sampling frequency, the number of appropriate channels will decrease - at 96 kHz it will be a maximum of 32 channels. The ES-Giga protocol offers greater possibilities where the total number of channels supported by the network increases to 256 (at 48 kHz).

There are three types of devices in the system, categorized according to the possibility of entering and/or outputting signals from and to the network: Master, Slave and Master/Slave. Master devices are equipped with analogue inputs and convert the analogue signal to the digital one. Slave devices behave reversely - they process a digital signal into an analogue one, sending a signal outside the system. Master/Slave devices are equipped with both inputs and outputs. The limitation to maximum 64 channels is not a disadvantage in this system. They work in both directions and there are 128 of them available.

Topology. The basic connection diagram in the EtherSound system is the daisy-chain. It is also possible to close the chain by connecting the last device in the chain with the first one, thus obtaining a redundant ring topology (Fig. 2). However, it should be remembered that if we close the daisy-chain topology, we will get a "ring", i.e. ring topology, which will be redundant by definition. It could cause the possibility of loops, in which the network packets somehow "get lost". To prevent this, one of the connections between the selected devices is programmed to be blocked. In the situation when one of the active connections is interrupted, it is restored in a short time, which allows further distribution of signals in the network.

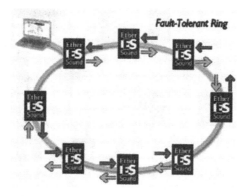

Fig. 2. Daisy-chain topology of Ethersound system

It is also possible to use star-topology combinations and mixed ones. Mixed combinations require switches. The limitation of this topology, using a standard, non-manageable switch, is that the signals can only go one way (unidirectional) (Fig. 3).

The first device in the chain is called Primary Master and it starts sending a 64-channel (at sampling rate of 48 kHz and 24-bit resolution) data stream "down" (i.e. from the OUT socket to the IN socket of the next device in the chain). Then, if the bidirectional mode has been selected, the device receives data back from other network elements (in this case the transmission is "up"). The bi-directional communication segments are programmed by setting the appropriate devices in the "Loop Back" or "End Of Loop" mode. In the daisy-chain connection, all channels are available in both directions for all connected devices.

In the application of the discussed system a mixed topology is predicted and the signals will properly flow "up", i.e. from the sensor to the network controller. The use of a managed switch, which enables to perform VLAN tunnelling, allows to get rid of the one-way inconvenience of the system by developing a virtual daisy-chain connection, and thus the use of full two-way communication (Fig. 4). If the application requires providing a signal to slave devices (e.g. broadcasting network), such a switch does not have to serve as VLAN tunnelling, but only as the distributor of the "downstream" signal.

Configuration and Control. Due to the fact that in the EtherSound system the control and monitoring data are "built-in" to the signal frame, no additional connections are needed. The configuration, control and monitoring of the network are done from one place, using a PC or a microcontroller (in the case of

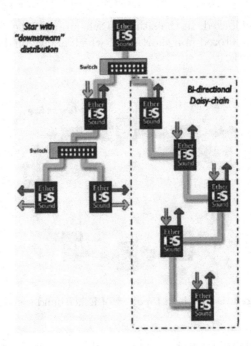

Fig. 3. Mixed topology of Ethersound system

less complicated solutions, the configuration is done by switches located in each interface). It is important that the computer with the control software should be connected with a cross-connected cable to the IN input of the device in the network. The remaining connections in the network should be made with a simple cable. In chain configuration, the problem does not exist, because in the Primary Master device the IN input is always free.

When the network is configured as a ring, any configuration and network control can be made only if at least one connection is physically disconnected. Of course, this does not affect the work of the network, which is still fully functional, but in the event of failure of any connection, the network redundancy may be lost. Therefore, if security is important, then in the ring topology - after configuration, it is necessary to disconnect the computer and close the network into a ring. All devices in the system will read information packets sent in both directions.

Latency. The advantage of the EtherSound network is its low latency (Fig. 5). Typically, these are 125 μs (6 samples) for bi-directional transmission of 64 channels with 24-bit resolution and 48 kHz sampling. However, knowing the size of the system, we can determine its latency almost in microseconds: on each of the EtherSound devices there is a delay of 1.5 μs (0.5 μs in the ESGiga network), a 5-sample buffer (about 104 μs) is used for time synchronization, and each 100-Mbit switch introduces a delay of about 22 μs.

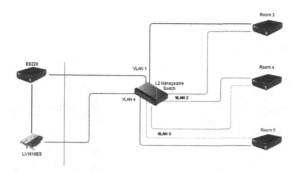

Fig. 4. Spreading-up the possibilities of upstream signal transmission via the switch using VLAN technology

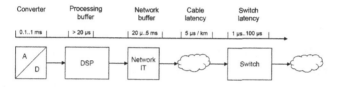

Fig. 5. Sources of signal latency in transmission path of Ethersound system

It should be noted that this delay does not have a significant impact on the operation of the tested system application. The full synchronization between channels is more important than the fact that the information about the measurement will reach the central station with a delay.

4   The Experiment

Studies on the synchronizing properties of the "one-sample accuracy" of Ether-Sound transmission system are not known. That properties were also not described in detail by the system authors and manufacturers. The reason for this is that these properties are not very important for system users in a conventional application. But these properties, as very important in seismic solutions, have been tested in the experiment.

The task of the experiment was to confirm the applicability of the Ethersound system for the recording of seismic phenomena with the preservation of synchrony of recordings using various transmission media and to confirm the unconventional VLAN tunnelling method for data transmission via a managed switch that is part of the network.

4.1   Research Set

The research stand, being the central part (headend) (Fig. 6) consisted of the following devices:

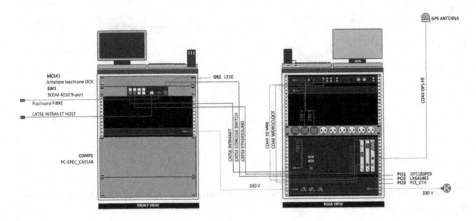

Fig. 6. Headend research stand for testing the transmission of the Ethersound system

- *COMP*1 - PC computer: Intel Core i7 3770 K 3.50 GHz processor, Giga-byte GA-Z77-DS3H motherboard, Corsair DDR3 2 GB 1600 MHz CL8 RAM memory, Western Digital GREEN 500 GB HDD, Intel Gigabit Pro/1000 CT Ethernet card, operational system Windows 7 Pro (32-bit);
- *PCI*1 - GPS card with antenna: Meinberg GPS180PEX;
- *MCLK*1 - Master Clock generator: Antelope Isochrone OCX;
- *PCI*2 - Ethersound card: Digigram LX6464ES;
- *SW*1 - Managed switch: CISCO SF200-24;
- *CV*1 - Media converter: Lanex SE37 SFP 1000BaseLX;
- *M*1 - xDSL modem: Proscend 5110H G.SHDSL.

The master clock signal in the form of a 10 MHz rectangular wave was taken from the appropriate output of the GPS card, and then connected to the input of the Antelope Isochrone OCX generator (as an alternative to the atomic clock). The OCX module produced a 48 kHz Word Clock signal that was connected to the synchronizing input of the Primary Master Ethersound LX6464 card.

The research kit, being an measuring - local part, consisted of the following devices:

- *ES*1, *ES*2 - Ethersound measurement module: Digigram ES220-L, 2 pcs.;
- *S*1, *S*2 - Geophone probe: EMAG-SERWIS SP-5.28/E (Geospace GS-11D geophone), 2 pcs;
- *M*2 - xDSL modem: Proscend 5110H G.SHDSL, 1 item;
- *CV*2 - Media converter: Lanex SE37 SFP 1000BaseLX, 1 item;
- *P*1..*P*4 - UTP patchcords: Cat5e, 1.5 m, 4 pcs.;
- *F*1 - Fiber optic patchcord: OPTRAL singlemode SC 1.5 m, 1 pair;
- *F*2 - Fiber optic patchcord: OPTRAL singlemode SC 100 m, 1 pair;

The Ethersound measurement card was configured as Primary Master accord-ing to the manufacturer's documentation. In the figure (Fig. 7), the ASIO driver settings for the experimental applications are presented.

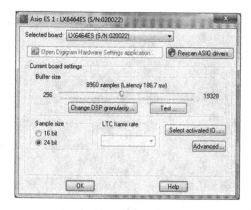

Fig. 7. ASIO driver settings for the Ethersound LX6464ES card

As a recording and analytical software, Adobe Audition CS6 Version 5.0 Build 708 was used.

4.2   Testing for the Correctness of Seismic Phenomenon Recording

The test system was arranged in accordance with the drawing (Fig. 8). The substrate was a concrete floor covered with PVC lining. The S1 geophone probe was placed vertically (the geophone construction forced the probe's vertical orientation) and connected to the audio input of the ES2 module. Ethersound modules ES1 and ES2 were connected in the daisy-chain topology with Ethernet P1 and P2 patch cords via IN-OUT connectors with the Primary Master card.

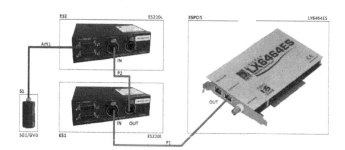

Fig. 8. Connecting a local kit for testing for the correctness of seismic phenomenon recording

As the inductor of the seismic wave, the Stanley FatMax demolition hammer weighing 4 kg was used. Excitation was performed as a free fall of a hammer from a height of 20 cm at a distance of 0.5 m from a geophone probe. With relatively low energy released, but with the total seismic silence kept, a high dynamics of

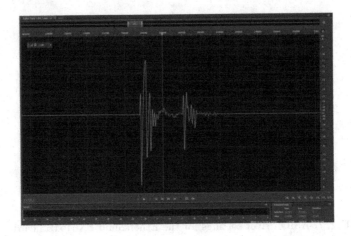

Fig. 9. Waveform of recorded seismic phenomenon

the recorded phenomenon had been expected, which was achieved (Fig. 9). The maximum value of the wave amplitude was about −9 dB, which in this case is very satisfactory.

4.3 Testing for Correctness Signal Synchronization Using Transmission via Ethernet UTP Patchcords

The test kit was arranged in accordance with the drawing (Fig. 10). The measurement and excitation conditions were identical to those described in Sect. 4.2. Of course it was impossible to expect identical shock energy, as the height of 20 cm was determined approximately, but in each of the following tests it was not important.

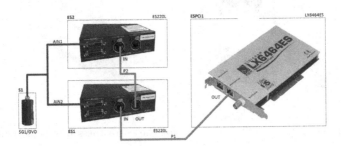

Fig. 10. Connecting a local kit for testing for the synchrony of the measurements transmitted via Ethernet UTP patchcords

The same geophone probe was connected to the analogue inputs of ES1 and ES2 converters (electrically it was a parallel connection of inputs). The shift of the seismic wave first-breaks caused by the different distance between the sensors and the source was irrelevant. It was important to check the synchrony of the measurements processed in different (electrically independent, but synchronized) devices. Two UTP Ethernet patchcords, 1.5 m long, were used to connect digital I/Os of the converters.

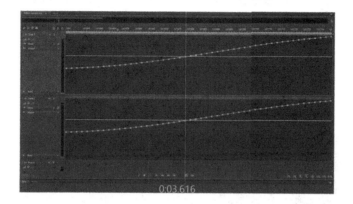

Fig. 11. The moment of waveforms transition by zero for test 4.3

In the figure (Fig. 11), the waveforms recorded by both modules are shown. Zooming in the waveform to watch individual samples, allowed to observe both waveforms the moment of transition by zero. It was found that this took place exactly in the same time (3.616 s), which confirmed the full synchrony of the measurements.

4.4 Testing for Correctness Signal Synchronization Using Mixed Transmission via Ethernet UTP Patchcord and Fibre Optics Path

The test kit was arranged in accordance with the drawing (Fig. 12). The measurement and excitation conditions were identical to those described in Sect. 4.2. Geophone probe was connected as described in Sect. 4.3. The difference was the replacement of the P1 patchcord with an F1 fibre optic consisting of a pair of 1.5 m singlemode fibre optics and two media converters CV1 and CV2. This allowed to introduce additional signal delays.

In the figure (Fig. 13), the waveforms recorded by both modules are shown. Both waveforms also had the moment of transition by zero in same time (1.267 s). It confirmed the full synchrony of the measurements despite increasing the signal delay.

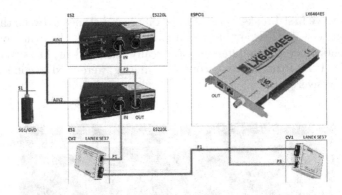

Fig. 12. Connecting a local kit for testing for the synchrony of the measurements transmitted via UTP patchcord and fibre optics path

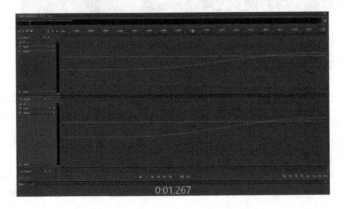

Fig. 13. The moment of waveforms transition by zero for test 4.4

4.5 Testing for Correctness Signal Synchronization Using Mixed Transmission via Ethernet UTP Patchcord and Long Fibre Optics Path

The test kit was arranged in accordance with the drawing (Fig. 14). The measurement and excitation conditions were identical to those described in Sect. 4.2. Geophone probe was connected as described in Sect. 4.3. Transmission connections was the same as described in Sect. 4.4. The difference was the replacement of the F1 1.5 m patchcord with an 100 m one. This allowed additionally increase signal delays.

In the figure (Fig. 15), the waveforms recorded by both modules are shown. Both waveforms also had the moment of transition by zero in same time (2.455 s). It confirmed the full synchrony of the measurements despite additional increasing the signal delay.

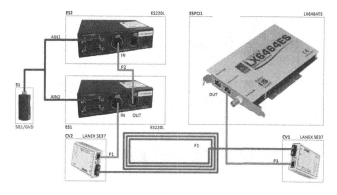

Fig. 14. Connecting a local kit for testing for the synchrony of the measurements transmitted via UTP patchcord and fibre optics long path

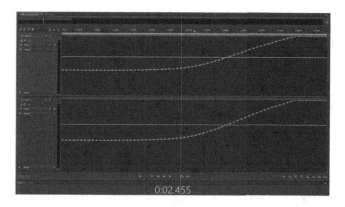

Fig. 15. The moment of waveforms transition by zero for test 4.5

4.6 Testing for Correctness Signal Synchronization Using Mixed Transmission via Ethernet UTP Patchcords and Managed Switch for VLAN Tunnelling

The test kit was arranged in accordance with the drawing (Fig. 16). The measurement and excitation conditions were identical to those described in Sect. 4.2. Geophone probe was connected as described in Sect. 4.3. Transmission connections was different. The signal from modules ES1 and ES2 was transmitted to the switch by two physical independent paths, which could suggest a star topology. However, the creation of the VLAN tunnel logically created a virtual daisy-chain topology, which enabled the transmission of an upstream signal to the Primary Master module. It should be noted that the switch additionally increased the signal delay.

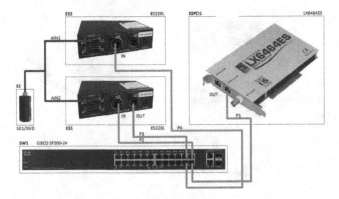

Fig. 16. Connecting a local kit for testing for the synchrony of the measurements transmitted via UTP patchcords and managed switch for VLAN tunnelling

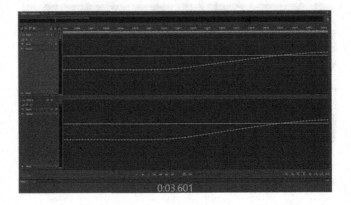

Fig. 17. The moment of waveforms transition by zero for test 4.6

In the figure (Fig. 17), the waveforms recorded by both modules are shown. And in this case also both waveforms had the moment of transition by zero in same time (3.601 s). It confirmed the full synchrony of the measurements despite increasing the signal delay and VLAN tunnelling.

4.7 Testing for Correctness Signal Synchronization Using Mixed Transmission via Ethernet UTP Patchcord, Long Fibre Optics Path and Managed Switch for VLAN Tunnelling

The test kit was arranged in accordance with the drawing (Fig. 18).

The measurement and excitation conditions were identical to those described in Sect. 4.2. Geophone probe was connected as described in Sect. 4.3. Transmission connections was the same as described in Sect. 4.6, only the length of fibre-optic cable has been increased to 100 m.

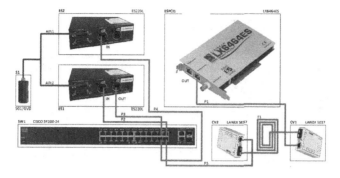

Fig. 18. Connecting a local kit for testing for the synchrony of the measurements transmitted via UTP patchcord, long fibre optics path and managed switch for VLAN tunnelling

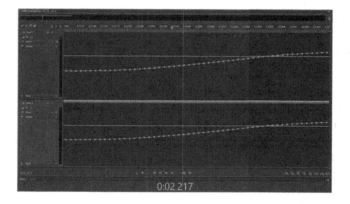

Fig. 19. The moment of waveforms transition by zero for test 4.7

In the figure (Fig. 19), the waveforms recorded by both modules are shown. In this case also both waveforms had the moment of transition by zero in same time (2.217 s). It confirmed the full synchrony of the measurements despite increasing the signal delay in long fibre-optics path and switch making VLAN tunnelling.

The change in the length of transmission cables as well as the transmission medium was aimed at investigating inter-channel desynchronization, because in the application of seismic systems, distances between sensors often differ by a few kilometres. Such desynchronization has no significance when using the standard Internet connections.

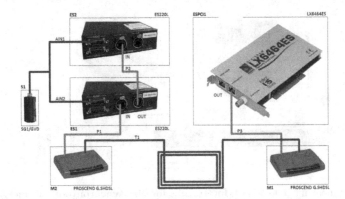

Fig. 20. Connecting a local kit for testing for the synchrony of the measurements transmitted via UTP patchcord and long xDSL modem path

4.8   Testing for Correctness Signal Synchronization Using Mixed Transmission via Ethernet UTP Patchcord and Long xDSL Modem Path

Sometimes in industrial applications there is a serious limitation of the possibility of using fibre optic transmission paths and the large required distances between sensors and the headend preclude the use of UTP Cat5e Ethernet patchcords. Therefore, it was decided to perform a test using xDSL modems, enabling connections up to 10 km with a bandwidth limit of up to 10 Mbit/s. Because the test system (2 measurement channels sampled at 48 kHz) used just over 3% of the bandwidth declared by the Ethersound system, it seemed reasonable. The test kit was arranged in accordance with the drawing (Fig. 20).

Modems were connected by a section of a 5 km standard telecom cable. Unfortunately, the Primary Master module did not connect to the ES1 and ES2 modules. The test was repeated with a few-meter section of the cable, but the effect was identical. Ethersound algorithms that check the availability of bandwidth for the transmission of its protocol are not known, but it is very likely that the ability to transmit packets at 100 Mbit/s is absolutely required, regardless of the practical need for bandwidth usage. It is also possible that the signal delay is too big for modem devices, which goes beyond the limits of the Ethersound control loop. In summary - it is not possible to use the Ethersound system with modem devices, and certainly converting 100 Mbit/s interface signals to lower speed systems.

4.9   Testing for the Time Shift Associated with the Different Distance of the Geophones from the Source of the Shock

The test kit was arranged in accordance with the drawing (Fig. 21). The S1 and S2 geophone probes have been connected to the analogue inputs of the converters - ES1 and ES2, respectively. Probe S1 was placed at a distance of 0.5 m, probe

S2 at a distance of 7 m from the source of the shock. Transmission connections and excitation conditions were identical to those described in Sect. 4.2.

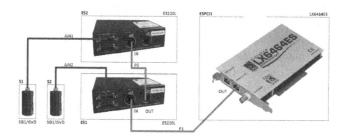

Fig. 21. Connecting a local kit for testing for the time shift associated with the different distance of the geophones from the source of the shock

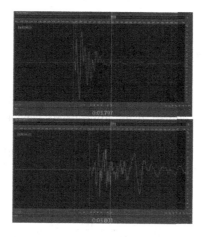

Fig. 22. The moments of wave first-breaks for test 4.9

In the figure (Fig. 22), the waveforms recorded by both modules are shown. The first-break wave time is shifted by 4 ms. Assuming the previously proven synchronism of the registration, it can be concluded that this effect is associated only with the phenomenon of the seismic wave moving in the solid body and depends only on the physical conditions of wave propagation (among others, distance, stress of the medium). Such measurement can be very useful for geophysical research.

5   Conclusions

The experiment has shown that the use of the Audio over Ethernet system (in this case the EtherSound solution) is highly justified in the application of the seismic measurement network. Although this was to be expected because the frequency spectrum of seismic and seismoacoustic signals is similar to the spectrum of audio signals, however, the advantages of a specific synchronization method required detailed testing of network elements, especially paying attention to time and phase shifts of signals using typical seismic sensors. It can be concluded that the use of the Audio over Ethernet system in seismic networks allows to achieve the following benefits:

- Lack of time and phase shift error between measurement channels – a virtually identical effect to analogue measurements;
- Lack of time and phase shift in measurement channels enables to use simple software algorithms for seismic noise reduction;
- 24-bit processing allows to achieve very high dynamics of measurements (>120 dB);
- Sampling, unfortunately, cannot be less than 48 kHz (in standard seismic systems 10 kHz is enough), which significantly increases the requirements for transmission channel capacity; still, thanks to this, it is possible to register higher harmonics of slow-changing seismic signal, which is a very useful research material;
- Sampling can be increased up to 192 kHz, which makes it possible to record ultrasound signals (of course, using appropriate sensors);
- For a sampling frequency of 48 kHz, the time between samples of 200 ns reduces the ambiguity of determining the distance to about 0.5 m, and in the case of 192 kHz – to about 12.5 cm (assuming the average speed of the carbon wave at 2,500 m/s);
- The system enables to use a fibre as a transmission medium, which significantly increases the transmission range (and thus the size of the system) to several dozen (or even several hundred) kilometres, while being completely immune to electromagnetic and electrostatic interference of the transmission path;
- The system has the possibility of two-way audio transmission, which allows simultaneous communication with the measurements (e.g. voice), and the communication does not interact with the measurement channels;
- Synchronization of measurements takes place only from one central source by broadcasting the Master Clock signal.

It should be noted that due to the characteristics of audio signals (especially the lower limit of the frequency band) of the Ethersound system (or any other AoE system), one cannot directly apply such a system in systems that monitor seismic phenomena. The lower limit of the seismic signal band starts from about 0.001 Hz. For acoustic signals it is 20 Hz.

Analog inputs of the EtherSound system are built according to the audio specification and are a problem when directly connecting seismometers (frequency 0.001 to 10 Hz), geophones (1 to 1000 Hz) or accelerometers (0.2 to over 20 kHz) [14], especially in the low range frequency. The reason is the standard capacitive decoupling input circuit, used primarily for protection of output electroacoustic transducers. It is necessary to redesign the input circuit, allowing the transmission of frequencies close to zero, which will ensure the possibility of using standard seismic sensors and will not damage electroacoustic transducers that are not present in the measuring paths of seismic systems.

The above problems are not a limitation of implementing AoE technology in seismic systems because parallel hardware-oriented work at the EMAG Institute is aimed at developing an analogue interface fully compatible with AoE systems. In addition, some work is underway on the distribution of the seismic signal through a wide area network, using a backbone Internet network, made available by one of the telecommunications operators.

Acknowledgements. The work was carried out within the statutory research project of the Institute of Innovative Technologies EMAG.

References

1. GeoSIG homepage. www.geosig.com/GMSplus--GMSplus6-id12557.aspx. Accessed 9 Feb 2019
2. Kinemetrics homepage. https://kinemetrics.com/post_products/quanterra-q330/. Accessed 9 Feb 2019
3. Ashjaei, M., Patti, G., Benham, M., Noite, T., Alderisi, G., Bello, L.: Schedulability analysis of ethernet audio video bridging networks with scheduled traffic support. Real-Time Syst. **53**, 526–577 (2017)
4. Cao, A., Dou, L., Cai, W., Gong, S., Liu, S., Jing, G.: Case study of seismic hazard assessment in underground coal mining using passive tomography. Int. J. Rock Mech. Min. Sci. **78**, 1–9 (2015)
5. Cianciara, B., Isakow, Z., Dworak, M.: Suggestion of a method for tremor prediction based on the ARAMIS-a measuring system. Mechanizacja i Automatyzacja Górnictwa **9**, 172–178 (2001)
6. Digigram: Ethersound, Overview - An introduction to the technology, Rev. 2.0c. Digigram S.A. (2004)
7. Digigram: Building Ethersound Networks, Rev. 2.0. Digigram S.A. (2005)
8. Hong-Jiu, Z.: First break of the seismic signals based on information theory. Inf. Eng. (IE) **4**, 38–43 (2015)
9. Isakow, Z., Dworak, M., Augustyniak, A., Sierodzki, P., Koza, J.: Sposób i układ do synchronizacji sejsmicznych i sejsmoakustycznych sieci pomiarowych, zwłaszcza kopalnianych sieci iskrobezpiecznych. Patent RP, PL 225485 (2016)
10. Isakow, Z., Juzwa, J., Kubańska, A., Siciński, K.: Nowoczesny system INGEO do monitorowania zagrożenia sejsmicznego i tąpaniami w kopalniach węgla kamiennego i rud miedzi. Zeszyty Naukowe Instytutu Gospodarki Surowcami Mineralnymi i Energią PAN **101**, 173–184 (2017)
11. Kornowski, J.: Energia umowna i energia emisji sejsmoakustycznej w przypadku obserwacji za pomocą aparatury ARES. Mechanizacja i Automatyzacja Górnictwa **6**, 18–28 (2009)

12. Mutke, G., Dubiński, J., Lurka, A.: New criteria to assess seismic and rock burst hazards in coal mines. Arch. Min. Sci **60**(3), 743–760 (2015)
13. Perol, T., Gharbi, M., Denolle, M.: Aging models and parameters of quartz crystal resonators and oscillators. In: Fall Meeting 2016. American Geophysical Union, San Francisco (2016)
14. Pilecki, Z., Harba, P., Adamczyk, A., Krawiec, K., Pilecka, E.: An overview of technical parameters of geophones used in seismic engineering. Przegląd Górniczy **70**(7), 12–21 (2014)
15. Wang, S., et al.: Aging models and parameters of quartz crystal resonators and oscillators. In: Symposium on Piezoelectricity, Acoustic Waves, and Device Applications (SPAWDA). Harbin Engineering University, Harbin (2015)

A Novel SQLite-Based Bare PC Email Server

Hamdan Alabsi[1], Ramesh Karne[2], Alex Wijesinha[2(✉)], Rasha Almajed[2], Bharat Rawal[3], and Faris Almansour[2]

[1] College of Business, Mathematics and Science, Bemidji State University, Bemidji, MN 56601, USA
[2] Department of Computer and Information Sciences, Towson University, Towson, MD 21212, USA
awijesinha@towson.edu
[3] IST Department, Penn State University, Abington, PA 19001, USA

Abstract. We describe a SQLite-based mail server that runs on a bare PC with no operating system. The mail server application is integrated with a *server-based* adaptation of the popular SQLite client database engine. The SQLite database is used for storing mail messages, and mail clients can send/receive email and share files using any Web browser as in a conventional system. The unique features of the bare PC SQLite-based email server include (1) no OS vulnerabilities; (2) the inability for attackers to run any other software including scripts; (3) no support for dynamic linking and execution of external code; (4) a small code footprint making it easy to analyze the code for security flaws; and (5) performance benefits due to eliminating OS overhead. We describe system design and implementation, and give details of the bare machine mail server application. This work serves as a foundation to build future bare machine servers with integrated databases that can support Internet-based collaboration in high-security environments.

Keywords: Bare machine computing · Bare PC · SQLite · Database engine · Web server · Email

1 Introduction

Private email servers are often used for collaboration and communication among small groups that do not want to use popular public email systems such as gmail that rely on external servers, or within an organization, where a select group of individuals do not want to use the organization's own email system for privacy reasons. Because of the importance of protecting sensitive content, private email servers require stronger security than ordinary email servers. However, private email servers that require support of an operating system (OS) are harder to secure since they are vulnerable to exploits that target the OS, and have a larger attack surface even if a lean kernel or customized OS is used (since the

© Springer Nature Switzerland AG 2019
S. Kozielski et al. (Eds.): BDAS 2019, CCIS 1018, pp. 341–353, 2019.
https://doi.org/10.1007/978-3-030-19093-4_25

server requires external code to support the email application). A private email server that runs as a bare machine computing (BMC) application with no OS, kernel or external code provides an alternative to using a conventional private email server.

As a first step towards building a complete private email BMC system, we implemented a bare machine mail server. The current prototype uses a novel *server-side* adaptation of the SQLite client database engine to support mail system operations. SQLite [18] is widely used as a client-side database in many applications, smartphones, operating systems, and web browsers. SQLite is also used as a database engine for web sites and as a database for application servers. In all such cases however, SQLite requires the support of a conventional operating system (OS) or some form of a kernel. BMC SQLite has no OS dependencies and it would be possible to replace BMC SQLite itself with a BMC database engine customized for private email. The current SQLite-based mail server is accessed via the Internet. It can also be used to build a private email system that runs on dedicated network with no external connections.

The BMC email server only requires a bare PC, which is an ordinary desktop or laptop, with no other software loaded except for the application itself. The server application integrates lean versions of the necessary network protocol code with the application code. The application also directly communicates to the hardware and the network interface controller (NIC). BMC applications are intrinsically secure due to the absence of features that attackers typically target. A BMC system could thus be viewed as a form of minimalist system, where the tradeoffs are less features versus reduced avenues for attack. The remainder of this paper is organized as follows. In Sect. 2, we discuss BMC systems, and in Sect. 3, we present related work. Sections 4 and 5 respectively provide design and implementation details of the bare mail server. Section 6 contains the conclusion.

2   BMC Systems

The bare machine computing (BMC) paradigm, previously called dispersed operating system computing (DOSC) [11], enables applications to run on a bare PC with no OS or kernel. Details of the BMC application development methodology, which differs from that in a conventional system, are given in [12]. The BMC programming paradigm may be viewed as a holistic approach, where one needs to consider all aspects of application and system programming including the execution environment. As there is no OS or centralized kernel running in the system only a user mode is supported, and the programmer writes BMC applications that consist of a sequence of events executing in an interleaved fashion. Each event code becomes a small thread of execution that runs without any interrupts or task switching. Since there is no OS, the BMC programmer has total control of both application and execution flow. For network communication, BMC applications have their own Ethernet driver and a lean TCP/IP stack that implements only those protocols needed by the application.

In addition to eliminating the overhead due to the OS, BMC applications do not have OS vulnerabilities. They present fewer attack vectors than conventional systems since each application only implements a minimal set of necessary features. Being smaller and simpler, the BMC code is also easier to analyze for security flaws. Furthermore, since the bare PC server application is statically compiled and runs as a single thread of execution, it is difficult to compromise. Furthermore, no exploit code can run on the server. While dynamic linking and script execution can be disabled on a customized minimalist OS-based server, it still has more avenues for attack than a BMC server application. These characteristics of BMC systems make the bare mail server especially suited for private communication in high security environments. If needed, the bare PC file system located on detachable physically secured USB flash drives can be used by the server for permanent storage.

3 Related Work

SQLite on OS-based systems provides database services for websites and application servers [3]. Email servers such as [4, 7, 10] store emails in a SQL database. Code for a Linux/free BSD-based mail server with account storage in a back-end database is given in [10]. A typical LAMP stack [13] consists of the Linux OS, Apache web server, MySQL database, and PHP, which could be used to build a customized mail server. NIST publications for securing servers include recent guidelines for secure hypervisor deployment [17]. Web, mail and database servers are frequently located in the cloud and virtualization is common. Conventional email servers have many external components, each of which would need to be secured in order to build a private email server. In contrast, the bare mail server runs without an OS or kernel and would have fewer components to secure. Alternative approaches to minimize OS impact include exokernel [5], bareMetal-OS [8], Linux kernel tinification [14], and the minimal Rust kernel [1]. Novel OSs, some of which are relatively small, are discussed in [16].

4 Design

4.1 Overview

Clients currently access mail and database services hosted on the BMC server using any Web browser (in future, BMC clients could be used for more security). The database client and database mail client data are stored on a USB flash drive. SQLite's command line interfaces are used for system administration and debugging. When the bare mail server is booted, any required files are transferred from a Windows system (using a lean version of trivial FTP), stored in memory, and used during server operation. If there is a pre-existing database, it will also be loaded. After the files are transferred, a dummy DB is created since SQLite needs a DB pointer to do any operation. A "mail" table defines the bare mail server's database. User profiles with username and password are also loaded before the bare mail server application starts.

The integrated bare mail server application is statically compiled and currently runs on any Intel x86 architecture based PC or laptop. When the application executes, it is not possible to run scripts, add new applications, or load dynamic libraries. The bare mail server only performs functions that are defined at development time. New profiles can be loaded to add or delete users as needed.

Figure 1 shows the system architecture for the BMC server. Server components include the BMC Web server [9], an integrated SQLite USB file system [20], and a novel task system. The tasks and their interactions are shown on the left in Fig. 2. When the system boots up from the USB, the user menu for selecting services is shown. The "Load" option loads the integrated server application, and then the "Run" option runs the application, which starts the Main Task (MT).

The server has an MT, which creates, initializes, and stores other tasks in their respective task pools, and a Receive Task (RT), which receives network packets. The four task pools are USB Task (UT), SQLite Task (ST), HTTP Task (HT) and Post Task (PT), which reside in their respective stacks. When needed, a task is popped from the stack, placed into a circular list for execution, and pushed back onto the stack upon completion. The current tasks in a circular list are processed in a first-in-first-out (FIFO) manner. When a packet arrives, the MT starts the RT, which runs as a single thread of execution without interruption until the packet is processed. If the packet has a GET request, an HT is started, and if it has a POST request, a PT is started. A USB event such as insert, remove, read, or write will invoke a USB task to process the event. Similarly, a client's new SQLite event will start a SQLite task. For simplicity, we only consider one SQLite event (i.e., a single SQLite client), although the number of SQLite events (clients) is unlimited in principle.

4.2   SQLite Server

The BMC web server runs after booting. The display then shows the SQLite prompt and other information that are used for diagnostics and debugging. An administrator can also enter command line queries and get a response. Loading user profiles and creating a dummy database is done before clients access the database. Reading the "t1.sql" file makes the database ready for client operations, and the DBR flag is set in memory so that POST commands will be processed when the client makes requests. The server screen now shows that ST and MT are running. The server runs until it is powered down. Wireshark traces to validate network packets were obtained using an HTTP GET request for the "barerocks530k.gif" file, and for read mail where POST has the user name and GET is used to obtain the results. The SQLite code integrated in this system is taken from the amalgamated package [19], which has two "C" files: shell.c and sqlite3.c. The file shell.c has "main()" and other user interfaces.

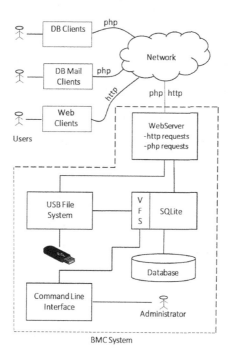

Fig. 1. System architecture

When this code was transformed to run on a bare PC, all system call interfaces were replaced with bare PC interfaces [15]. The right side of Fig. 2 shows the processing logic to integrate the bare PC SQLite system and tasks with the rest of the application. We changed the "main()" call in the shell.c file to "main_call()", so that there is only one main() in the system. The MT starts ST, which in turn calls "main_call()". The ST runs along with the other tasks in the integrated web server application. When ST starts, it prompts ">sqlite" and waits for the user requests. At the server side, all SQLite commands are run using this command line interface. The original SQLite is designed as a client. To build the bare mail server, we integrated SQLite with the bare PC web server by creating a dummy database. In "shell.c" there is an instance of a database, and appropriate initializations are done in the file "sqlite3.c". Otherwise, SQLite will not create any database instance. This approach allows us to use the SQLite code as is and simply tap into the database interface in the file "shell.c". A small script file t1.sql is used for this purpose. It includes the do-meta-command "read t1.sql" to create a dummy database and a dummy table t1. We can enter other SQL statements or meta-commands at the command prompt before running the script file. Once the script file is run, the SQLite interface at the command line interface is disabled as the ".DBR" meta-command sets a DBR flag in shared memory. The ".DBR" meta-command is added to the SQLite set of commands.

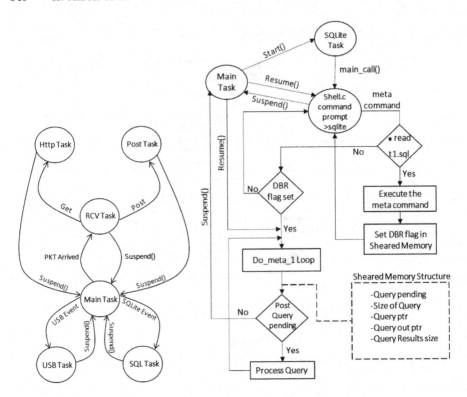

Fig. 2. Task interface (left) and task-based SQLite query processing (right)

When the DBR flag is set, the SQLite shell program will go into the "do_meta_1" loop. When necessary, ST suspends itself and waits for client requests. The suspend() function returns to the MT. The suspend() function in C++ was mapped to the suspend() function in C (as class instantiation cannot be done from our C program). Once the DBR flag is set, SQLite processes client queries via a POST command.

4.3   Mail Server Operations

A network event in the server application occurs when a TCP SYN packet arrives. Figure 3 shows control flow in the server including mail operations and the relevant TCP states and packets. Either an HTTP GET request or a HTTP POST request is sent by the client. When the client needs data from the server a GET request is used, and when it sends data to the server a POST request is used. For a GET request, the server knows how many packets to send and it can control the data transfer. For a POST request, the client controls the data transfer. In this case, the server keeps track of the number of packets and the total data size to verify that all packets have arrived. POST data processing requires a PHP parser. To avoid complexity, only the functionality needed

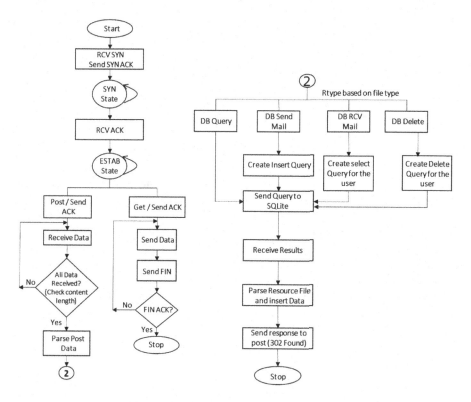

Fig. 3. TCP flow and mail operations

for the integrated application is implemented in the PHP parser. Since we are designing a server system, it is possible to control the design of the PHP files and their contents.

It should be noted that HTML files are static content and PHP files are dynamic content. When the client accesses the database, dynamic content is required due to the nature of queries and mail requests. POST data is received a packet of data at a time until all packets are received. Then the server processes the data according to the type of the request indicated by RType (circled value 2) in Fig. 3. There are five types of client-server interfaces, which result in five forms. Figure 4 shows the relation between functions, file names and operations (Rtype for login is not shown in Fig. 3). Each form corresponds to a different type of PHP file. In addition, the client fetches the results from the database using a separate form named "inbox.php". Thus a total of six PHP files are required in the system. Parsed attributes of each file are also shown in Fig. 4.

When the "login.php" file is received as POST data, the server parses the data for the username and password, validates them, and logs the information. This information is used throughout the session for a given user when logged in. Only necessary information is parsed using keywords; the file also has keywords to help the parser find the key reference points. Database queries from a client are

Function	File name	Operation
Login	Login php	Extract username, password
DB Query	Compose.php	Extract query
	Inbox.php	Put results
Send Mail	Sendmail.php	Extract mail parameters
		Form Insert query
Receive mail	Retrieve.php	Extract username
	Inbox.php	Form select query
		Insert results in file
Delete mail	Delete.php	Extract username
		Form Delete query

Fig. 4. Functions, file names and operations

received at the server via a "compose.php" file sent using the POST command. The server extracts the query data from the packet(s) and forms a query block for the SQLite server, which returns the results in a memory block. These results are inserted into the "inbox.php" file at the insertion points shown in Fig. 5. The insertion points "POINT1" and "POINT2" provide the references in the files to insert data into the file. When the result data is inserted, the "POINT2" reference will be moved down to increase the file size according the size of the data. This technique enables us to avoid writing a full PHP parser.

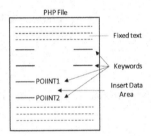

Fig. 5. Inserting results in the PHP file

When sending mail, the "sendmail.php" file is used to form a mail message. At the server, mail attributes such as from, to, subject, and body are extracted, and an INSERT query is formed. This query is sent to SQLite for execution and checked for a valid return code. Parsing of the "sendmail.php" file is similar to the other PHP files. In this case, a username is also parsed from the mail content and validated by comparing with the login name. Receiving mail is slightly more complex than other client requests. The "retrieve.php" file sent to the server via POST contains the username of the receiver. A SELECT query is formed to retrieve emails for this receiver from all other users (if any). This query is sent to SQLite and the results are inserted into the "inbox.php" file. The client sends a GET request to obtain the results. Inserting results into the file "inbox.php" is similar to inserting database query results. Finally, the "delete.php" file is received by the server as POST data. The server extracts the username from the message and the appropriate query is formed to delete the mail message from the database. For a given user, mail from a sender is deleted based on the sender's username. Only one mail message at a time can be deleted.

The server is a prototype that integrates client SQLite as a server component, and handles basic database queries and mail requests. While more features can be added to the server as it is designed in a modular fashion, this may increase server complexity and result in reduced security.

5   Implementation

SQLite integration is implemented in C/C++ with direct hardware interfaces. Communicating from C code to C++ code required the use of special pointer conversions and prototypes. The server application uses an Intel Gigabit Ethernet NIC (on-board chip) and its corresponding BMC driver. The executable size is 821,248 bytes. This includes the application code and the required execution environment code that makes it a self-contained, self-managed and self-controlled application. A user owned USB drive contains this executable including its boot code, which is used to boot up a bare PC and run the server application. The application runs on any Intel x86 architecture based PC or laptop. The SQLite and POST source code are 140K lines and 3K lines.

The SQLite DB application runs as a client server system. When a user clicks on the DB link to access the database or DB mail, the login interface appears. The user enters the username and password (which are validated), and the system menu is displayed as in Fig. 6. The page for composing a mail message is shown in Fig. 7. The results of a query are shown in Fig. 8. Selecting the DB Query option, a user can enter a SQLite query as in Fig. 9. Database queries can be typed in as text in the window or they can be included in an attachment. The system is tested for both options. A database application running on the system can be accessed by users and administrators. The bare PC web server has a debug mode, where some web pages can show internal server operations and database operations. Database queries are limited to the queries that can be performed by SQLite.

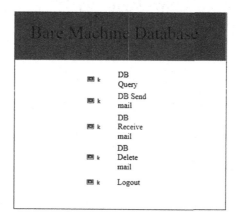

Fig. 6. Menu screen

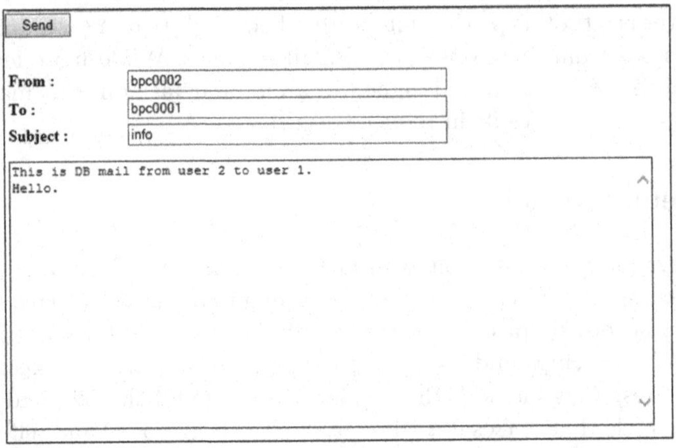

Fig. 7. Compose mail page

The bare mail server is designed to work with a limited group of users who share the SQLite database to send and receive email after logging in to the server. The server does not include TLS at present, but it can be added by integrating the existing BMC TLS code in [2]. An SMTP-based bare PC email server built previously [6] could also be integrated with the present system if needed to communicate with external mail servers. The SQLite database and user accounts are created by an administrator in a physically secure manner. An OS-based web browser is used to send, receive/read, and delete email messages. The corresponding database operations are send, receive, and delete. Sent messages are stored in the database. The send mail form requires the sender's name

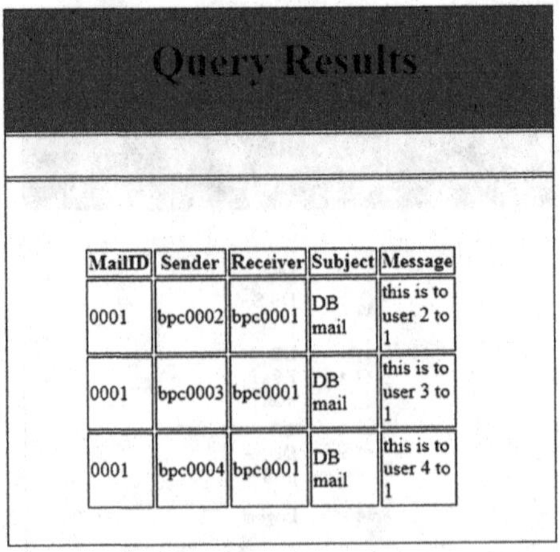

MailID	Sender	Receiver	Subject	Message
0001	bpc0002	bpc0001	DB mail	this is to user 2 to 1
0001	bpc0003	bpc0001	DB mail	this is to user 3 to 1
0001	bpc0004	bpc0001	DB mail	this is to user 4 to 1

Fig. 8. Query results page

address, receiver's name, subject, and message body. This form is sent to the server as a POST message. The server inserts this query into the mail database and returns. The receive mail form requires the username for the receiver. A page similar to that shown in Fig. 8 is used to display the results of receive mail. Currently, all emails received for a user are displayed. The server receives the username and password and forms a query. It submits the query to SQLite and collects the results, which are stored in a file and returned to the client via a GET request. The delete mail form requires the sender's username for the message to be deleted. A user must be logged in (as a receiver) to delete a sender's messages.

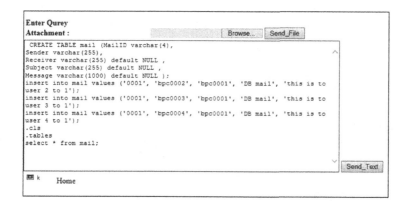

Fig. 9. Query entry page

6   Conclusion

We built a novel email server that provides mail services on a bare PC by running SQLite in a client-server environment. The server enables a small group of users to collaborate by sharing files or communicate privately without using a public email system. It is immune from attacks that require an underlying OS and is easier to secure than a conventional email server due to its reduced attack surface and smaller footprint. The code is currently written based on the Intel IA-32 CPU architecture, but can run in principle on other CPU architectures by developing the necessary hardware interfaces. The email server extends a bare PC webmail server built earlier in three significant ways: (1) it has the capability to store messages on USB-based external mass storage; (2) it uses the bare SQLite database engine and associated queries for message storage and retrieval; (3) it is integrated with a USB-based bare PC file system. The server is accessed by using any conventional OS-based Web browser. Since the browser/mail client can be compromised due to OS vulnerabilities, only the mail server retains the advantages of being resistant to OS-based attacks. For better security and enhanced privacy, the web browser should also be run on a BMC system. Such a web browser does not exist at present, but a text-based bare PC web browser is currently under development. While hardened lean OS-based private email servers can also be run on dedicated networks and limited to a select group of users, they are still vulnerable to OS-based attacks. This work serves as a foundation to build secure private bare PC email servers and clients that run with no OS support.

References

1. A Minimal Rust Kernel. https://os.phil-opp.com/minimal-rust-kernel/. Accessed 31 Jan 2019
2. Appiah-Kubi, P., Karne, R.K., Wijesinha, A.L.: A bare PC TLS webmail server. In: 2012 International Conference on Computing, Networking and Communications (ICNC), pp. 149–153. IEEE (2012)
3. Appropriate Uses for SQLite. https://www.sqlite.org/whentouse.html. Accessed 31 Jan 2019
4. Dbare mail: fast and scalable SQL based email services. http://www.dbaremail.org/. Accessed 31 Jan 2019
5. Engler, D.R., Kaashoek, M.F., et al.: Exokernel: an operating system architecture for application-level resource management, vol. 29. ACM (1995)
6. Ford Jr., G.H., Karne, R.K., Wijesinha, A.L., Appiah-Kubi, P.: The design and implementation of a bare PC email server. In: 2009 33rd Annual IEEE International Computer Software and Applications Conference, vol. 1, pp. 480–485. IEEE (2009)
7. Git-Hub – nodemailer/wildduck. https://github.com/nodemailer/wildduck. Accessed 31 Jan 2019
8. GitHub – ReturnInfinity/bareMetal-OS. https://github.com/ReturnInfinity/bareMetal-OS. Accessed 31 Jan 2019
9. He, L., Karne, R.K., Wijesinha, A.L.: The design and performance of a bare PC web server. Int. J. Comput. Appl. **15**(2), 100–112 (2008)
10. iRedMail-Freem Open Source Mail Server Solution. https://www.iredmail.org/. Accessed 31 Jan 2019
11. Karne, R.K., Jaganathan, K.V., Rosa Jr., N., Ahmed, T.: DOSC: dispersed operating system computing. In: Companion to the 20th Annual ACM SIGPLAN Conference on Object-Oriented Programming, Systems, Languages, and Applications, pp. 55–62. ACM (2005)
12. Khaksari, G.H., Karne, R.K., Wijesinha, A.L.: A bare machine application development methodology. Int. J. Comput. Appl. **19**(1), 10–25 (2012)
13. LAMP Stack. https://www.turnkeylinux.org/lampstack. Accessed 31 Jan 2019
14. Linux Kernel Tinification. https://tiny.wiki.kernel.org/. Accessed 31 Jan 2019
15. Okafor, U., Karne, R.K., Wijesinha, A.L., Rawal, B.S.: Transforming SQLite to run on a bare PC. In: 7th International Conference on Software Paradigm Trends (ICSOFT), pp. 311–314 (2012)
16. Operating Systems You May Not Have Heard Of (But Should). https://www.hongkiat.com/blog/lesser-known-operating-systems/. Accessed 31 Jan 2019
17. Security Recommendations for Hypervisor Deployment on Servers, NIST Special Publication 800-125A, January 2018
18. SQLite. https://www.sqlite.org/index.html. Accessed 31 Jan 2019
19. The SQLite amalgamation. https://www.sqlite.org/amalgamation.html. Accessed 31 Jan 2019

20. Thompson, W., Karne, R., Wijesinha, A., Chang, H.: Interoperable SQLite for a bare PC. In: Kozielski, S., Mrozek, D., Kasprowski, P., Małysiak-Mrozek, B., Kostrzewa, D. (eds.) BDAS 2017. CCIS, vol. 716, pp. 177–188. Springer, Cham (2017). https://doi.org/10.1007/978-3-319-58274-0_15

Building Security Evaluation Lab - Requirements Analysis

Dariusz Rogowski[✉], Rafał Kurianowicz, Jacek Bagiński, Roman Pietrzak, and Barbara Flisiuk

Institute of Innovative Technologies EMAG,
ul. Leopolda 31, 40-189 Katowice, Poland
dariusz.rogowski@ibemag.pl

Abstract. Physical protection of a laboratory must be built according to many stringent standards, criteria, and guidelines which are often general and difficult to apply in practice. This is a pity, because many organizations may save a lot of time, money and effort if they had a way of selecting the right security measures at the beginning of a process. Introducing a simple evaluation method of safeguards into requirements analysis can dramatically facilitate the designing phase of the lab's physical security. In the result, more institutions would decide to cope with the problem of fulfilling security requirements by choosing concrete solutions within the assumed budget.

Keywords: Common Criteria · Security requirements · Evaluation · Certification

1 Introduction

Building and organizing an Information Technology (IT) security evaluation laboratory is a very difficult and demanding task. It involves security specialists, technicians, analysts and even construction workers. At start you have to identify your lab's goal, future clients, and prospective IT products to be evaluated. And even though you have all these at place you still may face a main question what are the physical security requirements and how to deploy them properly to make your lab secure. The problem is even bigger when you have to build your lab from the scratch like in the case presented in this article.

We currently face such a problem in a research project "National schema for the security and privacy evaluation and certification of IT products and systems compliant with Common Criteria (KSO3C)" [18]. The project is financed from the Polish National Centre for Research and Development (NCBR) program "Cybersecurity and e-Identity". The project goal is to build a national schema for security and privacy evaluation of IT products compliant with ISO/IEC 15408 standard [7–9] (also known as Common Criteria – CC [3–5]). The project is a common initiative of three Polish research entities: NIT – National Institute of Telecommunications (Warsaw), NASK – Research and Academic Computer

© Springer Nature Switzerland AG 2019
S. Kozielski et al. (Eds.): BDAS 2019, CCIS 1018, pp. 354–365, 2019.
https://doi.org/10.1007/978-3-030-19093-4_26

Network (Warsaw), and Institute of Innovative Technologies EMAG (Katowice). The Polish schema will comprise of a Certification Body (CB) in NASK and two laboratories (ITSEF – IT Security Evaluation Facility) in NIT and EMAG.

According to the CC international arrangements of certificates authorizing countries and licensed laboratories, any new candidate schema and a lab have to fulfill a lot of strict organizational and security requirements. These requirements are spread out in many CC documents, other evaluation schemes, and international arrangements which we describe in Sect. 2.

One of EMAG's tasks in the project is to build, organize and secure the evaluation lab. After choosing the lab's premises you have to select proper physical security measures. In practice you have to select and apply minimum set of requirements listed by international arrangements and supporting documents. There are a lot of requirements to analyze and to decide which are the best to fulfill your lab's security needs. But unfortunately in most cases they are described in a general way without technical details answering questions like: How many cameras and with what resolution do I need for my surveillance system? How strong must be door locks or bars in windows? How many burglary-resistant doors must be included in an access control system? How many and what type of sensors are needed for an alarm notification system? All these details you have to adjust to your own needs. But how to do it objectively? How to assess whether the chosen security measures are good enough to protect the lab and clients' assets? How to find a trade-off between the level of security and costs of controls? In this paper we propose answers to these questions:

- First, in Sect. 3, we describe the problem of seeking for the proper criteria among dozens stored in the international arrangements, standards, and supporting documents. As an example we also present lists of basic requirements with references. In fact it is the first time in Poland when the criteria for securing the lab must consider the CC requirements. Thus the problem itself is new, and has not been addressed so far.
- Second, in Sect. 4, we suggest to use an approach supporting the selection of the lab's security measures based on the idea of combining the Polish regulation for physical protection of secret information [8] with the CC requirements.

The regulation itself is not something new or revolutionary but it helped us to choose adequate controls what we presented in Sect. 5. The regulation describes how to calculate a threat level to the critical information and how to select security controls relevant to that level. In Sect. 5 we also discuss the results and compare our idea to the standard way of choosing security controls. In Sect. 6 we conclude our research and state the future work which can help to adjust the proposed approach for new applications.

2   Common Criteria Methodology

In this section we briefly outline the Common Criteria methodology and its main goals. It will set the scene for this paper and help the reader to understand the gist of the problem presented in Sect. 3. Common Criteria (CC) is a multi-part international standard for the security evaluation and certification of IT products which was published in 1999. The standard is the result of common work of six countries: The United States of America, Canada, France, Germany, The Netherlands, and Great Britain. They adjusted and integrated their former security standards into one common CC standard.

CC parts 1–3 [3–5] became an International Organization for Standardization (ISO) standard (ISO/IEC 15408) and it is supported by the Common Evaluation Methodology (CEM) [2]. CEM is also the international standard ISO/IEC 18045 [12]. CC methodology is based on three main processes [15,16]: (1) IT security development process in which a Security Target (ST) document is created. ST describes a product's security features countering threats. (2) IT product development process – it is building the product and implementing its security features. (3) Security evaluation process – it is verification whether the product's security features work well or not. In this section we will focus only on the last process.

In CC the assurance of security features (i.e. security functions) is measured by Evaluation Assurance Levels (EAL) from EAL1 to EAL7. The higher level the more stringent rules and requirements must be fulfilled during the development of a product. Minimum Site Security Requirements (MSSR) document [13] defines a set of minimum controls that a developer shall meet at the development site.

A common sense let us assume that at least the same requirements must also apply to the evaluation lab. And in fact, all ITSEFs use MSSR to determine the proper security controls what we describe in Sect. 3. MSSR was used as a reference document for the lab's security requirements in the KSO3C project.

Next, evaluation results are validated in a certification process run by Certification Body (CB). In the end, a certificate is issued and published on the Common Criteria website [17]. The CC website provides statistics that list a grand total of 3779 products have been certified so far. The CC certificates are issued by different countries using different CC schemes. It raised a question whether certificate issued by one country is still valid in other countries? That is why countries decided to create and sign two international arrangements that solved this problem:

1. SOG-IS MRA (Senior Officials Group – Information Systems Security Mutual Recognition Agreement) [14] – in this European agreement 15 countries decided to mutually recognize CC certificates up to EAL4. Within the group there are 8 countries – certificate producers and 7 countries – certificate consumers (thanks to the KSO3C project Poland joined that group in March 2017 and is represented by NASK);
2. CCRA (Common Criteria Recognition Arrangement) [1] – in this international agreement 29 countries established mutual recognition of evaluated IT products at EALs 1-2. Within the group there are 17 certificate authorizing members and 12 certificate consuming members (and likewise in SOG-IS, in November 2018 Poland became a certificate consuming member thanks to the KSO3C project).

Both arrangements put conditions on the Certification Body and Evaluation Facility for mutual recognition of certificates. Article 5 of the arrangements says that ITSEF must: perform evaluations impartially; apply CC, CEM, and CCRA supporting methods correctly and consistently; and the most importantly – adequately protect the confidentiality of sensitive information. Article 5 also says that CB and ITSEF must be accredited to ISO/IEC 17065 [11] and ISO/IEC 17025 [10] standards respectively.

In this section we described in a great short the Common Criteria standard and accompanying arrangements and supporting documents. We focused on the possible sources of requirements we must follow to build the physically secure lab. In the next section we list the most important requirements taken from these sources and show they are described in a very general manner. This is the main problem cause general statements about security measures must be somehow adjusted to the concrete lab's use case.

3   The Problem – General Requirements

In previous section we described many sources of requirements necessary for building a secure lab. These sources comprise of many guidelines, rules and advice how to implement security measures. The problem is that there are plenty of such documents and requirements. Even though you completed all necessary controls then you have to adopt them to your specific lab context. Beneath we present the examples of requirements sets with their references.

But the problem is that requirements are described in a general way. This is still good cause it allows to gather all necessary prerequisites and prepare preliminary budget, equipment and staff. But in the end, this broad and general set of conditions must be narrowed down to the concrete needs and financial capabilities, and must assure adequate level of physical security. It is an interesting problem to solve how effectively achieve this goal. We currently face and try to solve this problem in the KSO3C project.

3.1   CCRA and SOG-IS Requirements

The CCRA and SOG-IS arrangements determine conditions of the mutual recognition of evaluations and certificates. Main conditions are described in Article 5 of both arrangements. But additionally they also demand the use of criteria, methods and requirements based on Common Criteria, CEM, ISO/IEC 17025, and supporting documents issued by the Common Criteria Management Committee (CCMC).

The CC and CEM standards describe security assurance requirements concerning development security in a special assurance family Development Security (DVS). But as we wrote in Sect. 2, security measures typical for development site can be used also for an evaluation site. The DVS family deals with controls measures which help to counter the threats existing at the developer's (here also evaluator's) site. The family comprises of two components:

– First says that a site shall have all the physical, procedural, personnel, and other security measures to protect confidentiality and integrity of the TOE;
– Second says that the documentation of a site shall justify the security measures provide the necessary level of protection to maintain TOE security.

Article 5 says that any ITSEF must be accredited in accordance to ISO/IEC 17025. The standard includes requirements for the competence of testing and calibration laboratories enabling them to demonstrate they operate competently and generate valid results. The standard focuses on quality, documentation, and testing methods management. That is why we will not find there any physical security requirements.

Article 5 also mentions about CCMC supporting documents. Among them there is one that has got all we need at the moment – it is the MSSR (Minimum Site Security Requirements) supporting document issued by Joint Interpretation Working Group (JIWG). The group provides guidelines to facilitate application of the requirements and methods by the members of SOG-IS arrangements. In the next section we present some examples of requirements from this document.

3.2 Minimum Site Security Requirements

MSSR defines the necessary security measures based on the best practices from developers, ITSEFs, and CBs. Every section of the document consists of security goals and measures. Each control is described by its objective and security measure. The objective defines the mandatory target of control. The security measure describes possible controls to protect the assets of the lab.

Among others, MSSR consists of the following sections concerning security: Development Security System (DSS) and Documentation, Internal DSS audits, DSS Improvement, Control Objectives and Controls. The last section includes subsection of Physical and Environmental Security which is the most of our interest because it describes basic physical means preventing unauthorized physical access to the lab's premises.

Thus we are now very close to find out what physical controls we shall implement to make our lab compatible with CCRA and SOG-IS physical security requirements. Below there are some examples of security measures [13].

1. Physical security perimeter – the objective is to protect the premises by at least two line of defense, a detection layer and a stop layer. These layers shall separate authorized from unauthorized people.
2. Physical entry controls – the objective is to protect secure areas by appropriate entry controls to ensure that only authorized personnel are allowed access.
3. Securing offices, rooms and facilities – the objective is to implement measures to ensure detection and prevention of unauthorized access.
4. Media handling – the objective is to protect confidentiality of TOE related data and information, all media shall be protected against unauthorized disclosure, modification, removal, theft, destruction or damage.

The requirements sets in MSSR and schemes are generic and they are intended to be applicable to all organizations, regardless of type, size, and nature. These security measures can be modified, replaced, or enhanced by the owner of the lab. Here comes our idea to adjust general requirements taken from CC, CEM, MSSR, and evaluation schemes by using the Polish regulation for the physical protection of secret information.

4 The Idea - Adjusting Controls to the Requirements

Polish regulation for the protection of secret information [6] determines:

– Basic criteria and the way to define a threat level,
– Selection of security measures relevant to the defined threat level,
– Types of threats that must be considered,
– Application scope of physical security measures.

The main assets of the lab are the TOE parts and its documentation that must be physically protected against unauthorized disclosure. Thus we can assume that all information concerning an IT product can be treated as confidential information in the same way like in the regulation. This is why we decided to use the regulation guidelines for the selection of security measures. But at the same time we have to fulfill conditions of CCRA and SOG-IS arrangements. Later in this section we will trace MSSR requirements to the regulation requirements to show that conformance.

Our main idea is to change the general sets of conditions from MSSR into the detailed set of physical security measures ready to be applied in the lab. The whole idea includes only two steps: (1) Determining a threat level for the lab's premises and assets; (2) Selecting security measures according to the determined threat level. In the following subsections we describe how the idea works.

4.1 Threat Level Calculation

The threat level is calculated by using factors which impact on disclosure or loss of confidential information. The evaluation says how significant the given factor is for the organization. Each factor is evaluated in points within 3-stage scale: Very substantial importance (8 points); Substantial importance (4 points); Low importance (1 point). There are the following seven factors:

– Classification of confidential information – it shows whether the information in the organization is top secret (8 pts), secret (4 pts), or confidential (1 pt).
– The number of sensitive documents – the more sensitive documents in the organization the higher score.
– The form of secret information – electronic form is the most important so it gets the highest score (8 pts).
– The number of people – this factor pertains to all the people in an organization who have access rights to confidential information. The more people having access the more points.

- Location – the factor value is higher when the building is used together with other entities or is placed near by another publicly available buildings.
- Access to a building – the importance of this factor is higher when more people who are not the employees have access to a building.
- Other – this factor considers such events as sabotage, terrorist attacks, or force majeure.

We calculate the sum of factors' points and assign a threat level for our sensitive information according to a key presented in Table 1.

Table 1. Threat levels values

Threats level		
Low	Medium	High
7–16 pts	17–32 pts	Above 32 pts

Next, for the calculated threat level we must select a proper set of security measures which effectively counter the threats.

4.2 Security Measures Selection

The right combination of security measures depends on the calculated threat level and the confidentiality clause of sensitive information. In the regulation [6] there are four tables for confidentiality clauses: top secret, secret, confidential and restricted. Below we present the table for the secret information (Table 2). Other tables are built in the same way but of course with different values.

Table 2. Minimal values of security measures categories for secret information

The highest confidentiality clause in the organization	Threats level		
	Low	Medium	High
SECRET			
Mandatory: categories C1+C2+C3	8	9	10
Mandatory: categories C4+C5[a]	4	5	5
Additionally: category C6	4	5	5
Total in points	**16**	**19**	**20**

[a]All values different from 0

The regulation [6] defines the following six security measures categories: C1 (Safes for secret information storage), C2 (Rooms), C3 (Buildings), C4 (Access control), C5 (Security personnel and burglar alarm systems), C6 (Perimeters).

Let us assume we calculated the medium level of threats (cf. Sect. 4.1). According to Table 2 it means that the total points for all security categories should be 19. Further, for categories C1, C2, C3 the sum is 9, for categories C4, C5 the sum is 5, and for additional category C6 the sum is 5. In the next step we select specific security measures within these categories.

Security measures must be selected in a way which allows mandatory categories to reach minimal values from Table 2. If the total points for mandatory categories is still less than 19 (Medium threats level) then the additional category C6 must considered. The regulation describes the following security measures for each category (CM), Table 3. Measures' values vary from 1 (low) to 4 (high).

Table 3. Enumeration of security measures

		Value ranges	Calculation
C1	C1M1	Class of a safe construction safety (1..4)	C1M1 × C1M2
	C1M2	Class of a safe locks safety (1..4)	
C2	C2M1	Class of a room construction safety (1..4)	C2M1 × C2M2
	C2M2	Class of a room locks safety (1..4)	
C3	C3	Class of building construction safety (1..4)	C3
C4	C4M1	Class of an access control system (1..4)	C4M1 + C4M2
	C4M2	Methods of checking the people without permanent authorization to enter the building (values: 1 or 3)	
C5	C5M1	Restrictions quality level of security personnel (1..4)	C5M1 + C5M2
	C5M2	Class of burglar alarm systems (1..4)	
C6	C6M1	Class/quality of fence safety (0 or 1)	C6M1 + C6M2 + C6M3 + C6M4 + C6M5 + C6M6
	C6M2	Class/quality of security check at control points according to fence classification (0 or 1)	
	C6M3	Methods of security checking system of people and goods at entry/exit (0 or 1)	
	C6M4	Quality of fence trespass detection system (0 or 1)	
	C6M5	Quality of lighting of perimeter and security area (0 or 1)	
	C6M6	Quality of perimeter surveillance system (0 or 1)	

Using Table 3 we are able to select the measures very precisely. Each measure has its own range of values. These values depends on how the measures are effective. Efficiency of a measure relies on the implemented security features which can be more or less stringent. These security features are described in standards and guidelines. They are divided into classes or are rated in points. In the regulation there are tables with security categories and measures. The stronger (better) security feature the higher value of a measure. Below we present Table 4 with an example of the "C2M2 – a room lock" measure which strength

Table 4. A room lock security measure values

Points	Functions or features
4	A lock fulfills at least class 7 of Polish standard requirements
3	A lock fulfills at least class 5 of Polish standard requirements
2	A lock fulfills at least class 4 of Polish standard requirements
1	A lock fulfills at least class 3 of Polish standard requirements

depends on the lock's class. Values for other measures are selected in the same way by using dedicated tables from the regulation [6].

We additionally traced the regulation's categories to MSSR to check whether all MSSR's requirements are met. It revealed we could assign every category to at least one of the MSSR"s security requirements listed in Sect. 3.2 in the following way:

- Physical security perimeter – C3, C6,
- Physical entry controls – C2, C4,
- Securing offices, rooms and facilities – C2, C5
- Media handling – C1.

It means that by using Polish regulation we can still select and evaluate security measures which are conformable with Common Criteria requirements. In the next section we present the results of selecting security measures for our lab.

5   Results and Discussion

We assumed that we will process "Secret" information in the lab. According to Sect. 4.1 we assessed the importance of factors in the following way:

- Classification of confidential information – 4 pts (for "Secret" information),
- The number of sensitive documents – 4 pts (not so many),
- The form of secret information – 8 pts (the highest because the most of documents will be in digital form),
- The number of people – 1 pt. (only a few people will have access rights to the "Secret" information),
- Location – 4 pts (a building will be partially separated from other facilities),
- Access to a building – 1 pt. (only lab workers will have an access to the lab),
- Other – 1 pt. (we do not consider atypical events).

The sum of all factors is 23 points. According to Table 1 we should apply security measures countering the medium threat level.

For the secret information and medium threat level, the minimal total points value of security measures should be 19 (Table 2). The minimum value of security measures should be 9 pts (for categories C1, C2, C3) and 5 pts (for categories C4, C5). If required 19 pts are not achieved then additional security measures from category C6 should be used. In the first iteration we assumed the following security measures depicted in Table 5. The points were assigned to each measure according to the dedicated tables in the regulation [6] as it was described for Table 4. The sum of all security measures categories is 34 points – far more than requested 19 points. It means that in the first approach we chose too strict security measures. It means there is still a room to choose less stringent measures and reduce the costs.

Table 5. Values of main security measures categories

Category		Justification	Points	Result
C1	C1M1	Safe structure fulfills S2 class of Polish standard PN-EN 14450	3 pts	C1M1 × C1M2 = 6 pts
	C1M2	A lock of type 2 fulfills A class of Polish standard PN-EN 1300	2 pts	
C2	C2M1	A room structure with high resistance	3 pts	C2M1 × C2M2 = 12 pts
	C2M2	A room lock in accordance with class 7	4 pts	
C3	C3	Medium resistant lab's building	3 pts	3 pts
C4	C4M1	Access control system fulfills class 2 of Polish standard PN-EN 50133-1	2 pts	C4M1 + C4M2 = 5 pts
	C4M2	Visitors are accompanied by authorized workers	3 pts	
C5	C5M1	Security personnel	3 pts	C5M1 + C5M2 = 6 pts
	C5M2	Burglar alarm systems fulfills class 3 of Polish standard PN-EN 50131-1	3 pts	
C6	C6M1	2 m high fence without any sensors	1 pt.	C6M1 + C6M2 + C6M3 + C6M4 + C6M5 + C6M6 = 5 pts
	C6M2	A guard at reception	1 pt.	
	C6M3	Security checking system of people and goods	1 pt.	
	C6M4	No fence trespass detection system	0 pt.	
	C6M5	Lighting of secure area	1 pt.	
	C6M6	Perimeter surveillance system	1 pt.	
		Total score		34 pts

As outlined in the introduction, building the physical protection of the lab according to the Common Criteria standard relies on the international CC arrangements and many supporting documents. But these security requirements are described in a general way. This is why it is very difficult to find an objective means of selecting security measures matching CC requirements at the beginning of building the lab.

To solve this issue we decided to implement the Polish regulation which is used for the evaluation and selection of security measures. The regulation helped us to select controls and evaluate them in the objective manner.

We verified the conformance of physical security measures proposed by the regulation with measures described in the MSSR document. It turned out the regulation's measures match MSSR's requirements so they can be used for the lab protection.

We also presented the example of security measures evaluation. In the first iteration we chose too strong security measures rated with 34 points in total. It was far too many than demanded 19 points for the medium threats level. In the next iteration we should adjust security measures so that we could achieve demanded level of protection at the lower costs.

6   Conclusions

The goal of this paper was to describe the problem of selecting security measures for the physical protection of the IT security evaluation lab. The paper includes the review of the Common Criteria methodology and accompanying arrangements, and documents which are the potential sources of requirements for physical security controls.

The CC standard includes general requirements stating how to protect the lab and its assets. A developer must follow these requirements and adjust them to the specific needs of a lab. In the paper we proposed the solution based on the Polish regulation which allows to select the proper security controls.

The idea uses the method of security measures evaluation based on Polish technical standards which impose concrete conditions on security features. The more strict conditions are fulfilled by a security measure the bigger score it gets during evaluation. In this way we can objectively assess security measures and select the best one to protect the lab.

Of course the method still needs to be supplemented by other security measures which are not included in the Polish regulation. We plan to improve the method according to the European Union or NATO (North Atlantic Treaty Organization) requirements for the protection of confidential information. Then the method could be used for the selection of security measures required for the laboratory where some special IT products must be highly protected.

Acknowledgements. I would like to thank my co-author colleagues for their help in preparing this paper. The article is financed within the project "National schema for the security and privacy evaluation and certification of IT products and systems compliant with Common Criteria (KSO3C)". The KSO3C project is financed from the National Centre for Research and Development (NCBR) national programme "Cybersecurity and e-Identity", Contract ID: CYBERSECIDENT/381282/II/ NCBR/2018.

References

1. CCRA – Arrangement on the Recognition of Common Criteria Certificates. In the field of Information Technology Security. http://www.commoncriteriaportal.org/ccra/. Accessed 6 Dec 2018
2. Common Methodology for Information Technology Security Evaluation – Evaluation Methodology. CCMB-2017-04-004. CCMB-2017-04-004, Version 3.1, Revision 5, April 2017
3. Common Criteria for Information Technology Security Evaluation. Part 1: Introduction and general model, CCMB-2017-04-001, Version 3.1, Revision 5, April 2017
4. Common Criteria for Information Technology Security Evaluation. Part 2: Security functional components. CCMB-2017-04-002, Version 3.1, Revision 5, April 2017
5. Common Criteria for Information Technology Security Evaluation. Part 3: Security assurance components. CCMB-2017-04-003, Version 3.1, Revision 5, April 2017
6. Dziennik Ustaw Rzeczypospolitej Polskiej, Poz. 683, Rozporzadzenie Rady Ministrow z dnia 29 maja 2012 r. w sprawie srodkow bezpieczeństwa fizycznego stosowanych do zabezpieczania informacji niejawnych. Warszawa, dnia 19 czerwca 2012 r., z późniejszymi zmianami
7. IEC/ISO 15408-1:2009(E) Information technology – Security techniques – Evaluation criteria for IT security – Part 1: Introduction and general model (2009)
8. IEC/ISO 15408-2:2008(E) Information technology – Security techniques – Evaluation criteria for IT security – Part 2: Security functional components (2008)
9. IEC/ISO 15408-3:2008(E) Information technology – Security techniques – Evaluation criteria for IT security – Part 3: Security assurance components (2008)
10. IEC/ISO 17025:2005 General requirements for the competence of testing and calibration laboratories (2005)
11. IEC/ISO 17065:2012 Requirements for bodies certifying products, processes and services (2012)
12. IEC/ISO 18045:2008 – Information technology – Security techniques – Methodology for IT security evaluation (2008)
13. Joint Interpretation Library. Minimum Site Security Requirements, Version 2.1 (for trial use), December 2017
14. SOG-IS MRA – Mutual Recognition Agreement of Information Technology Security Evaluation Certificates, version 3.0. Final version 8 January 2010. https://www.sogis.org/uk/mra_en.html. Accessed 6 Dec 2018
15. Rogowski, D.: Identification of information technology security issues specific to industrial control systems. In: Zamojski, W., Mazurkiewicz, J., Sugier, J., Walkowiak, T., Kacprzyk, J. (eds.) DepCoS-RELCOMEX 2018. AISC, vol. 761, pp. 400–408. Springer, Cham (2019). https://doi.org/10.1007/978-3-319-91446-6_37
16. Rogowski, D.: Software support for common criteria security development process on the example of a data diode. In: Zamojski, W., Mazurkiewicz, J., Sugier, J., Walkowiak, T., Kacprzyk, J. (eds.) Proceedings of the Ninth International Conference on Dependability and Complex Systems DepCoS-RELCOMEX. 30 June–4 July 2014, Brunów, Poland. AISC, vol. 286, pp. 363–372. Springer, Cham (2014). https://doi.org/10.1007/978-3-319-07013-1_35
17. The Common Criteria Homepage. http://www.commoncriteriaportal.org/. Accessed 5 Dec 2018
18. The KSO3C project web page. https://www.kso3c.pl/. Accessed 6 Dec 2018

Correction to: Serialization for Property Graphs

Dominik Tomaszuk, Renzo Angles, Łukasz Szeremeta,
Karol Litman, and Diego Cisterna

Correction to:
Chapter "Serialization for Property Graphs"
in: S. Kozielski et al. (Eds.): *Beyond Databases,*
***Architectures and Structures*, CCIS 1018,**
https://doi.org/10.1007/978-3-030-19093-4_5

In the originally published version of the chapter 5, the grant number in the acknowledgments section was missing. The grant number has been added.

The updated version of this chapter can be found at
https://doi.org/10.1007/978-3-030-19093-4_5

Author Index

Printed in the United States
by Baker & Taylor Publisher Services